Ecology and Management of Coastal Waters

The Aquatic Environment

Springer
London
Berlin
Heidelberg
New York
Barcelona
Hong Kong
Milan
Paris
Santa Clara
Singapore
Tokyo

Gilbert Barnabé and Régine Barnabé-Quet

Ecology and Management of Coastal Waters

The Aquatic Environment

Professor Gilbert Barnabé, Université de Montpellier II, Station Méditerranéenne de l'Environnement Littoral, Laboratoire Écologie Marine, Sète, France

Régine Barnabé-Quet, Technicienne CRNS (UPR 9060), Université de Montpellier II, Station Méditerranéenne de l'Environnement Littoral, Laboratoire Écologie Marine, Sète, France

Translator: Dr Jennifer Watson, Department of Zoology, University of Aberdeen, UK
Translation Editor: Dr Lindsay Laird, Department of Zoology, University of Aberdeen, UK
Original French edition: *Ecologie et aménagement des eaux côtières*
Published by © Technique et Documentation, 1997
This work has been published with the help of the French Ministère de la Culture - Centre National du Livre

---

SPRINGER-PRAXIS BOOKS IN AQUACULTURE AND FISHERIES
SUBJECT *ADVISORY EDITORS*: Lindsay Laird, M.A., Ph.D., F.I.F.M., University of Aberdeen, UK
Selina Stead, B.Sc., M.Sc., Ph.D., Land Economy, University of Aberdeen, UK

ISBN 1-85233-647-1 Springer-Verlag Berlin Heidelberg New York

British Library Cataloguing in Publication Data
Barnabe, Gilbert
Ecology and management of coastal waters: the aquatic environment. - (Springer-Praxis books in aquaculture and fisheries)
1. Coastal ecology
I. Title II. Barnabe-Quet, Regine
577.5′1
ISBN 1852336471

Library of Congress Cataloging-in-Publication Data
Barnabé, G. (Gilbert)
Ecology and management of coastal waters: the aquatic environment/Gilbert Barnabé-Quet
p. cm. - (Springer Praxis books in aquaculture and fisheries)
Includes bibliographical references
ISBN 1-85233-647-1 (alk. paper)
1. Coastal ecology. 2. Marine ecology. 3. Coastal zone management.
I. Barnabé-Quet, Régine, 1942- . II. Title. III. Series.
QH541.5.C65 B37 2000
577.7—dc21 00-037371

Printed by MPG Books Ltd, Bodmin, Cornwall, UK

Cover design: Jim Wilkie
Project Copy Editor: Rachael Wilkie
Typesetting: Originator, Gt. Yarmouth, Norfolk, UK

Printed on acid-free paper supplied by Precision Publishing Papers Ltd, UK

1

Contents

List of colour plates

The colour plates section appears between pages 192 and 193.

10. At other sites, the flat oyster *Ostrea edulis* constitutes the dominant population of filter feeders. Jetty at Sète harbour, depth 8 m.
11. This 50 cm diameter buoy submerged at a depth of 5 m in May, was colonised by flat oyster (*Ostrea edulis*) spat at the beginning of July. Shellfish culture zone at Agde (Hérault), photograph taken in August.
12. The same buoy photographed 10 months later. Mussels have now also colonised the substrate. The oysters are smothered by this competition for attachment and access to open water which is their source of food and oxygen. The populations succeed one another in time and are in competition for the same substrate.
13. Even in tropical waters, all objects which are submerged in mid-water, such as this buoy and especially its mooring, are covered in flora and fauna which in turn attract fish. Caribbean coast of Martinique, Saint Pierre, depth 3 m.
14. Coral reefs provide habitats for juvenile fish in tropical waters. Nouméa lagoon, New Caledonia, depth 6 m.
15. This car wreck (front wheel axle unit) attracts juvenile fish in the absence of a coral reef on this sandy seabed. Caribbean coast of Martinique, Fond Corré, depth 7 m.
16. Coral reefs are frequented by the multitude of adult fish (here, soldier fish). Caribbean coast of Martinique, Prêcheur, depth 12 m.
17. Plankton harvesting equipment, filter bag containing the harvest (copepods), achieved in a few hours from a productive salt marsh.
18. Artificial reef frequented by the sea bass, *Dicentrarchus labrax*. (Photograph by Gy F. Bachet.)
19. Artificial reefs are often made up of very large structures, such as this 158 m^3 module (the diver shows the scale). (Photograph by Gy F. Bachet.)
20. When waters are rich in nourishing seston, the biomass and production of mussels in suspended culture become enormous—each 10 m long rope produces more than 100 kg in nine months. Shellfish culture zone at Agde, depth 22 m.

Introduction

Management of coastal waters: Why?

I.1 OBJECTIVES OF THIS BOOK

This book has been produced primarily in response to our own needs with regard to university education, but we also wanted it to fill in the gaps in information on the management of lagoon and coastal waters which currently exist in Europe. It therefore involves scientific, environmental and economic activities which are rapidly expanding in many countries, but which are relatively undeveloped in Europe. Thus there was not actually a lack of information, but a lack of synthesis of the subject, whatever the language of the references.

In order to overcome this lack, we have attempted to analyse and then to synthesise a large amount of data spread throughout a multitude of documents, in order to make them comprehensible to all those concerned with the management and development of coastal waters: administrators, social science specialists, ecological scientists, divers or underwater fishermen, decision makers or even simply the enlightened, self-taught man in the street.

The development and ecology of coastal waters constitutes a vast subject involving knowledge belonging to a wide range of disciplines including: oceanography, hydrology, biology, ecology, fisheries science, aquaculture, civil engineering, geography, economics, law and social sciences. Without pretending that this is an exhaustive review, we have tried to achieve a compromise, a sorting out which is balanced and which allows the reader to understand the great stakes involved in the development and management of coastal waters without having to refer to numerous other books.

The specialists may not be satisfied, but we hope that the references found in the bibliography will alleviate this dissatisfaction. In addition to classical scientific sources, we have used many of our own research results, some of which are published here for the first time. The mass of data forced us to make choices: it was important to maintain a broad view and not become too specialised, without being superficial when the subject required more in-depth discussion.

Our thought processes have remained those of the scientific ecologist, essentially global, attempting to seize on relationships between functional phenomena, to the detriment of being descriptive, but also involving ecology in terms of a science of synthesis, using methods and knowledge from modern science and its technical progress (Frontier and Pichot-Viale, 1991). To this we have added our own experiences of diving in lagoons and coastal environments, constituting another original approach. We also hope that our approach is distinct from a political type of ecology.

The physicochemical and naturalist-type approach of the ecological scientist leads to description and understanding of natural phenomena, but goes no further; this was the aim, for example, of the systematics scientists who proposed in "Systematics, Agenda 2000" (initiated in the USA and taken up elsewhere), to discover, describe and classify all species on the planet. This initiative is commendable, but will a simple inventory, however big, stop the disappearance of endangered species, for example? The answer must be a resounding no.

Therefore we propose to challenge this aim, which only constitutes one aspect of the problem; because of its implications for the economy, the liaison between science and technology is now recognised as the principal source of productive resources. The ecologist must therefore become a partner of the players in the economy and propose real ecological management which integrates all the relationships between man and the environment. This is the new dimension which we wish to add to our approach.

I.2 POSITION OF THIS BOOK

There are certainly many general books, in French, on land management (Lamotte, 1985), ecological management and fish culture in freshwaters (Arrignon, 1976), or the degradation of the coastal environment (Lacaze, 1993), but none which, strictly speaking, concern the development or management of lagoons or coastal waters. Aubert's book *Cultiver l'océan* (1965) and those on the theme of aquaculture (Barnabé, 1990, 1991) are not in the same context, since these activities are only one aspect of management; more recently, Chaussade (1994) discussed the future of fishing, without really proposing new solutions for a traditional activity in which development is limited.

The English language literature is certainly richer and, without pretending to be exhaustive, one can cite the works on coastal environments by Carter (1991), the ecology of sandy shores (Brown and McLachlan, 1990), the study of ecological engineering (Mitsch and Jorgensen, 1989), the book on managed aquatic ecosystems by Michael (1987) or that of Adey and Loveland (1991) dealing with the physical modelling of different types of aquatic ecosystem. Recent aquaculture books, once again, are devoted to intensive aquaculture and are far removed from management, except perhaps fish culture in ponds, but this involves freshwater. In the past, the prospects for aquaculture in the open sea (Hanson, 1974) or the rearing of fish and shellfish in coastal waters (Milne, 1972) included a raised awareness aimed partly at development; the progress which has occurred since has rendered these works obsolete and they have had no recent successor. The term "management" has been applied to large marine ecosystems (Sherman and Lewis, 1985) and the Club of Rome became interested in the problem of management with the report by Mann Borgese (1986), although more in economic terms. More in-depth reflection, or even "ecological philosophy", can be found in the works of Frontier and Pichod-Viale (1991) and Goldsmith (1994); but this is theoretical in nature.

The work of Mann and Lazier (1991), which is devoted to the dynamics of marine ecosystems, deserves a mention in this review since for the first time the interactions between hydrology and the production of living matter constituted the objective of an entire study; previously, it would have been necessary to search in separate publications for data whose correlations had never before been so clearly demonstrated. Now, physical processes and biological productivity form the starting points for management in ecology and in the role played by the oceans; in terms of the world's climate, for example, the integrated physical and biological functioning of marine ecosystems must be considered.

I.3 DEFINITIONS OF COASTAL DEVELOPMENT

In ecological terms, development is defined as the organisation of a space by modifying an ecosystem in order to exploit it, or the creation of habitats with a view to encouraging reproduction or settling of particular species (Parent, 1990). Thus, pastoral, rural or forestry developments have become well known on land and involve the development and careful exploitation of natural resources with respect to the environment, since development also requires judicious organisation.

The development of European coastal waters which is aimed first at animal and, to a lesser degree, at plant resources can be defined in the same way. This consists of the application of scientific and technical principles and concepts to animal or plant populations, as well as to their habitats, with a view to encouraging the reproduction or settling of particular species or ensuring their healthy survival.

These aims are of no direct benefit to man, whereas those of land management concern human ecology and are defined as the organisation of space so as to improve living conditions for populations, to develop economic activities and to develop natural resources while avoiding disturbance of natural ecosystems (Parent, 1990). According to Lamotte (1985), the concept of development consists of the transformation by man of a system—extending to the land, productivity, or some complex combination of these—with a view to more rational or efficient utilisation of resources; he added that it involves an activity which is essential to human society ... man's objective is to free himself from the constraints of the functioning of natural ecosystems in which he evolves and which he transforms.

The coastal domain is not very well defined; in France, it is considered to be a zone approximately 40 km on either side of the water's edge (Anon., 1994). Coastal waters thus extend 40 km out from the coast.

Development of coastal waters must aim to compromise between man's requirements (in terms of food, leisure, etc.) and safeguarding aquatic ecosystems. There is no contradiction between these two objectives, even from the point of view of strict human utilitarianism; this goal can also be attained by conservation of ecosystems (since reserves or wildlife parks constitute developments) as well as by modifications. It has been said that it has now become essential to save the whales and the elephants ... not for the whales and elephants themselves, but in order that we develop the qualities that will save them and which will save ourselves (Panneau, 1990).

This compromise between strict human utilisation and conservation of natural ecosystems can also be shown by sustainable development*, which consists of management and development that are limited with respect to nature but are permanently renewable, as advocated by the Rio Conference in 1992.

Man is present in ever-increasing numbers and has an impact on nature; development is required to counteract this impact. As stated by Quignard (1994), of all the animal species, man is among the most gregarious and lives in concentrations only equalled by termites.

While resources are limited, development has the dual goal of increasing these resources and avoiding pollution.

I.4 GENERAL LAYOUT OF THIS BOOK

Oceanic and continental living resources in aquatic environments are already exploited by fisheries and various forms of aquaculture, but the shore is also a very sought-after area for many uses (dwellings) and for many leisure activities (tourism, sailing, diving, angling, shellfish harvesting, etc.). We will set ecological development alongside these current uses. The book is divided into several complementary parts.

The first part constitutes a review of the basic knowledge required for an understanding of the physical, chemical and biological phenomena that are acting in coastal waters, from the waves and currents to the food chains which explain the abundance or scarcity of fish to the increasing sea level which causes the regression of the coastline.

We have stressed the biological and physical interactions in coastal ecosystems: first, because they involve data which do not feature in the classic books; second, because these processes are determinant in biological production (see *Coastal upwellings: physical factors feed fish*, Hartline, 1980); and lastly, because they are required to play an increasingly important role in development.

It is now possible to add another dimension to marine ecology; instead of placing organisms at the centre of problems and considering them in relation to other organisms and the environment, it is possible to work with marine ecosystems in which the physical, chemical and biological components are equally important in the

* The definition of words and terms followed by an asterisk can be found in the glossary.

definition of the entire system: the biology cannot be considered separately (Mann and Lazier, 1991). The understanding of this interaction is vital for the proposal of future development solutions, which justifies the importance we have accorded to it.

Part C presents, for the first time, an exhaustive review of the technology available and the various types of development achieved around the world. These data and current experiments lead to the proposal of new solutions to often ancient problems.

We have reserved a significant amount of space for future prospects and emphasis is placed on the equilibrium between economic development based on an ecological approach and environmental protection.

This book should thus constitute a succinct and up-to-date summary of the functioning of coastal ecosystems and a thought-provoking document on the development and management of coastal seas; at least this is what we hope.

Only functional terms are used; descriptions, categories and classifications are avoided.

I.5 BIBLIOGRAPHY

Adey W.H. and Loveland K. *Dynamic aquaria: Building living ecosystems.* Academic Press, San Diego, 1991.

Anon. Les disciplines marines: situation des stations marines. *C.R. Acad. Sc.*, Série *générale* 1994; 11(1): 19–58.

Arrignon J. *Aménagement écologique et piscicole des eaux douces.* Gauthiers-Villars Éd., Paris, 1976.

Aubert M. *Cultiver l'océan. La science vivante.* PUF Éd., Paris, 1965.

Barnabé G. (Ed.). *Aquaculture: Biology and ecology of cultured species.* Ellis Horwood Ltd, Chichester, 1994.

Barnabé G. (Ed.). *Bases biologiques et écologiques de l'aquaculture.* Lavoisier Tec & Doc, Paris, 1991.

Brown A.C. and McLachlan A. *Ecology of sandy shores.* Elsevier, Amsterdam, 1990.

Carter R.W.G. *Coastal environments.* Academic Press, London, 1991.

Chaussade J. *La mer nourricière, enjeu du XXIst siècle.* Institut de Géographie, Nantes, 1994.

Frontier S. and Pichod-Viale D. *Écosystèmes: Structure, fonctionnement, évolution.* Masson Éd., Paris, 1991.

Goldsmith E. *Le défi du XII siècle: une vision écologique du monde.* Éd. du Rocher, Monaco, 1994.

Hanson J.A. Open sea mariculture in utilitarian perspective. In: Hanson J.A. (Ed.) *Open sea mariculture.* Dowden Hutchinson & Ross Inc., Strousburg, 1974.

Hartline B.K. Coastal upwellings: physical factors feed fish. *Science* 1980; 208: 38–40.

Kesteven G.L. and Deacon G. Les recherches océanographiques et leur contribution à l'étude des pêches. *Bull. Pêches*, FAO 1955; 8(2): 1–13.

Kullenberg G. La science doit s'atteler à la gestion intégrée de la zone cotière: que faire. *Bull. Int. Sc. Mer* (UNESCO) 1994; 2–3.

Lacaze J-C. *La dégradation de l'environnement côtier*. Masson Éd., Paris, 1993.
Lamotte M. *Fondements rationnels de l'aménagement d'un territoire*. Masson Éd., Paris, 1985.
Mann Borgese E. *The future of the oceans*. A report to the Club of Rome. Harvest House, Montreal, 1986.
Mann K.H. and Lazier J.R.N. *Dynamics of marine ecosystems*. Blackwell Scientific, Boston, 1991.
Michael R.G. *Managed aquatic ecosystems*. Ecosystems of the world no. 29. Elsevier, Amsterdam, 1987.
Milne P.H. *Fish and shellfish farming in coastal waters*. Fishing News, London, 1972.
Mitsch W.J. and Jorgensen S.E. (Eds). *Ecological engineering: An introduction to ecotechnology*. John Wiley & Sons, New York, 1989.
Panneau J-M. *La Nouvelle République*. 6 April 1990.
Parent S. *Dictionnaire de l'environnement*. Broquet Inc., Ottawa, 1990.
Quignard J-P. *Séance d'accueil*. Okéanos 94, Montpellier.
Sherman K. and Lewis M.A. (Eds). *Variability and management of large marine ecosystems*. Westview Press Inc., CO, 1986.

Part A

Ecology of coastal waters

1

Water, the oceans and the atmosphere

1.1 PHYSICAL AND CHEMICAL CHARACTERISTICS OF MARINE WATERS

Marine water, which makes up the seas and oceans, is a complex mixture of a large number of substances. Water is the main constituent, but almost all the chemical elements known on earth are present in sea water. The physical and chemical characteristics of sea water determine its biological properties, as well as many other aspects.

1.1.1 Temperature

The surface waters of the oceans in contact with the atmosphere are subjected to sunlight, evaporation and transfer by convection into the air as it is moved around by winds and other phenomena. As a result, their temperature varies from the poles to the equator (Table 1.1; Guilcher, 1979). Annual variations in temperature are between 1–3 degrees for polar and intertropical waters, but this range is wider in temperate zones (8°C in the English Channel and 13°C in the Mediterranean, for example). In contrast, diurnal fluctuations rarely exceed 1 degree owing to the high specific heat of water (see Section 2.6. below) and only affect the superficial layers.

Very shallow littoral waters and especially confined waters (bays, lagoons, lakes and rearing ponds) whose thermal inertia is lower, reflect variations in atmospheric temperature more quickly and intensely than large water masses, especially when winds facilitate heat exchange through water mixing such as by waves and currents (Chapter 2); currents can also bring waters of different temperatures to the surface. Rapid changes in temperature can thus occur, despite the high specific heat of water. These shallow zones follow the air temperature and are warmer (in summer) and colder (in winter) than the coastal waters of seas or oceans.

Latitude, and also altitude, plays a role in the temperature of freshwaters. In the sea, dominant currents take warm or cold water far from their origin, which can allow the rearing of exotic species (thus the Gulf of Mexico receives warm water

Table 1.1. Mean temperatures of surface waters (°C) at different latitudes (Guilcher, 1979).

	Atlantic	Indian	Pacific
Latitude north			
70°–60°	5.6		
60°–50°	8.7		5.7
30°–20°	24.2	26.1	23.4
10°–0°	26.7	27.0	27.2
Latitude south			
0°–10°	25.2	27.4	26.0
20°–30°	21.2	22.5	21.5
50°–60°	1.8	1.6	5.0
60°–70°	1.3	1.5	1.3

from the south, while at the same latitude the Californian coast receives cold water from the north). The thermal equator does not coincide with the geographic equator; the former is further north.

Depth must also be considered, as deep oceanic waters are rich in nutrients and, although rarely utilised, have shown potential; while these waters are cold (~4°C at depths greater than 1000 m), three-quarters of the oceans' volume have a mean temperature of 3.25°C (between 0–6°C). Another factor which is at a relatively advanced stage of investigation by American workers (project OTEC), is the use of the sea's thermal energy; they also envisage the use of cold, nutrient-rich waters being brought to the surface.

This variation in water temperature with depth does not occur gradually but, most often, by a sudden change known as the thermocline*. Its formation is the result of the surface layers of water being heated by the sun and the actions of winds and swell at depths less than 40 m. This heating of surface, superficial water is accompanied by a decrease in its specific density; thus the lighter, warm waters do not mix with colder, denser waters immediately below, resulting in stratification. The thermocline is a real physical, and sometimes biological, barrier (tuna do not traverse a thermocline). An example of the development of this phenomenon is illustrated in Fig. 1.1. In Norway, such artificial stratification (the introduction of freshwater which spreads without mixing across the surface of the salt water in a lake), allows solar energy to be harnessed by the salt water, which is thus isolated from the cold atmosphere. Temperatures of 30°C have been recorded: ongrowing of the flat (or European) oyster, *Ostrea edulis* has been achieved (Korringa, 1976).

The study of the distribution of living organisms both on land and in the water has become a scientific discipline: zoogeography. In the aquatic environment, this distribution is largely regulated by temperature. There are warm water species (penaeid prawns, fish such as tilapias, some species of carp and mussels), temperate water species (bass, sea bream, catfish, common carp, oysters, mussels, etc.) and cold water species (salmonids, cod, halibut). This classification is not very satisfactory: a

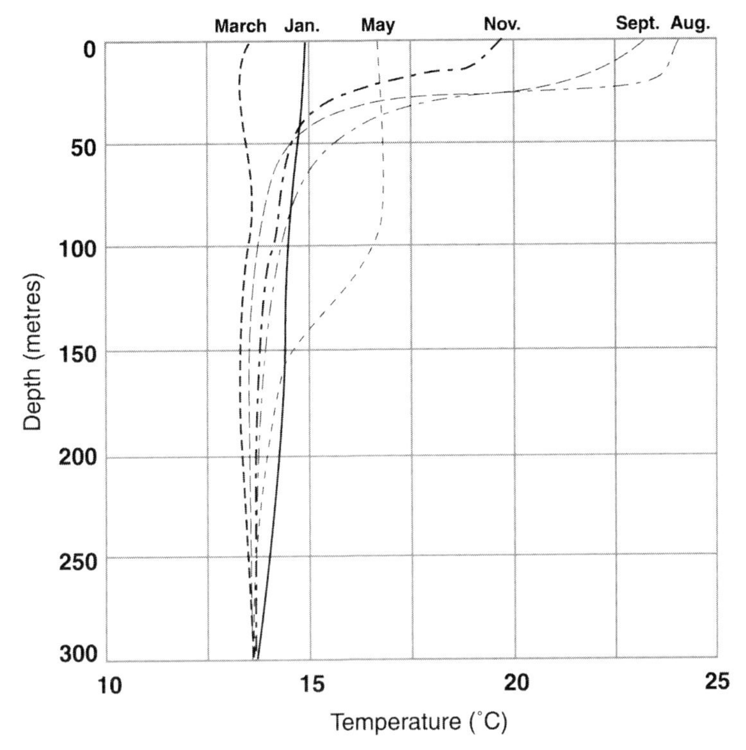

Fig. 1.1. Vertical temperature profiles in the Mediterranean at different times of year (from data collected by the Zoological Station at Villefranche-sur-Mer).

temperate water fish such as the bass spawns at temperatures between 10 and 14°C, but its optimum growth is at 23–25°C and it can survive in water temperatures of 1–30°C.

The effects of temperature on living organisms will not be discussed here, but marine species (other than mammals and birds) are poikilotherms*, their metabolism thus being dependent on the ambient temperature. Optimal temperatures are often different for species which are reared together (plankton and plankton-eating fish, for example), which makes the problem even more complex.

1.1.2 Salinity and dissolved salts

1.1.2.1 Primary constituents

Marine water is saline and salinity is expressed in parts per thousand (S‰), representing the "mass (in grams) of solid substances contained in sea water when iodine and bromine ions are replaced by their equivalent in chlorine, carbonates are converted into oxides and all organic matter excluded".

Table 1.2. The percentage of salts making up the composition of oceanic waters.

Cl^-	(chlorine)	55.200	(18.98 g/l)
Na^+	(sodium)	30.400	(10.56 g/l)
SO_4^{--}	(sulphates)	7.700	2.64
Mg^{++}	(magnesium)	3.700	1.27
Ca^{++}	(calcium)	1.160	0.40
K^+	(potassium)	1.100	0.38
Br^-	(bromine)	0.100	0.06
Sr^{++}	(strontium)	0.040	0.01
H_3Bo_3	(boric acid)	0.070	
HCO_3–CO_3	(carbonic acid and carbonates)	0.035	

In addition to these constituents, sea water contains small amounts of many other elements, whose concentration does not exceed 0.025% of the main constituents.

Chlorinity (Cl‰) represents the mass (g) of chlorine equivalent to the total amount of halogens in a kilogram of sea water. Salinity and chlorinity are related by the equation:

$$S‰ = 1.805Cl‰ + 0.030‰ \qquad \text{(Eq. 1.1)}$$

Mean salt content is about 35 g/l in the ocean (33–37‰) but the northern part of the Red Sea contains 42‰, while certain gulfs of the Baltic Sea have salinities of 4‰.

1.1.2.2 Secondary constituents

These constituents (0.02–0.03%) are also significant, as they hold mineral elements (nitrogen [N] and phosphorus [P]) which are vital to the production of living organisms in the aquatic environment. These nitrogenous (nitrites, nitrates and ammonia salts) and phosphoric (various phosphates) compounds are used by the phytoplankton, together with carbon dioxide (see dissolved gas, below) in order to make living tissues, under the action of sunlight (photosynthesis*). Their concentration varies as biological processes can utilise them entirely.

The concentration of essential mineral salts in surface waters in the open ocean is too low (0.001–0.08 mg N per litre, 0.002–0.003 mg P per litre) to allow good primary production (0.05 mg N and 0.005 mg P per litre are required). Much higher levels are found in deep waters (0.2–0.5 mg/l N and 0.05–0.09 mg/l P). This fact is essential for fisheries planning and management perspectives and will be returned to at a later stage.

Oligo-elements (substances present at very low levels) such as iron (1–50 mg/m^3), copper (4–10 mg/m^3), manganese (1–10 mg/m^3), zinc (5–14 mg/m^3) and selenium (4 mg/m^3) are vital, while certain vitamins and other organic substances (amino acids, carbohydrates, etc.) are present in sea water; their role is not always known. Silicon (10–1000 mg/m^3) is a constituent of the skeleton (theca) of diatoms; low or non-existent levels constitute a limiting factor for the development of diatom populations.

Many dissolved organic substances such as amino acids, simple and complex carbohydrates, vitamins, etc. are also found in the sea. They result from the activity of living organisms or from the decomposition of their carcasses.

These substances alternate between a dissolved and a particulate state, interacting together, sometimes having the capability to stimulate or inhibit the development of other species. Thus diatoms and dinoflagellates secrete "telemediators"* which can impede the development of other, potentially competitive, species. Aubert (1994) reviewed the information on these transmitter substances which are very diverse molecules: phenols, tannins, terpenoids, polysaccharides and fatty acids.

These organic substances can also act as growth factors: this is true of certain amino acids such as arginine, histidine, tryptophan or vitamins such as B12 or biotin. The culture media for all micro-algae contain these vital components.

1.1.2.3 Substances introduced by man

Man discharges a range of substances into the seas; these can either be incorporated into biological cycles (C, N, P) or accumulate passively in food chains or sediments. Hydrocarbon discharges from the cleaning out of oil tankers at sea are of the order of 1.5 million tonnes per annum. Pesticides (the ocean contains 1.5 million tonnes of DDT), herbicides and radioactive waste often end up in the oceans. Concentrations of N and P have increased by a factor of 2.5 over the past 20 years in the Atlantic Ocean and the English Channel; in contrast, O_2 has disappeared from depths below 80 m in the Baltic.

According to Guillemau-Drai (1991), around one million synthetic chemical substances exist, most of which have been discharged into the sea. This author reports the conclusions of international organisations (OMS) which indicate that, at certain sites, the concentration of organochlorides is such that the consumption of fish once a week can increase the risk of cancer to above the permissible level.

1.1.3 Dissolved gases

Atmospheric gases are found dissolved in sea water at relatively low concentrations, but they are in equilibrium with atmospheric concentrations at the surface of the oceans; they follow the law of dilution, or Henry's Law.

Transfer occurs across a superficial film a quarter of a μm* thick, molecules moving by molecular diffusion; the layers of water and air appear to be areas of rapid mixing (see Copin-Montégut, 1993 for further details).

1.1.3.1 Oxygen (O_2)

Oxygen is essential for all living organisms (aerobes). It is used for oxidation of food, liberating energy necessary for all vital activities (swimming, hunting, reproduction, growth, etc.). This oxygen is taken from the surrounding environment by respiration, expressed by the overall equation:

$$CH_2O + O_2 = H_2O + CO_2 \qquad \text{(Eq. 1.2)}$$

The aquatic environment contains relatively little oxygen (less than $10\,cm^3/l$ in contrast to $200\,cm^3/l$ in air); an adequate concentration of oxygen in water is thus a crucial factor for all types of biological management.

In natural environments, this concentration is close to saturation (sometimes even supersaturation) and it varies from $8\,cm^3/l$ in cold waters to $4.5\,cm^3/l$ in tropical waters, at least in marine or surface continental waters, as a result of the photosynthetic activity of plants. In eutrophic zones, too great an O_2 demand can exhaust the environment (anoxia): this is called dystrophy. In high-density rearing (intensive rearing such as that of salmonids), a stoppage in the water turnover can lead to the exhaustion of dissolved oxygen from the environment and death of the cultured fish.

For a chlorinity of 20‰ at atmospheric pressure, the maximum number of cm^3 which can be dissolved (saturation) in one litre of sea water varies with temperature:

5°C: 7.07 10°C: 6.35 15°C: 5.79
20°C: 5.1 25°C: 4.86 30°C: 4.46

This solubility varies similarly, but less markedly, in relation to water salinity; thus for a chlorinity of:

0‰: $6.57\,cm^3$ 5‰: 6.26 10‰: 5.95 15‰: 5.63 20‰: 5.31

The curves in Fig. 1.2 (Alzieu, 1989) show the relationship between O_2 saturation, salinity and temperature.

Decomposition of organic matter by numerous micro-organisms living in the water column or on or within the seabed (see Chapter 2) also consumes oxygen. This requirement for oxygen and respiration constitutes the Biological (or Biochemical) Oxygen Demand (BOD); it can become significant when animals or plants are decomposing in a closed environment. Using herbicides to treat areas of water may therefore present problems. In certain managed ecosystems, such as the Étang de Thau (France), oxygen concentrations are limiting.

Lack of oxygen occurs particularly in summer, in natural shallow or stagnant environments (stratified waters) and is due to high temperatures, abundance of vegetation, short day length, high food consumption by reared animals and high respiration and excretion rates. Nocturnal oxygen consumption exhausts the environment towards dawn and the most sensitive species die, further increasing the BOD (as a result of their decomposition by micro-organisms), at a time when no photosynthesis is taking place.

1.1.3.2 Nitrogen (N)

This gas is present in natural aquatic environments at around 12 ml/l, with supersaturation of about 2% close to the surface. Supersaturation exceeding 3–4% in aquaculture ponds can lead to a non-infectious disease known as gas bubble disease, symptoms of which include exophthalmia. This supersaturation is due to the

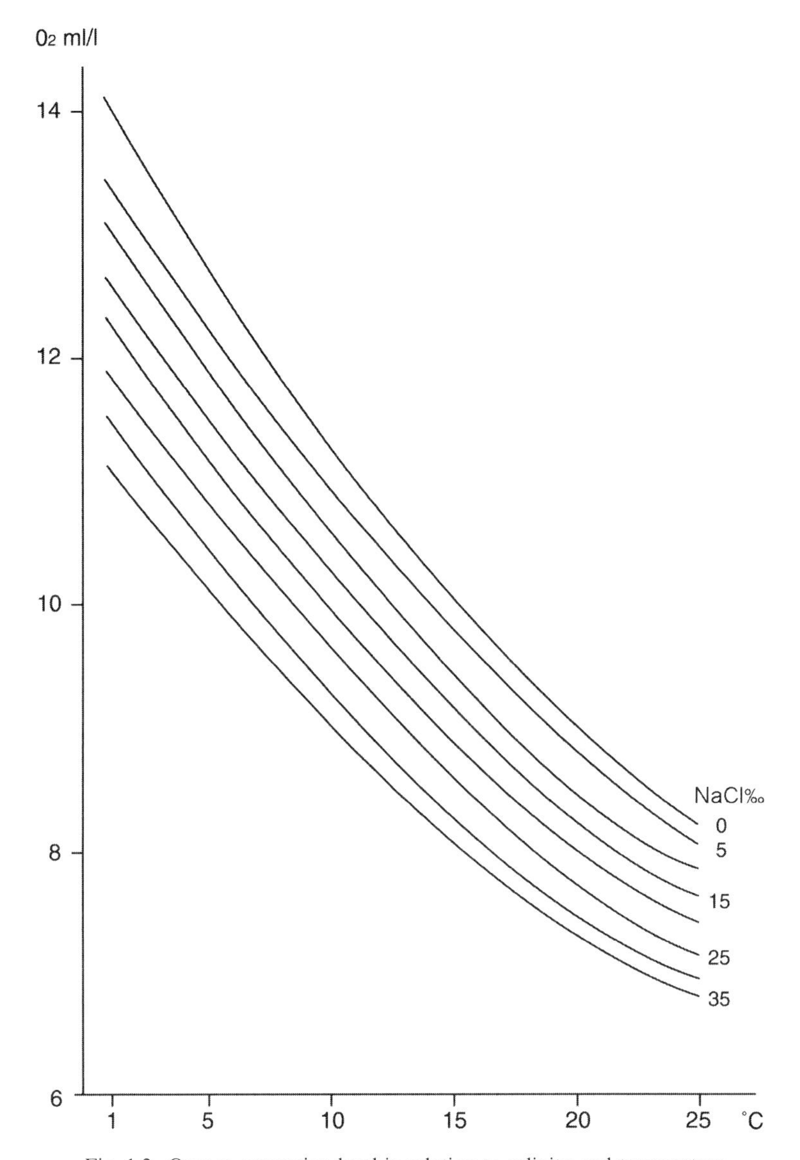

Fig. 1.2. Oxygen saturation level in relation to salinity and temperature.

dissolving of nitrogen at the air intakes during the uptake of water by pumps, or the warming of water, which can affect its solubility.

1.1.3.3 Hydrogen sulphide (H_2S)

This is produced from anaerobic decomposition and results in an anoxic environment; thus it cannot be present in water destined for aquaculture. It is often found in organic-rich sediments, below rearing cages or below suspended mollusc culture systems; such sediments are toxic to the reared species, if resuspended in the water column.

1.1.4 Carbon dioxide gas and pH

1.1.4.1 The carbon dioxide gas system

Carbon dioxide (dioxide of carbon made up of non-dissociated molecules of CO_2) is dissolved and is in equilibrium with that in the atmosphere. As in the air, it is not very abundant in the sea, as it participates in a reversible chemical equilibrium (buffer system) with carbonate and bicarbonate ions and carbonic acid, as shown in the equation :

$$\underset{\text{carbon dioxide gas}\uparrow}{2H_2O + 2CO_2} = \underset{\uparrow\text{carbonic acid}}{2H_2CO_3} = H^+ + \underset{\uparrow\text{bicarbonate}}{2HCO_3^-} = 2H^+ + \underset{\uparrow\text{carbonate}}{2CO_3^{2-}} \quad \text{(Eq. 1.3)}$$

The "total available" concentration of CO_2 varies between 40 and 50 mg of CO_2/l. The molecules of carbonic acid only represent a few percent of the total amount of non-dissociated molecules of CO_2. The main part of total carbon dioxide gas is thus represented by the carbonate and bicarbonate ions.

The concentration of carbon dioxide gas stabilises in well-buffered sea water in relation to temperature; according to Frontier and Pichod-Viale (1991), at a salinity of 35‰:

Temp (°C)	0	10	15	20	30
CO_2	0.44	0.31	0.27	0.25	0.21

1.1.4.2 pH

pH expresses the concentration of hydrogen ions: pH = $-\log 10\ [H^+]$. It measures the acidity or alkalinity of water, on a scale from 0–14. Between 0 and 7, the water is acidic; it is neutral at 7 and alkaline above that. The pH of sea water varies between 7.9 and 8.3 depending on area.

The chemical environment is greatly influenced by pH; for example, the equilibrium between NH_4^+ and NH_3 in water is displaced towards the formation of NH_3, which is very toxic to fish, when pH increases (Fig. 1.3). Water which is not toxic at pH 6.7 can kill animals at pH 8.

pH equilibrium is dependent on other interactions, mainly on the CO_2-carbonates system and pH governs the carbonate richness of waters. pH modifies various forms of CO_2 (see Frontier and Pichod-Viale, p. 71, 1991).

1.1.5 Specific density

Oceanic water has a specific density 800 times greater than that of atmospheric air. The temperature and salinity of a body of water determines its specific density: the presence of dissolved salts results in the higher density of sea water (1024 g/l) than

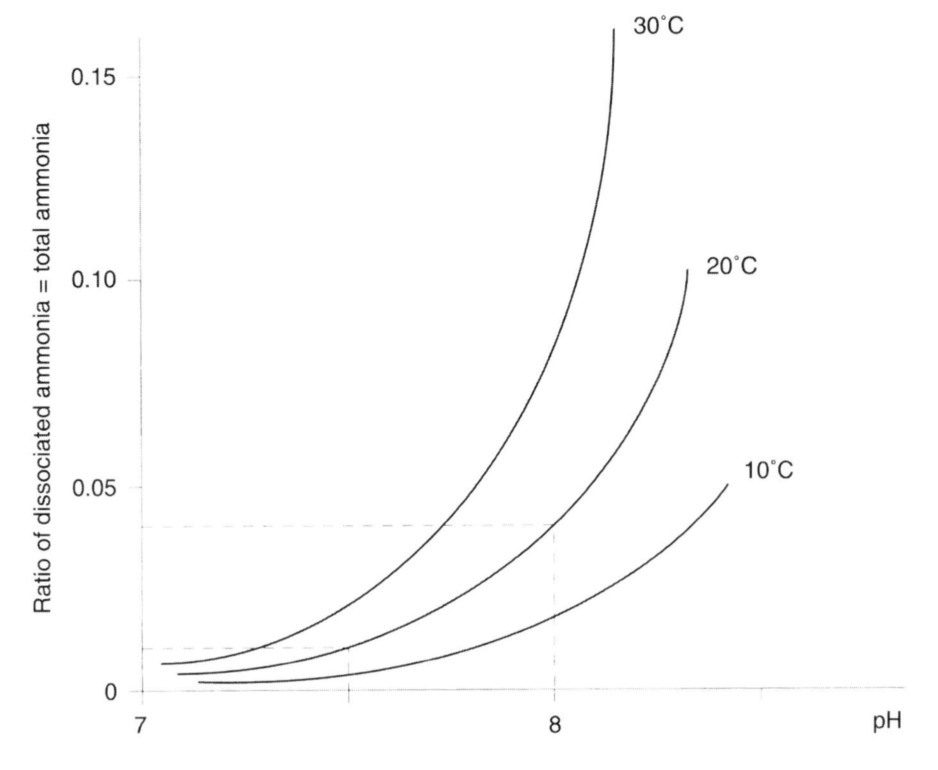

Fig. 1.3. Variations in the proportion of dissociated ammonia in relation to pH (data from Frontier and Pichod-Viale, 1991), and temperature.

that of freshwater (1000 g/l), which in turn affects the buoyancy of planktonic animals and also water stratification; when large water masses are involved, waters of different salinities do not mix; the denser water tends to lie below the less dense. Water is densest at 4°C; cold water therefore tends to lie at the bottom, below warmer waters. The result is a layering effect, a stratification of water layers which do not mix (see colour section, plate 1).

In the sea, in summer, the surface waters are warmer (and thus lighter) than the deeper waters, from which they are separated by the thermocline, a very thin, intermediate layer of water ($\leq$1 m). This is most often situated at a depth between 20 m and 50 m (but sometimes 200 m). Its presence is of great interest: it can separate water layers whose temperatures differ by 5–10 degrees. The thermocline sometimes constitutes a real barrier for aquatic animals which do not cross it. In winter, movement caused by storms, currents and tides leads to mixing, at least to a depth of 40–50 m; this is the "mixed layer" which sometimes extends as far as 200 m with a temperature similar to that at the surface.

In estuaries, freshwaters spread out over salt waters, which can ascend the water course for several kilometres resulting in specific phenomena such as a saline wedge.

1.1.6 Specific heat

This represents the amount of energy required to increase the temperature of a mass of substance by 1 degree (the calorie is the energy required to increase the temperature of 1 g of water from 14.5–15.5°C). Specific heat is thus 1 cal/g/°C or 4.186 J/g/°C. This value rises in particular circumstances, but hardly varies with the salinity or temperature of the water.

In practice, this shows that water is one of the best known accumulators of heat and, for this reason, it is used both in dwellings and vehicles as a means of heating as well as cooling. This characteristic allows the ocean to play the role of "heat sink" in the climate (see Section 1.4 below).

Sudden changes in temperature in aquatic environments are thus cushioned. The thermal stability (and also the physical and chemical stability) of the marine environment in relation to the biological environment is the reverse, for example, of terrestrial habitats. For the manager, this indicates that a large amount of energy will be required to alter the temperature of an aquatic rearing system.

1.1.7 Hydrostatic pressure

An increase in depth of 10 m is accompanied by an increase in pressure of 1 bar (or atmosphere). This characteristic does not affect aquatic species although anomalies in stability (probably linked to the swim bladder), have been noted in trout placed in submerged cages at depths of several tens of metres. In contrast, human diving is very dependent on this phenomenon, which influences the build-up of gases in the tissues and limits dive durations.

1.1.8 Methods of analysis

Analytical methods used for sea water are described in the work of Aminot and Chaussepied (1983). Note that the measurement of concentration of certain constituents must be carried out using different methods (e.g. ammoniacal components) to those used for freshwater. Various manufacturers produce diverse and sophisticated equipment for the assessment of the main parameters in marine ecology and development: while these tend to avoid long and complex analytical methods, the tendency is to measure an increased number of parameters.

1.2 EARTH, OCEAN AND PLANET

Using the many physical, geological and geographical oceanography atlases and other works, we have put together data on the general characteristics of the oceans which, in view of their size, can be taken as characteristics of our planet itself.

1.2.1 General characteristics of oceans

The total surface area of the earth is estimated to be 510 million km^2, the oceans covering 361 million km^2 of that, i.e. 70%. The volume of marine waters is estimated at 1.33 thousand million km^3 or 1.33 billion km^3, i.e. 1/800 of the total volume of the earth and 97.5% of the total water present on earth. Salomon (1994) calculated that, on average, each inhabitant of the planet has at their disposal 0.2 billion m^3 of marine water, corresponding to a cube with 625 m sides.

The ice caps represent 1.8% of the total water (25×10^{18} litres). Underground reservoirs represent 0.6%, i.e. 8.4×10^{18} litres, while lakes and rivers with 200×10^{15} litres, represent only 0.015%. Water in the ground (66×10^{15} litres) and atmospheric water vapour (1.3×10^{15} litres) represent only 0.005% and 0.009% respectively.

From the shore, the seabed usually starts by sloping gently (less than 1°) in the form of a continental shelf, often several tens of kilometres wide (70 km on average). These shelves cover 5.7–8%, even 9.9% of the surface of the world's surface, depending on author. They thus represent 30–40 million km^2. In its simplest form, we imagine a shelf's surface to be bare, like that of the moon. Figure 1.4 shows the main subdivisions of the marine environment.

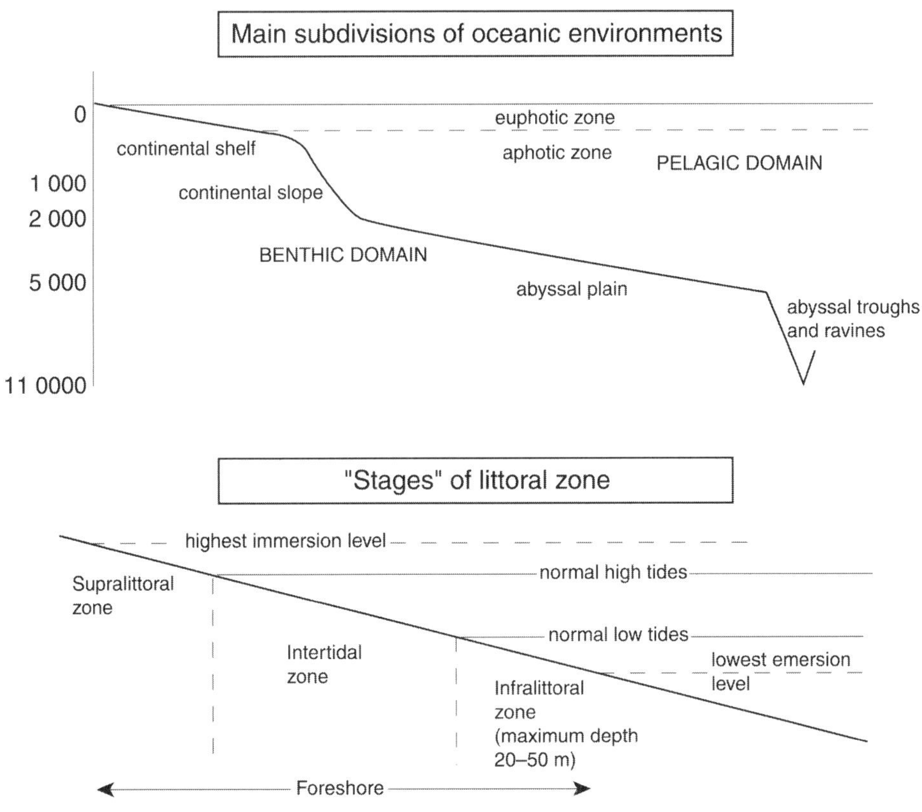

Fig. 1.4. Subdivisions and stages.

Continental shelves, potential areas for developmental activities, correspond to portions of continents which were submerged during recent geological periods; they extend the continental form into the sea and are therefore narrower when mountainous areas plunge straight down into the sea than in gently sloping continents where they constitute vast underwater plains. Tropical seas generally have narrow plateaux, while those of Europe are wide.

Waters which extend around the land to a depth limit of 200 m are called coastal or neritic waters.

Ninety percent of the exploitation of the marine environment is concentrated on continental shelves for several reasons: being close to land makes them more easily accessible and they are biologically rich, as the factors controlling production of living matter are favourable (shallow, mixing of waters, nutrient salts, see Chapter 4). One-quarter of the primary production of oceans occurs in coastal zones, on the continental shelf.

At a depth of 150–200 m, a steeper continental slope leads to deeper zones. In ecological terms, this continental slope separates the neritic waters (corresponding to the continental shelf) from those of the open ocean.

The ocean has a mean depth of 3800 m. The relationship between mean depth of oceans and their mean width, i.e. 4000 km, is a factor of 1000. In fact, deep zones (abyssal troughs and plains) situated below this depth represent the majority (64%) of the world's surface. The general profile of mean sea depths is shown in Fig. 1.4.

In oceanography, however, "averages" should be avoided, as situations such as the scale of a continent are often very different; the northern hemisphere, for example, is represented 40% by land and about 60% by sea, while these proportions change to 20% and 80% respectively in the southern hemisphere. Considering Salomon's (1994) estimate of the quantity of sea water "available" per inhabitant of the planet, an inhabitant of northern Europe would only have a ten-thousandth of the mean predicted value, i.e. 40,000 m^3, although such estimations are only illustrative.

Table 1.3. Global data relevant to aquatic environments.

	Surface area (millions of km^2)	Volume (millions of km^3)
Salt water		
Atlantic Ocean	106	354
Pacific Ocean	180	720
Indian Ocean	74	291
Total	360	1365
Fresh water		
Lakes	0.70	0.10
Glaciers	18.00	30.00

1.2.2 Spatial and temporal scales in the ocean

On the one hand, the marine ecologist must consider bacteria, which can be smaller than one micron and, on the other, the oceanic basin, the width of which can be greater than 10,000 km. In terms of time scale, the circulation of large thermohaline currents which circumnavigate the world can take 1000 years; on the basin scale, main currents can take several years, while small disturbances can have a duration of seconds. The size of the organisms roughly determines the time scale of their life cycle: whales can reach 100 years old, teleost fish generally between 1 and 10 years and zooplankton several days or weeks. Phytoplankton double in numbers in a few days, bacteria in a few hours. This results in small organisms with greater fluctuations in abundance than larger ones. Phytoplankton and the smallest animals constitute about one-half of the biomass of the ocean.

1.3 THE OCEAN AND THE ATMOSPHERE

1.3.1 Solar radiation

Solar radiation is the basis for all life on earth and in aquatic environments, since it determines the process of production of living organisms by photosynthesis. Apart from this function, solar radiation plays a major role in the thermal budget of the seas and thus on the climate.

The calorific action of the sun exerts itself on the atmosphere, then on the oceans or continents; the upper part of the atmosphere receives around 340 watts/m^2 (7.2×10^6 cal/m^2/day), but only 173 watts reach the level of the sea, the rest being absorbed by atmospheric gases or reflected into space. In practice, the numbers vary according to latitude: the sun provides 0–256 cal/cm^2/day to the poles, 379–447 cal/cm^2/day at the equator, 157–480 cal/cm^2/day at a latitude of 30° and 0–410 cal/cm^2/day at a latitude of 60°. Note that the intertropical zone represents two-fifths of the planet.

The upper atmosphere absorbs most ultraviolet radiation as a result of the ozone layer, which plays an important role in blocking UV; without it, animals and plants would not survive. The ozone layer is thinner in the tropics than elsewhere and so ultraviolet radiation is stronger there.

Total solar energy received by earth is enormous, of the order of 5×10^{20} kcal (1.73×10^{17} W) per year. The oceans receive 71% of this radiant energy; at average latitudes it is estimated that this represents 9–10 billion kcal/ha/year, i.e. 1 million kcal/m^2/year (Duvigneaud, 1981). Only around 42% will be utilisable, as absorption by the atmosphere, reflection at the surface and, above all, the differential absorption of active photosynthetic radiation (APR) by the waters, lead to the restriction of the potentially productive water layers to about 15–100 m, depending on turbidity. Only half of this 42% is utilisable for photosynthesis, as photosynthetic pigments mainly absorb blue and red wavelengths.

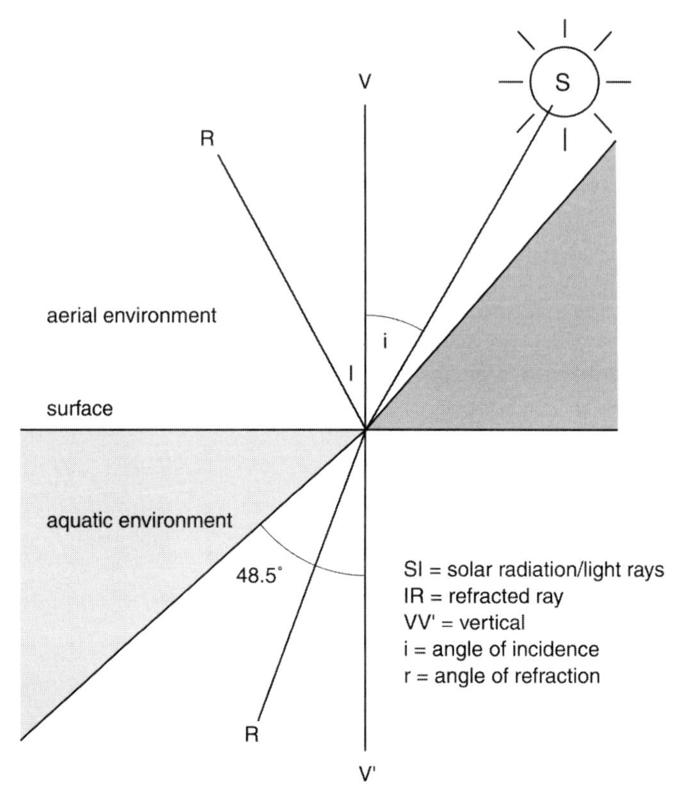

Fig. 1.5. Light refraction at the water's surface.

The route followed by a ray of light reaching the surface of an area of water is shown in Fig. 1.5; if the sun is at an angle of less than 48.5° in relation to the horizon, it will be reflected at the surface without penetrating into the water. At a greater angle, it will penetrate the water, but its direction is altered; this angle of refraction (r), is calculated from the angle of incident light rays (i) with the vertical: $\sin r = \frac{3}{4} \sin i$. Day length is thus much shorter in water than on land.

The surface of aquatic environments is rarely flat; waves may reflect most of the light rays. While it has been proved that variations in light levels lead to increased photosynthetic efficiency, there is nevertheless a significant loss in quantitative terms.

As on land, radiation intensity also varies with latitude, season, time of day, cloud cover, etc., but its penetration is also subject to other parameters:

- the transparency of water is very variable in relation to particles in suspension (including plankton). Relative transparency can be estimated using a Secchi disc (white, 30 cm diameter disc); the depth at which it is no longer visible is measured. Alternatively, a spectrophotometer can be used to measure absorption of light travelling through a sample of water. Chinese aquacultur-

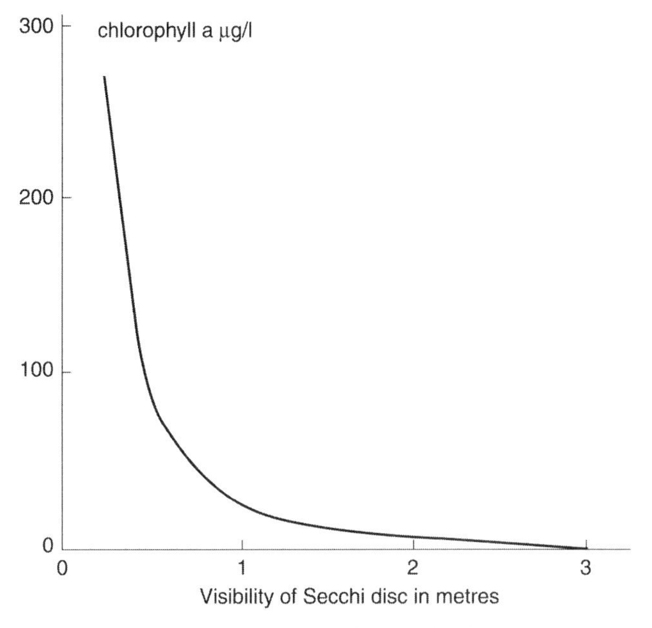

Fig. 1.6. Relationship between visibility of Secchi disc and chlorophyll a concentration.

ists dip their forearms into the water in their ponds to assess the density of algal plankton which colour the water green. Figure 1.6 (after Almazan and Boyd, 1978) relates the disappearance depth of the Secchi disc to the plankton chlorophyll content of the water.

In the Vigo Ria, Fraga (1979) demonstrated a 26% decrease in productivity (assessed by plankton concentration), linked to a decrease in transparency resulting from particulate material in suspension created by human activity (these particles absorb and/or reflect light), constituting an effective filter. Oceanic waters are much more transparent and are described as blue waters.

- Water coloration, either through import of material from the land, or by planktonic organisms, reduces the penetration of solar radiation. In coastal waters "yellow-green substances", related to humic acids, slightly colour the water, but this phenomenon must be distinguished from planktonic blooms which produce a very marked red or brown colour, and are known as "red tides". This "marine humus" is in fact made up of organic molecules (dissolved organic matter). Figure 1.7 (after Pérès, 1976), demonstrates the absorption of light in relation to the depth of coastal waters and of the oceans. In eutrophic zones or in algal cultures, absorption of light by phytoplankton is much more rapid, as their presence limits the transparency of the water to less than 1 m.

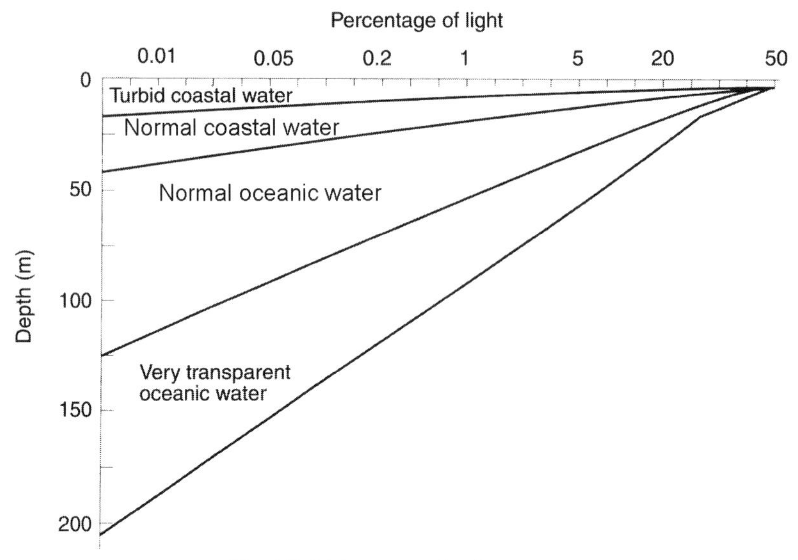

Fig. 1.7. Light penetration into water.

- The differential absorption of various colours and wavelengths of radiation results in an underwater domain dominated by blue, the radiation which penetrates furthest (Fig. 1.8).

In relation to light penetration, one can define a euphotic zone (water layer extending to depths receiving less than 1% of the light intensity at the surface) in which photosynthesis occurs. It extends from the surface to a depth between 20 and 100 m. The oligophotic zone (400–600 m) is only penetrated by blue-violet radiation, deeper zones remaining dark (aphotic). Note that littoral waters fall within the euphotic zone.

Ultraviolet radiation which has not been absorbed by the ozone layer in the upper atmosphere is absorbed in the first centimetres of water, to which it gives up part of its heat. Infrared radiation (rays with the highest calorific value) is absorbed partly by the atmosphere, but mostly by water. As a result of its high specific heat, water has a great capability for absorbing this type of radiation.

1.3.2 Heat gains and losses

Heat gains and losses vary in relation to latitude and season. The absorption of light radiation also varies depending on season (Fig. 1.9): it occurs in the first few metres of oceans, but heat losses occur across the first centimetre; they occur through evaporation, infrared radiation and conduction (all these phenomena are increased by wind). The ocean stores 90% of the heat accumulated by the climate. Heated to 30°C in the tropics, surface water, transported by currents, transfers this heat to high latitudes.

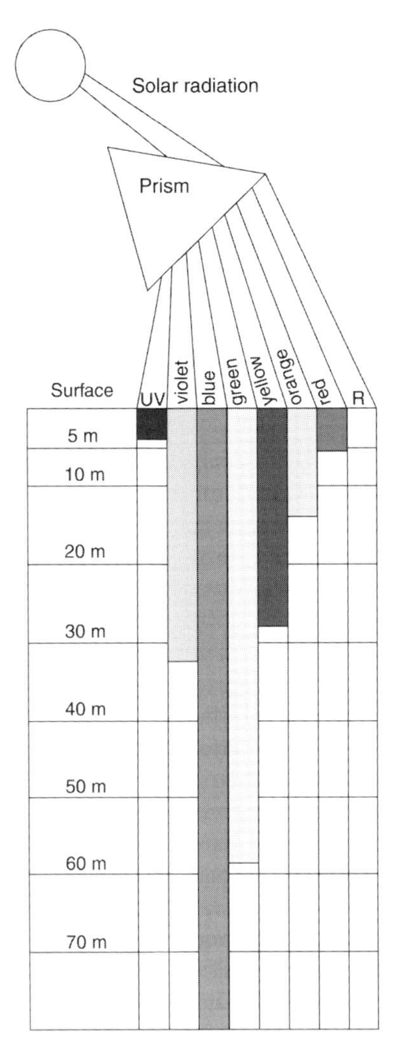

Fig. 1.8. Absorption of different wavelengths of solar radiation.

Mann and Lazier (1991) provide the following example: during one day in July in the North Atlantic (40°N and 0°W), around 22,500 kJ/m^2 of energy was absorbed by the ocean, losses being about 10,400 kJ. If the mixing layer is 5 m deep, 76% or 17,100 kJ would be absorbed by this layer (in relation to the differential absorption of the components of the solar spectrum, the remainder being absorbed below 5 m in depth). In deducing the losses, the energy gain δQ in this layer is 6700 kJ/m^2. This gain must be divided by the mass of water, that is 5000 kg, and by the specific heat (4.2 kJ/°C). The gain is about 0.3°C. The temperature does not decrease gradually because mixing due to waves only affects the water to a certain depth, which is a function of this agitation: it is the "mixed layer" (see Chapter 2).

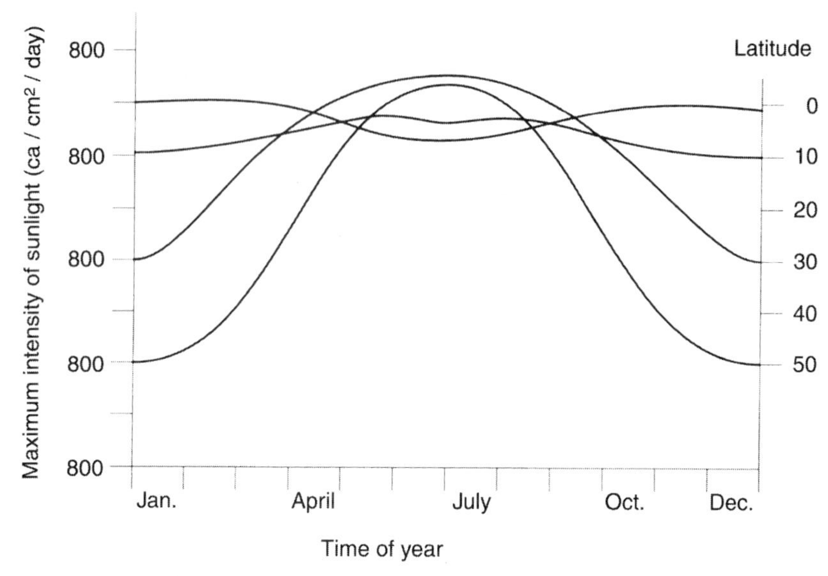

Fig. 1.9. Maximum intensity of sunlight in relation of latitude and month.

1.3.3 Consequences for ocean–atmosphere interactions

Thermal radiation results in the differential heating of air, land and water. Thus, the atmosphere and the ocean do not have a constant relationship because they are subjected to numerous phenomena: the incidence of solar rays means that the atmosphere, and thus the solid planet, receives more heat at the equator than at the poles. This leads to the formation of masses of air of different densities which move relative to each other, leading in turn to either steady or irregular winds (trade winds, monsoon, etc.) This radiation varies according to the angles of both the sun and the earth (which determine the seasons) and also between day and night, but also depending on cloud cover, abundance of tropospheric aerosols, etc.

As a result of the high specific heat of water, the ocean absorbs the majority of the heat in the hot equatorial zones. The volume of the oceans (1320 million km^3, 1/800 of the volume of the planet), confers upon them a prodigious thermal capacity and they play a guiding role in the regulation of the climate: its temperature only varies between -2 and 30°C, whereas on land it can vary by 80°C. The ocean stores 90% of the heat accumulated by the climatic system. It also plays a role as transporter and it is estimated that it contributes at least half of the transfer of heat moved from the equator to the poles, via currents (Chapter 2), of which the Gulf Stream is the best-known example. Warm water from the tropical west Atlantic is moved towards northern Europe. This water warms up the west winds, which thus make regions of Europe milder in relation to American regions at the same latitude.

Transport of energy by atmospheric and oceanic circulation is similar; Morel (1992) talked of the "regulated system" to describe this interaction. This regulation is never perfect; there are oscillations around the equilibrium, which results in

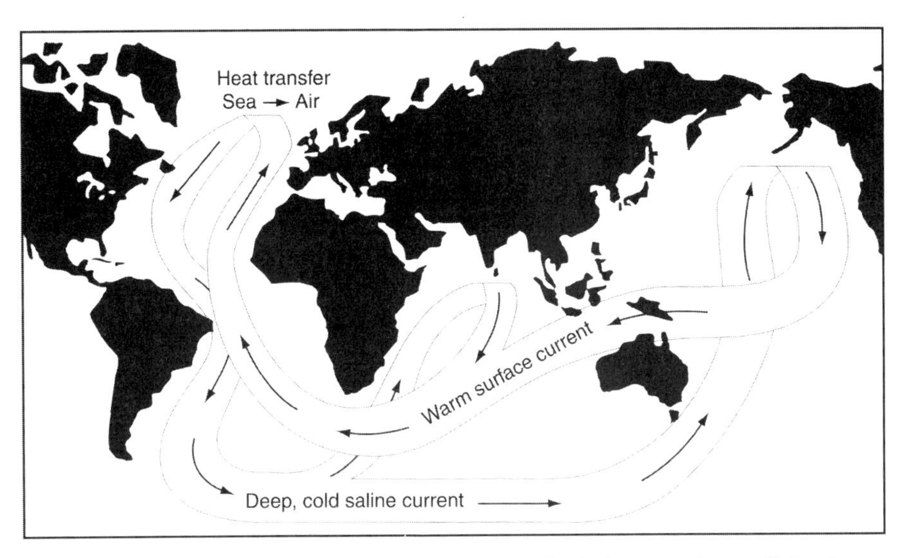

Fig. 1.10. Diagram explaining the concept of thermohaline circulation, sometimes called a "conveyor belt".

interannual variations. Figure 1.10 represents, in diagrammatic form, the general circulation of the main marine currents in the oceans (Anon., 1990): we have mentioned the Gulf Stream and its effect on the European climate, but the waters which constitute it, after being cooled while heating the atmosphere, become denser and descend to the depths, resulting in a deep, cold current (5°C) which circulates from north to south in the Atlantic. The same current passes close to Antarctica. Thus a sort of conveyor belt transports warm (and less dense) waters at the surface and denser, colder waters in the depths.

Sometimes, fluctuations in the ocean–atmosphere interactions in the intertropical zone become irregular and result in interannual climate variations (for example El Niño and the Southern Oscillations designated by the name ENSO); it is similar in the polar regions (interaction between ocean–atmosphere and ice) which in turn has an effect on the climate.

1.3.4 Precipitation, evaporation and atmospheric transport

It has been estimated (Morel, 1992) that the oceans receive an average annual rainfall of 1.03 m, while 1.11 m is restored to the atmosphere through evaporation. These figures differ noticeably for the continents (rainfall 0.60 m, evaporation 0.40 m). Evaporation, caused by the warming of the ocean in a warm region, generates water vapour which is taken via convection into the atmosphere; the ocean is the main source of atmospheric water vapour. This vapour can recondense into droplets which, in turn, warm the atmosphere. This succession of interactions, the turbulent character of the atmosphere or marine currents, demonstrates their complexity and that of the resultant climatic fluctuations. Each cubic centimetre of evaporated water represents a thermal loss for the sea of 2400 joules.

The warming of the atmosphere increases the temperature of the superficial layers of the ocean by 1–2°C, to which ecosystems respond: coral reefs or mangrove swamps can thus extend and the flat oyster has extended its reproductive area on the north coast of Brittany in France. In contrast, too great an increase in temperature appears to be the root cause of bleaching of corals.

Winds blowing across the oceans have very diverse effects, notably at a climatic level; at the hydrodynamic level, their main action is to create the waves and currents studied in Chapter 2.

In addition, winds transport very fine particles ($<50\,\mu m$) over long distances, constituting loess. Thus sand lifted from the Sahara by the *harmattan* fertilises the Atlantic as far as the Caribbean. Similarly, industrial lead released into the atmosphere in northern Europe is found several days later in the Mediterranean (Frontier and Pichod-Viale, 1991); what illustration can these be of the global nature of these phenomena?

Marine aerosols, produced from the ocean's surface, fertilise continents in the same way: over a 10-year period, a poor soil will acquire all the necessary elements. French forests located on poor soils are fertilised in this way: aerosols made up of particles less than $1\,\mu m$ in diameter are carried by winds and clouds to the hearts of continents. Tens of thousands of m^3 of water fall on one hectare of land each year, representing about 20 kg of elements (Labeyrie, 1994). Aerosols can form at sea when wind speed exceeds 11 km/h. We shall see (cf. Section 1.5, below) that aerosols can also influence the climate.

Rain can have many repercussions for the ocean: as well as the desalination of the waters, which can eliminate marine species from normally salt, shallower waters, it can lead to the proliferation of pathogenic bacteria produced in freshwaters, which do not survive in salt waters (the sale of cockles was prohibited from the Étang de Thau in 1989 because of the presence of *Salmonella* bacteria). Acid rain leads to nitrogenous and sulphuric compounds of aluminium, etc. (Solbé, 1988); no water body is therefore "virgin". Rains can also cause plankton blooms and lead to the deposition of fertilisers in the coastal zone which, in turn, has an influence on fisheries and aquaculture as they can induce better larval survival or faster growth of shellfish. This is an example of the interactions between climate, human activity and marine life.

1.3.5 An example of ocean–atmosphere interaction: "El Niño" and "ENSO"

The Peruvians habitually fish the Pacific waters off their coasts for huge quantities of anchovies (around 15 million tonnes, i.e. 18% of the total world fisheries catch). This amazing fishery is the consequence of a hydrological phenomenon created by the climate: in Peru, the trade winds from the east blow strongly and carry the coastal waters westwards with them, towards the centre of the Pacific. These surface waters, which drift towards the open sea, lead to the upwelling of deep, colder waters, which are also richer in fertilising elements (nitrogen and phosphorus) than the surface waters which they replace. These rich waters, subjected to strong sunlight at the surface, form the basis of high planktonic production (see Section 4.4), which is exploited by planktonivores such as the anchovy. The proliferation of anchovy in these waters is thus a result of both hydrological and climatic events.

These surface waters are gradually reheated by the intense Pacific sun and by the trade winds which carry them eastwards. They concentrate in an immense mass of warm water—a sort of "reservoir" more than 100 m deep and with a temperature over 28°C, close to Indonesia (Fig. 1.11a).

Thus, there are warm waters in the western Pacific Ocean and cooler waters in the east; these thermal contrasts reinforce the trade winds, since the mild air above the reservoir of warm water is associated with lower barometric pressures than the mass of cooler air. Crossing the Pacific, these air masses will also warm and their density will thus decrease. Gradually, and especially above the reservoir of warm water close to Indonesia, this air lightens and fills with water vapour, creating cumulo-nimbus clouds which rise to a height of 18 km. The reciprocal influence of the atmosphere

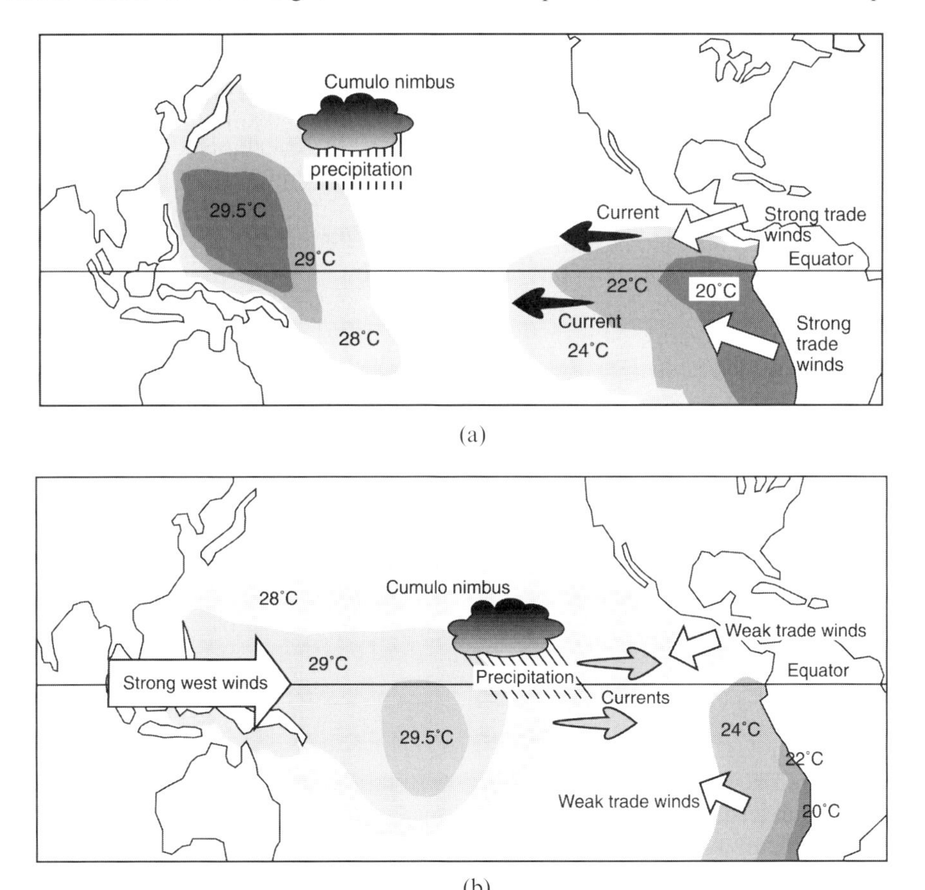

Fig. 1.11. (a) Normal situation (strong trade winds and upwelling). (b) Situation during El Niño (weak trade winds and strong west winds).

and surface waters constitutes a coupling of the ocean and atmosphere, which maintains the trade winds and the thermal contrasts between the East and West Pacific. This convective system is called "Walker's circulation".

In some years, the trade winds become weaker, the temperature of surface coastal waters increases by 3–5°C and the direction of surface currents is reversed (Fig. 1.11b); soon afterwards, the anchovy catches collapse. These phenomena usually begin around Christmas and the Peruvians have associated them with the infant Jesus, whence the name El Niño (infant Jesus in Spanish) is attributed to these occurrences. This periodic anomaly occurs on average every 3–7 years.

The increase in surface temperatures and the reversal of marine currents which characterise the El Niño phenomenon originate from the abnormal weakening of the east trade winds (usually they blow quite strongly across the Pacific): warm waters are no longer carried to the west towards Indonesia, but stagnate in the middle of the Pacific; this completely changes the climatic process. There is no longer an upwelling of cold water along the coasts of Peru, no more rich waters; hence there is less plankton production and so reduced anchovy production: the fishery can fall by 80%.

El Niño, an oceanographic phenomenon, is thus coupled with a climatic phenomenon called the Southern Oscillation. The two, El Niño and the Southern Oscillation, are known collectively by the abbreviated name ENSO. The term El Niño is used in the media but it actually defines the whole assembly of ENSO.

The weakening of the trade winds is accompanied by an increase in the winds from the west that originate from Asia, which blow from the other side of the Pacific, pushing warm waters back to the east. The explanation of this anomaly of climatic function is still only hypothetical; given that convective clouds forming in the western Pacific mostly release their water there in the form of rain (2 m/year), this mass of freshwater, less dense than salt water, will float for a long time on top of marine water, forming a sort of isolated and super-heated mattress (temperature greater than 30°C). This huge mass of warm water transmits its heat and humidity to the atmosphere, creating convective clouds (cumulo-nimbus type), which modify the atmospheric circulation and the normal pattern of rains.

In the absence of east trade winds, the west winds drive back these superheated waters and this zone of convective clouds to the east, towards the centre of the Pacific. Although still only a hypothesis, it is thought that the salinity of the ocean is the determining factor of a climatic phenomenon. However, the important effect of the ocean on climate and the multiplicity of interactions (and doubtless retrospective effects) which link all these phenomena are clearly demonstrated.

El Niño and ENSO have consequences not only for the anchovy fishery. The weakening of the trade winds, the cause of which is unknown, has many other consequences for the climate, as all climatic "cells" (water and air masses) interact and climatic disorders are felt across the entire planet: the monsoon is upset in India, there are cyclones in Polynesia, a drought in the Caribbean, India, Australia and rains in Southwest USA.

In health terms, El Niño favours the spread of disease as a result of the proliferation of mosquitoes due to the high precipitation and warming: this is the case with dengue fever (tropical fever caused by an arbovirus) and also viral encephalitis transmitted by mosquitoes in Australia. The cholera vibrio, which can survive for 50 days in sea water spreads in the same way.

The El Niño phenomenon from 1997–98 was the 22nd this century and the most devastating, as it lasted longer and was more severe than its predecessors: a high rise in sea water temperature from spring 1997 was a first indicator of its severity. The resulting drought in Indonesia led to devastating fires at the end of 1997 which in turn created atmospheric pollution over the whole of Asia.

The warming of surface waters in the Pacific, created by El Niño, seemed to spread across the entire planet, particularly in southern Europe. The year 1997 was the warmest known in France, with a mean temperature of 17.3°C, 1.5°C greater than the mean recorded for the years 1961–1990. In New York, the temperature in Central Park was 19.8°C in the middle of January.

There are also other climatic fluctuations, one of which can be detected in the north Atlantic: its periodicity is around 10 years and it is due to an interaction between the atmosphere and the north Atlantic. Temperature anomalies can thus be followed over 10 years, in the course of their slow progression away from the main current of the Gulf Stream, from American to European coasts (Reverdin, 1998). This is known as the North Atlantic Oscillation (NAO).

1.4 THE GREENHOUSE EFFECT AND THE OCEAN

The greenhouse effect is a natural phenomenon without which the surface of the earth would have a mean temperature of −18°C; some atmospheric gases absorb thermal radiation and, as a result, the mean global temperature is around 15°C. Modification of the composition of the atmosphere would simply accentuate the greenhouse effect. "Global change" is the term used to describe actual climate modifications which cannot be attributed with absolute certainty to man's activities or to natural changes, although each passing day brings new proof of the effects of human activities.

As a result of these activities, the composition of atmospheric air is modified and its carbon dioxide content has increased by 0.5% per year (i.e. 1.5 ppm) since 1958, the year in which these measurements were started. In one century, the emissions of CO_2 have increased by 30%, those of methane (CH_4) by 100%, those of chlorofluorocarbons (CFC) by 25%, the same increase as for nitrous oxide (NO). This increase in atmospheric CO_2 is inexorable and is predicted to double in the course of the next 70 years. Human activity will produce 20 billion tonnes (gigatonne: Gt) of CO_2 per year, i.e. 5.5 Gt of carbon. According to various sources, the USA has contributed to 17% of the greenhouse effect, just ahead of the European Union (EU) (14–15%), CIS (12%), Brazil (10.5%), China (7%) and India (4%). Brazil's position is explained by the annual destruction of 4.8 million hectares of Amazonian rainforest. Total deforestation, more significant than had been thought, is estimated at between 160,000 and 200,000 km^2/yr. Since the start of the century, one-fifth of the forested areas on earth have disappeared.

A report on the development of climate, based on a large number of scientific studies on the subject, was published by Greenpeace (Leggett, 1994).

Simultaneously, the mean temperature of the atmosphere is slowly increasing, although not all authors agree on the amplitude of this increase in the future (from 1.7°C to 4.4°C from now until the year 2050). Although mean global temperature was lower between 1991 and 1993, this was attributed to the eruption of the Pinatubo volcano in the Philippines which released thousands of tonnes of dust and aerosols into the upper atmosphere, slowing down the warming-up process. However, the long-term trend is inescapable and the mean temperature has increased by one degree since 1850, the warmest years being 1990 and 1994, and especially 1997. Eight of the ten warmest years in the past 140 years (period over which data are available) have been recorded since 1980. The increase in temperature is greatest on the continents and especially in winter. Rains and droughts have become more severe. In Siberia, the increase in temperature has been 2°C over 100 years, but reading the annual rings of spruce trees which are several centuries old shows that the additional increase in carbon dioxide does not increase their growth.

The Norwegian Centre for the Environment confirms that 1.4% of the ice cap has melted in 10 years. An ice-cap loss of 4.3% has been measured in the Antarctic; Alpine glaciers have decreased by a third since 1850. British fishermen have caught hammerhead sharks, normally a tropical species, and trigger fish, warm water fish which usually arrive in the Bay of Biscay in the summer, are now fished in the North Sea!

The meteorological disturbances have resulted in severe floods since 1990 (The Netherlands, Germany, lower reaches of the Mississippi), severe droughts (Sydney surrounded by flames), wild storms in Northern Europe (Braer ran aground in Scotland) and cyclones in the Pacific; in the Pacific, corals are bleaching in waters that have become too warm, while in the Antarctic, a $2600\,km^2$ iceberg detached from the ice field, thus proving the accelerated thaw.

A link between CO_2 concentration in the air and mean atmospheric temperature has been established for the past 1,600,000 years and the relationship has been proved to be reliable and long-term. The scientific community attributes the temperature increase and climatic disturbances to the greenhouse effect, created by the increase in the concentrations of CO_2 and other gases (Fig. 1.12). This state of affairs is new, although the experts do not all agree on the details of the causes. Certainly, the increase in temperature during the 20th century has only been 0.5°C, but this has been the fastest rate of increase in the last 10,000 years.

This correlation between CO_2 concentration and temperature is linked to the greenhouse effect: it is known that this gas, along with water vapour, inert gases, methane, nitrous oxide and CFCs, absorbs infrared heat radiation passing through the atmosphere. These gases are present at extremely low concentrations in comparison with CO_2 ($1.7\,cm^3/m^3$ of air for methane), but absorb much more thermal radiation and could be responsible for up to 40% of the greenhouse effect.

Mean production of CO_2 per human is about 1.1 tonne/yr (0.4 for an African, 5 for a North American), i.e. close to 6000 billion tonnes/yr for mankind. CO_2 concentration increases by 0.5%/year. There are also about a billion cattle on Earth,

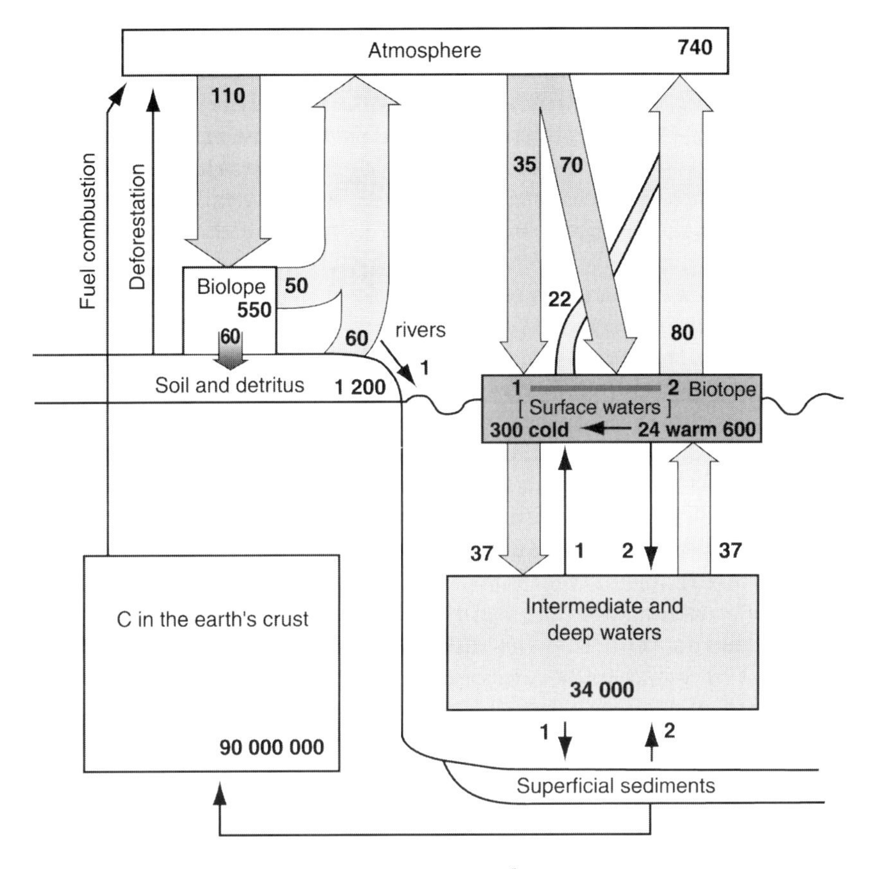

Fig. 1.12. Global carbon cycle (values in 10^9 t of carbon) (after BOFS).

each one producing 40 litres of methane daily! The heat thus trapped by these gases increases the mean air temperature.

Huge masses of marine water are warmed much more slowly, but Bethoux *et al.* (1990) noted an increase of 0.12°C between 1959 and 1988 in the temperature of deep Mediterranean waters, the temperature of which had been stable since the start of the century. They attributed this phenomenon to the warming of the air and the superficial layers of this sea by infrared radiation from the atmosphere, to the benefit of the sea (greenhouse effect): the surface and deep waters are mixed together over a period of 15 years, which explains the cumulative trend of the increase. The warming of surface waters could be between 0.6–0.7°C (Boudouresque, 1995).

A study carried out on the long-term (60 years) changes in the intertidal fauna of a rocky shore in California was made by Barry *et al.* (1993). They noted an increase in the abundance of eight of the nine southern species and a decrease in abundance of five of the eight northern species. The mean annual temperature of the ocean increased by 0.75°C and the means of the maximum temperatures were 2.2°C higher between 1983 and 1993 than during the period 1921–1931.

These increasing trends are not universal. Colbourne *et al.* (1994), studying climatic variation and environmental conditions in the north-west Atlantic (Canadian continental shelf) between 1970 and 1993, noted a trend towards cooling in this region, which has been very marked since 1990. It resulted in greater deposits of ice and a reduced thaw, a thinner and colder mixed layer between the waters, and a lower heat content of the water column. These phenomena were associated with colder winter air temperatures, 1991 being the coldest year for 30 years.

The greenhouse effect is not the only result of man's impact on the atmosphere: we also contribute to freon production, including chlorofluorocarbons (CFCs) produced by human industrial activity (used in refrigerators) and spread out in the atmosphere where they play a role in the destruction of the ozone layer which absorbs UV at high altitudes. The concentrations of these gases, despite international agreements limiting their use, continue to grow by 5–7% per year. In November 1994, data from the space shuttle Atlantis confirmed the theory that the ozone layer is very thin and that the winter of 1994–1995 would break the record for having 30% less ozone.

The passage of UV through the "ozone holes" which appear above continents could be fatal for terrestrial vegetation and phytoplankton, but its main effect will undoubtedly be on the more frequent development of cancers in man exposed to sunlight. The increase in ultraviolet radiation has already reduced the amount of phytoplankton by 6–12% in the area situated below the Antarctic hole, according to data reported by Weber (1994).

This development is not inevitable. The use of methyl alcohol (methanol) in car engines would slow the formation of ozone (a model achieved by the city of Los Angeles) and a new Canadian process allows the extraction of sulphur dioxide from the smoke from power stations, which is known to be responsible for acid precipitation.

Another way in which the greenhouse effect affects both the climate and the oceans is reported by Bakun (1990) on the west coast of the United States. The greenhouse effect intensifies the impact of winds on the ocean along the length of the coast, which leads to an acceleration in the rate of coastal upwelling: thus the data suggest that large upwellings have increased in intensity while the greenhouse effect gases have accumulated in the atmosphere. As a result, fresh and misty summers, which characterise the coasts of northern California and other regions with similar resurgence, are becoming increasingly pronounced and marine breezes have become stronger. The effects on the oceanic ecosystem are not clear, but the distribution of oceanic temperatures may have been affected, while an increase in primary production would result in accelerated circulation of atmospheric CO_2, counterbalancing the greenhouse effect.

The greenhouse effect is also partly slowed down by aerosols and dust in the high atmosphere, not all of which originate from the ocean:

- As well as dusts, volcanoes emit aerosols, mainly formed from sulphuric acid. While the effect of aerosols on the transmission of sunlight is not entirely understood, it is known that they constitute a diffusing layer which reflects part of the solar radiation back into space. After the eruption of Krakatoa in 1883 and the Agung in 1963, a cooling of the atmosphere was noticed over the subsequent 1–2 years.

- Human industrial activity also emits many aerosols, mainly sulphur dioxide (SO_2), produced from fossil fuels which oxidise into sulphates in the atmosphere. According to Hansen and Lacis (1990), this phenomenon could explain the fall in temperatures in the northern hemisphere between 1940 and 1970 (purification of fumes has since progressed). These sulphates play a role in the formation of acid precipitation.

The greenhouse effect, as described above, results in an increase in mean sea level, which has been estimated using various models. The increase in this level has been 15 cm since 1900. It is actually estimated at 1.5 mm/year, but has not shown any recent signs of acceleration (Huet, 1994). The Topex-Poseidon satellite indicated an increase of 2.6 mm in mean sea level between 1993 and 1994, simply from the expansion of the waters as a result of warming. As waters become warmer, their capacity to fix CO_2 decreases. In effect, 80% of the world's coasts are receding. In Aquitaine (France), the coast loses an average of 1.5 m/year, i.e. nearly 75 m since 1946. The rise in the oceans is estimated at between 15 and 95 cm, with a mean of 50 cm for this new century.

Other phenomena are undoubtedly involved and the increase in solar activity has been proposed as an explanation for global warming.

1.5 CONSEQUENCES OF GLOBAL WARMING

The increase in the relative level of the sea can have numerous consequences for coastal zones (Kjerfve, 1991):

- flooding of the littoral zone (an increase of 1 cm corresponds to the loss of 1 m of beach);
- more rapid erosion of beaches;
- increase in salinity of ground water, rendering it unfit for human consumption;
- ingress of salt water into estuaries;
- destruction of terrestrial vegetation and its replacement with halophilic species.

Large numbers of coastal agricultural areas and also entire archipelagos (Bahamas, Maldives) are under threat: already in the Maldives, where the highest ground does not exceed 2 m, beaches have had to be protected using sacks filled with coral debris to prevent erosion. The concern of the Maldivians resulted in the demonstrations held at the Conference of the Environment held there in November 1989 (Pugh, 1990).

In 1995, 35 small island states in the Pacific formed an association, AOSIS, to put pressure on industrialised countries to reduce their emissions of CO_2 by at least 20%

of 1990 levels by the year 2005. Global warming was the subject of passionate debates at the Conference on the Climate in Berlin (1995) and an inhabitant of western Samoa declared there: "While you spend time in discussion, we are being subjected to climatic changes, warming of the oceans, massive loss of coral reefs; we don't want more assurances or investors, but simply more hope". It is true that four inhabited islands had to be evacuated as a result of erosion by waves and the salination of ground water.

On the continents, the low land of the coastal plains represents 3% of the planet, but these areas are very densely populated, very fertile and represent a quarter of the cereal-producing areas in the world. Thus the Camargue (France) and certain coastal zones in the Languedoc would be submerged by an increase of 50 cm in sea level; this phenomenon also poses a great threat to coastal zones on all continents.

The greenhouse effect could potentially result in real terms in a general warming, varying between 2–5.5°C from now until 2100, but certain zones could become colder. Instead of melting as a result of the greenhouse effect, Greenland glaciers have increased in thickness by 23 cm/year since 1970. However, this is also due to the effects of warming (other authors predict a regression of the ice field): increased evaporation of the seas leads to an increased freezing of water vapours. Aquatic ecosystems are constantly evolving as they are already subjected to fishing pressure. The warming also results in an increasing difference in temperature between land and sea, which increases the transport of precipitation towards land. This increase is estimated at between 3–15% for the years to come.

Knowledge of the oceans becomes clearer with technological advances. The Topex-Poseidon satellite sends data every 10 days which allow sea levels to be determined to the nearest centimetre. This has allowed a seasonal pattern of variation to be constructed, and thus it has been discovered that there is seasonal expansion of 7 cm in the oceans of the southern hemisphere towards 40°S, and a greater expansion (15–50 cm) around the Gulf Stream in the northern hemisphere. The Gulf Stream flow varies by 10–15% depending on season; its mean flow represents about 100 million m^3, that is 100 times the total discharge of all rivers in the world.

The estimation of the amount of CO_2 absorbed and its discharge into the oceans by physical or biological processes, is therefore central to current research on the ocean. It is true that the world's oceans contain an enormous amount of CO_2 (estimated at 350,000 billion tonnes of C), i.e. three-quarters of all carbon that exists on earth. Seven billion tonnes of carbon are liberated into the atmosphere by man each year in the form of CO_2: 5 billion tonnes result from the combustion of hydrocarbons and coal and around 2 billion from deforestation. It is estimated that the ocean also absorbs 105 Gt of carbon and only produces 102 Gt (Jacques, 1991); as only 4 billion tonnes of excess are to be found in the atmosphere, it is considered that the "missing" 3 billion tonnes are absorbed by the ocean, although the mechanisms by which this occurs have yet to be explained. This difference could be due to errors in biomass estimations, as has been found for boreal forests : two researchers estimate that it has been overestimated by 200–300%. Figure 1.12 shows the various routes of the carbon cycle and puts quantitative estimates on the fluxes. Thus it is suggested that surface water phytoplankton play the role of a carbon pump.

Jeffries (1994) showed from 35 years of statistics from identical weekly trawl samples in Narrangasset Bay (USA), that the population of winter flounder (*Pleuronectes americanus*) had fallen while those of other species increased. This was not attributed to a direct temperature effect, but to the predation of the young by a shrimp (*Crangon septemspinosa*) whose abundance could reach $80/m^2$. Laboratory studies of shrimp predation on larval flounder revealed that they could completely destroy an age class. Normally, they are not active during cold winters. During successive mild winters, they are able to start feeding earlier and earlier, therefore overlapping with the reproductive season of the flounder. Mean winter temperatures more than 1.5°C higher than normal were sufficient to set this phenomenon in action. The flounder populations can recover if the warming continues, as predators of the prawns can also start feeding earlier in winter. On the basis of calculations using the same statistics, the author indicates that a mean increase in temperature of 1°C increases the stock of clams (*Mercenaria mercenaria*), lobsters, the crustacean *Squilla mantis* and crabs (by 73%) and decreases the stock of the northern shrimp, *Pandalus borealis* by 75%.

The reality of the greenhouse effect, while very controversial, is taken to be proven, despite the paucity of data concerning the role of the oceans. Figure 1.13 (adapted from Morel, 1992) demonstrates the uncertainties in evaluating mean sea levels, and also their convergence.

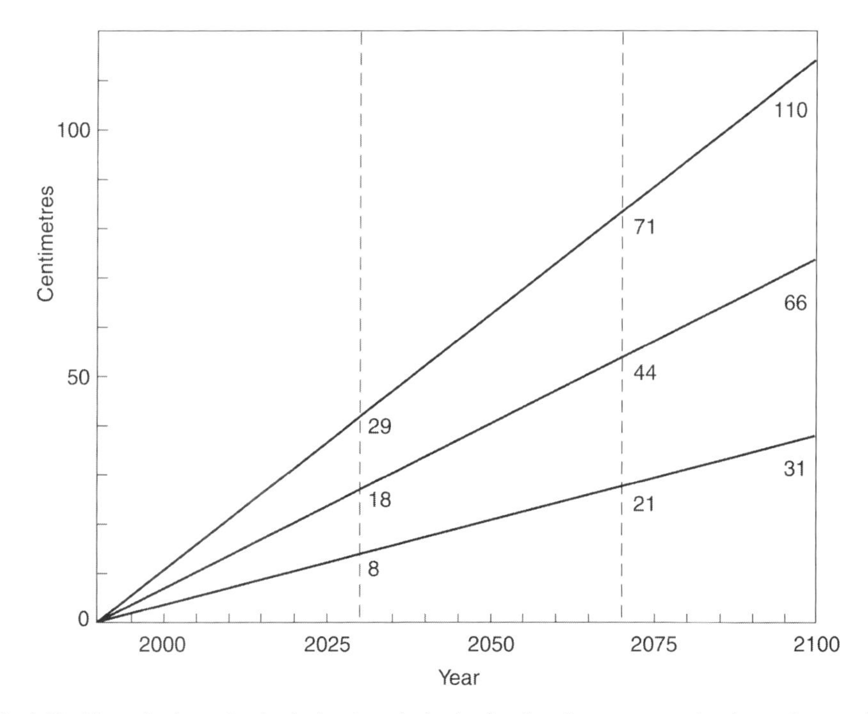

Fig. 1.13. Mean rise in sea level calculated on the basis of various increase scenarios due to the greenhouse effect (Morel, 1992).

1.6 THE ROLE OF ECOSYSTEMS IN A DYNAMIC CLIMATE

Without revisiting Lovelock's "Gaia" hypothesis, i.e. the earth considered as a living organism, it must be emphasised that living organisms play a role in climatic mechanisms.

Examples of adaptation to the physical environment have been described by Frontier and Pichod-Viale (1991), notably the adaptation to a lack of nitrogen in assimilable forms by rapid recycling. In the oceans, however, the abundance of plankton dictates the absorption of solar radiation, which is absorbed by chlorophyll. A high abundance of phytoplankton thus increases the absorption and, as a result, the speed of warming of the ocean's surfaces. Data collected for the Arabian Sea by remote sensing show that the development of plankton exerts control over the development of surface temperatures. A biological mechanism thus creates a regulatory influence on ocean–atmosphere interactions.

In coral reefs, excessively high temperatures lead to their bleaching, but an increase in atmospheric CO_2 also results in an increase in the waters (half of the CO_2 dissolves there). We have seen (Section 1.1) that an increase in carbon dioxide acidifies the waters and deprives them of carbonates, but that an equilibrium exists between the various forms of carbon: researchers have thus calculated that the formation of corals could fall by 40% if the CO_2 content of the atmosphere doubled.

According to Charlson *et al.* (1987), the main source of condensation nuclei for clouds over the oceans seems to be the dimethylsulphate produced by planktonic algae, which is oxidised into a sulphate aerosol in the atmosphere. The reflection of solar energy is sensitive to the density of these condensation nuclei for clouds and a biological regulation of climate could equally be possible through the effects of temperature and sunlight on the phytoplankton and thus on the production of dimethylsulphate. Thus to combat warming due to the increase of CO_2 in the atmosphere, one could double the number of cloud-condensing nuclei by altering plankton densities. Effectively, aerosols make clouds and satellites have shown increases in clouds in the wake of ships (their fumes are full of aerosols). According to Kienne and Bates (1990), the amount of dimethylsulphate in the ocean is regulated by biological processes, mainly by the microbial cycle (see Chapter 3), by microbial activity and not by exchange between ocean and atmosphere.

We also know that photosynthesis absorbs CO_2 and produces oxygen. Pérès (1976) reported that if the phytoplankton no longer carried out its photosynthetic function, the level of oxygen in the atmosphere would fall by about 10% in 25 years.

1.7 CONCLUSION

The relationships between the oceans and the atmosphere are many and complex; they are studied by climatologists and oceanographers using increasingly sophisticated techniques including specialised satellites, networks of buoys, etc.

Regarding the greenhouse effect, the role of carbon sink played by the ocean is important: it may be that the ocean could limit the impact of the greenhouse effect, but quantitative estimates are lacking. However, the ocean could not carry out this function beyond certain limits; there are other fertilising elements which limit photosynthesis, even where CO_2 and light are not limiting. These are nitrates, phosphates and oligo-elements such as iron (Martin *et al.*, 1990). Certain secondary constituents of sea water are therefore of vital importance in terms of biological marine production, even when they are present in reduced quantities in aquatic environments.

With regard to global climate change, the experts of the Intergovernmental Group of the Evolution of the Climate concluded their report in 1996 by stating that a host of facts suggests that man has a perceptible influence on climate.

1.8 BIBLIOGRAPHY

Almazan G. and Boyd C. An evaluation of Secchi disk visibility for estimating plankton density in fish ponds. *Hydrobiologia* 1978; 61(3): 205–208.

Alzieu C. L'eau, milieu de culture. In: Barnabé G. (Ed) *Aquaculture*. Lavoisier Tec & Doc, Paris, 1989; 17–45.

Aminot A. and Chaussepied M. *Manuel des analyses chimiques en milieu marin*. CNEXO Éd., Brest, 1983.

Anon. Schema expliquant la notion de circulation thermohaline dite du "tapis roulant". *Bull. Intern. Sciences de la Mer* (UNESCO) 1990; 53/54: 6.

Anon. La commission océanographique et intergouvernementale de l'UNESCO et l'étude de l'océan dans notre environnement. *Bull. Intern. Sciences de la Mer* (UNESCO); 59/60: 3–6.

Aubert M. La Méditerranée, la mer et les hommes. *Rev. Int. Oceanogr. Med.* 1994; 109–112; 488.

Bakun A. Global climate change and intensification of coastal ocean upwelling. *Science* 1990; 247: 198–201.

Barry J.P., Baxter C.H., Sagarin R.D. and Gliman S.D. Climate related long-term faunal changes in a rocky intertidal community. *Science* 1993; 267: 672–675.

Bethoux J-P., Gentilly B., Ranet J. and Taillex D. Warming trend in the western Mediterranean deep water. *Nature* 1990; 347: 660–662.

Boudouresque C.F. *Impact de l'homme et conservation du milieu marin en Méditerranée*. Univ. Mer, Fac. Sc. Luminy. Gis Posidonie Éd, Marseille, 1995.

Brown A.C. and McLachlan A. *Ecology of sandy shores*. Elsevier, Amsterdam, 1990

Charlson R.J., Lovelock J.E., Andrew M.O. and Waren S.G. Oceanic phytoplankton, atmospheric sulphur, cloud albedo and climate. *Nature* 1987; 326(6114): 655–661.

Colbourne E., Narayanan S. and Prinsenberg S. Climatic changes and environmental conditions in the Northwest Atlantic. In: *Cod and climate change*. ICES Mar. Sci. Symp. 198, 1994; 331–322.

Copin-Montégut G. Chimie marine. *Océanis* 1993; 19(5): 197.

Crépon M. Initiation à la dynamique de l'océan. *Océanis* 1993, 19(2): 110.

Duvignaud P. *La synthèse écologique*. Doin Éd., Paris, 1981.

Fraga F. La profundidad de visión del disco de Secchi y su relación con las concentrationes de fitoplancton y arcilla. *Invest. Pesq.*, 1979; 43(2): 519–528.
Frontier S. and Pichod-Viale D. *Écosystèmes, structure, fonctionnement, évolution.* Coll. Écologie 21, Masson, Paris, 1991.
Guilcher A. *Précis d'hydrologie marine et continentale.* Masson, Paris, 1979.
Guillemau-Drai C. Pollutions chimiques et conséquences sanitaires. *Rev. Int. Oceanogr. Med.* 1991; 101: 205–213.
Hansen J.E. and Lacis A.A. Sun and dust versus greenhouse gases. *Nature* 1990; 346: 713–719.
Huet S. L'océan au centimètre près. *Science et Avenir* 1994; 565: 42–47.
Jacques G. La mystérieuse disparition du carbone. In: *La vie dans les océans.* Science et Vie 1991; 64–71.
Jeffries J.P. The impacts of warming climate on fish populations. *Maritimes*, The University of Rhode Island Marine Program 1994; 37(1): 12–15.
Kienne R.P. and Bates T.S. Biological removal of dimethyl sulphide from sea water. *Nature* 1990; 345: 702–705.
Kjerve B. L'évolution future du niveau de la mer. *Bull. Intern. Sciences de la Mer* (UNESCO) 1991; 57–58: 4–5.
Korringa P. Farming the European flat oyster (*Ostrea edulis*) in a Norwegian poll. In: Korringa P. (ed) *Farming the flat oyster of the genus Ostrea.* Elsevier, Amsterdam, 1976; 187–204.
Labeyrie J. Échange océan-atmosphère, le pétillement des océans. *Science et Avenir*, hors série 98: 48–51.
Leggett J. (Ed). *Climat: une bombe à retardement.* Greenpeace International Publishers, Brussels, 1994.
Mann K.H. and Lazier J.R.N. *Dynamics of marine ecosystems.* Blackwell Scientific, Boston, 1991.
Martin J.H., Gordon R.M. and Fitzwater S.E. Iron in Antarctic waters. *Nature* 1990; 345: 156–158.
Morel P. Introduction à la dynamique de l'atmosphère, des océans et du climat. *Océanis* 1992; 18(3): 265–369.
Pérès J. M. *Précis d'océanographie biologique.* PUF, Paris, 1976.
Pérès J. M. and Devèze L. *Océanographie biologique et biologie marine. Vol. II: La vie pélagique.* PUF, Paris, 1963.
Pugh D. À bas la montée du niveau de la mer! *Bull. Intern. Sciences de la Mer* (UNESCO) 1990; 53/54: 3–4.
Salomon J.C. Gérer l'environnement marin par la modélisation mathématique. *Équinoxe* 1994, 50: 13–18.
Solbé J. Water quality. In: Laird L. and Needham T. (eds) *Salmon and trout farming.* Ellis Horwood Ltd, Chichester, 1988; 68–86.
Weber P. Sauver les océans. In: *L'état de la planète 1994.* Worldwatch Institute; Éd. La Découverte, Paris, 1994; 65–92
Weller R.A. Mixing in the upper ocean. *Nature* 1987; 328: 13–14.

2

Water currents

2.1 WATER CURRENTS: THE BASIS OF OCEANIC PRODUCTIVITY

Waves, currents, tides and upwellings* move waters on diverse spatio-temporal scales, from the smallest whirlpool with a duration of about a minute affecting a few plankton*, to large planet-wide currents which can displace the mass of water in the oceans over one millennium. These currents and the eddies which accompany them not only mix the waters, but also play a role in the displacement of planktonic species, nekton* migrations and influence encounters between predators and prey. All this movement renews the waters at the air–water and water–sediment exchange surfaces and also the entire volume of water around the organisms living in it. On this scale, water movements alter the layers surrounding organisms, transporting nutrients and excretory products, whence their biological importance.

These physical phenomena create the optimal conditions for many important biological processes but, before concentrating on waves and currents, it is vital to examine the basic physical processes.

2.2 FLUID DYNAMICS: THE BASIC FACTS

2.2.1 Turbulent flow

Particles constituting a fluid can move in a way which is regular and homogeneous and is hence called laminar flow; this is what occurs in very slow flows. It is a theoretical rather than real concept as, above a certain speed, the various particles have different, irregular, disorganised trajectories: the system is said to be turbulent.

Thus, in a water current, variable speed indicates the presence of patches of turbulence. These patches can be of variable speed (of the order of a second with variations of intensity of speed of several millimetres, or tens of minutes in the case of variations of speed intensity of several metres).

The speed vector of a current is not a constant, varying most often around a mean position. Fluctuations in speed are very rapid (occurring over several seconds or several hundredths of a second). Turbulence characterises the movement of a fluid and favours the diffusion of quantities of movement and matter by exchange of fluid elements (Crépon, 1993).

These turbulent patches are three-dimensional and extend equally in all directions, but this approximation can only be used up to the depth of the water. Beyond this, homogeneity gives way to two-dimensional eddies. The best-known instability is the breaking of waves when they become too steep: the movement of the waves is transformed into random turbulence. At the molecular level, the speed of each water particle can vary, which makes these phenomena very complex. The simplification is, of course, to return to mean flows, but this does not resemble reality at all. Various mathematical approaches can be used, but problems of representation of movements on a small scale mean that using numerical models, and taking into account the memory size of computers, it is too difficult to refine the representation; the spatial scales of such models are of the order of tens of kilometres in the open sea and of a kilometre in the coastal zone, while what is happening between these zones is ignored.

In currents, energy is continually being transferred from movements on a large scale to movements on a small scale, in a less violent way than the breaking of waves. There is an "energy cascade" which neither changes the total amount of energy, nor transforms the nature of the energy (Mann and Lazier, 1991).

The widest eddies are tens of metres wide in the surface layer and several metres in depth. A decrease in eddy size is accompanied by an increase in speed gradient. When the eddies are quite small, the shearing point is quite high, thus the molecular viscosity, a kind of internal resistance of the water, acts to slow the water molecules and absorb the speed gradient. This "absorption" by viscosity is the route by which the turbulent energy is eventually transformed into heat. This viscosity is very weak in large-scale terms (see below).

Mann and Lazier report that when wind speed varies from 15 to 5 m/s, small scale turbulences, varying between 6 and 20 mm, occur close to the surface, as the viscosity limits the size of the smallest turbulent eddies to 6 mm in a very high-energy environment. At greater depths, or when the wind is light, the smallest eddy has a diameter of 35 mm.

Eddies greater than a few millimetres in diameter mix waters the most efficiently. The significance of turbulences generated by water movements in relation to molecular diffusion (simple contact, without movement) is that the propagation of turbulence is much more rapid: this is termed turbulent diffusion.

The transfer of physical and chemical properties of water by eddies is the same for all parameters (temperature, salinity, dissolved salts) whereas the situation is completely different for simple molecular diffusion (see below).

The specialist literature shows that many laboratory studies do not take into account the part played by turbulence, which has an even more significant role on a small scale. Similarly, academic studies frequently separate hydrology and biology.

2.2.2 Molecular diffusion

Turbulent energy is efficient for the transport of nutrients and waste for organisms of about 1 mm in size; below this size, micro-organisms must rely on the flow arising from molecular diffusion, which corresponds to slow mixing caused by the random movement of molecules. This is a very slow process in water since it takes around 10 seconds to produce an effect at a distance of 100 μm. Horizontal diffusion, typical of an eddy, is 1 billion times faster than molecular diffusion, which demonstrates the differences in scale.

Diffusion varies depending on certain parameters; considering viscosity as the parameter which absorbs and diffuses the acquired speed, diffusion of heat is 10 times less rapid than that of speed, that of salinity or nitrates being 100 times less rapid. As a result, small fluctuations persist for longer before being eliminated by diffusion.

Mann and Lazier thus calculated that the smallest scale for a temperature fluctuation is 2–13 mm while that for nitrates is 0.2–1 mm.

2.2.3 Boundary layer

Solid layers, such as the surfaces of organisms, are associated with a boundary layer in which water movement is reduced. There is also a boundary layer at the bottom of the oceans. A fundamental property of this layer is the condition of friction: molecules in contact with a solid surface adhere to its surface and are stationary. The speed of displacement of the molecules increases as one moves away from this layer.

Two distinct layers are recognised in the boundary layer: the part in which turbulent fluctuations can still be sensed (but are gradually reducing) and a viscous layer, adjacent to the surface. From 10 cm to 0.2 cm away from the solid substrate, the speed profile in relation to open water is logarithmic. Below this layer, the speed profile is linear, speed decreasing linearly in this thin viscous layer (see Fig. 4.1).

The limiting layers are thus layers where turbulence is reduced; this can reduce the exchange of nutrients but, the larger the organism, the thinner the layer in relative terms; also, as speed increases, the layer becomes thinner. For sedentary organisms, fast-moving water thins the boundary layer and facilitates the exchanges of substances; for actively-swimming organisms, rapid swimming acts in the same way.

The rate of exchanges thus relates not only to the size of the object, but also to the current speed and mode of flow around the object; it can be calculated from the Reynolds number (R) which characterises the viscosity of a fluid.

2.2.4 Viscosity and Reynolds number

We have seen that, in nature, flows are turbulent, while the other type of flow, laminar flow, has been studied mainly in the laboratory; the passage from a laminar regime to a turbulent regime was analysed by Reynolds while considering the viscosity of fluids.

The dynamic viscosity corresponds to the property of a fluid to present resistance or internal friction to relative displacements of two adjacent layers (for example, two water layers of different densities superimposed on one another). It is expressed in poises (1 poise = 1 dyne/s/cm^2) according to the CGS system and in SI units it is expressed in Newtons per seconds per metre squared (N/s/m^2); kinetic viscosity (v) is the quotient of the dynamic viscosity over the density and is expressed in square metres per second (m^2/s). On an oceanic scale, viscosity has little effect: a change in current speed of 1 m/s over a distance of 1000 m creates a change in viscosity* of 10^{-6} N/m^2.

This theoretical approach has been used to calculate the reduction in flow in pipes; this is greatly influenced by whether flow is laminar or turbulent. Reynolds showed that if U represents the fluid speed in metres/second, D the diameter of the pipe in metres and v the kinematic viscosity, the relationship UD/v, called the Reynolds number (Re or R), could be used to describe the type of flow (Re = UD/v).

Below a threshold value of about 2500, flow is laminar, indicating that the forces of viscosity regularise the flow, while the forces of inertia are 2500 times greater than the forces of viscosity; above this, and especially for values >10,000, flow becomes turbulent.

In a channel, and also in a lagoon or the sea, the Reynolds number R can be applied, with depth being used instead of tube diameter. Crépon (1993) reported the example of an ocean 1000 m deep influenced by a very slow-moving current of 1 mm/s, with a kinematic viscosity (v) of the water of 10^{-6}SI, giving:

$$R = \frac{0.1 \times 10^5}{10^{-2}} = 10^6 \qquad \text{(Eq. 2.1)}$$

The system is therefore turbulent even at this very low-speed current. The dimensions of the ocean are such that the Reynolds numbers are very large and exceed the critical value of 2500.

Forces of viscosity become important only for small dimensions: for a mass of water wider than 1 cm (in a pipe, for example), moving at 1 cm/s, R is ≈102; in such a current, the forces of viscosity become significant in relation to the forces of inertia and oppose turbulent perturbations. R can also be used to described solid bodies in a fluid, such as living aquatic organisms, by using the size of the animal around which flow is being studied. For a 5 cm diameter fish swimming at 1 m/s, R = 50,000. As stated by Mann and Lazier (1991), the forces of inertia and viscosity can be ignored in the animal world.

On a microscopic scale, however, things are different; a planktonic organism* of 50 μm moving at 10 μm/s has a Reynolds number of 10^{-4}. It lives in world where viscosity dominates: Mann and Lazier (1991) compare this to a man swimming in honey. They also list several Reynolds numbers as calculated by Vogel (1981).

2.2.5 Drag

Movement of a body in water generates a drag force which must be overcome by the

Table 2.1. Estimation of Reynolds number of different organisms (after Vogel, 1981).

	R
Large whale swimming at 10 m/s	300,000,000
Tuna swimming at same speed	30,000,000
Duck flying at 20 m/s	300,000
Large dragonfly flying at 7 m/s	30,000
Copepod in a current of 20 cm/s	300
Smallest flying insects	30
Invertebrate larva 0.3 mm long moving at 1 mm/s	0.3
Sea urchin spermatozoon swimming at 0.2 mm/s	0.03

body in order to maintain this movement. Drag force is caused by two things. The first is the adherence of water molecules to the body at the boundary layer; this force is a function of the area and type of surface of such an object (smooth or rough). If R is low, it is the only component of drag (e.g. for small organisms). The second cause relates to the mass of water which must be deformed in front of the body and replaced behind it.

In the sea, submerged objects in moving water with horizontal currents (the most common case) are subjected to horizontal stress from drag. This is true for stationary structures used for rearing in the open sea, for fixed fishing nets and marker buoys, etc. This stress is expressed by Morrison's formula:

$$Ft = \frac{1}{2} D.T.S. \ v2$$

where Ft = drag force (in Newtons); D = density of water, i.e. 1025 kg/m^3; T = drag coefficient, varying between 0.5 (sphere) to 1.2–1.5 for a flat surface (depending on author); S = contact surface in m^2 (surface of the projected volume subjected to current action); v = current velocity in m/s.

Observations show that the current tends to tilt pliable structures towards the bottom in open water: seaweed, nets, moorings and their buoys are thus keeled over by strong currents. This effect can be interpreted using the preceding formula: by aligning parallel to the currents, all these structures essentially reduce the area of the contact surface S and often the drag coefficient T (which expresses the hydrodynamic profile of the structure). A position of equilibrium is thus reached, especially close to the bottom, where the current often decreases very rapidly (this is noticeable when diving).

2.3 THE MIXED LAYER AND WATER STRATIFICATION

Many studies, including those of Thorpe and Hall (1987), concerning large, open, water bodies have shown that, under the action of windspeeds greater than 3 m/s, waves breaking at the surface carry clouds of air bubbles to deeper water (<10 m),

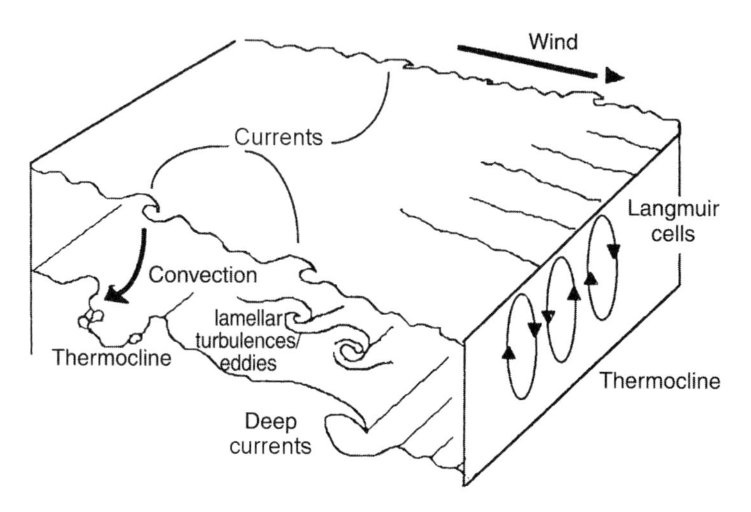

Fig. 2.1. The superficial or "mixed layer" of waters.

which in turn carries warm surface water to deeper levels. These waves also inject kinetic energy to the deeper waters in the form of turbulent movements. Other processes also intervene, such as the formation of convection cells (Langmuir cells). This collection of phenomena is shown in schematic form in Fig. 2.1 (adapted from Weller, 1987). The surface of the oceans is, in practice, a "mixed layer", but identical relationships characterise the surface of all water surfaces, with intensity depending on factors such as size, depth, etc.

Passive accumulation of organisms (plankton) or inert particles occurs in the calm zones between the turbulences; thus between convection cells, floating elements accumulate at the surface around superficial convergences forming surface "slicks". The thin superficial layer of the slicks is moved much more quickly by wind than the rest of the water.

Because of their raised temperature, dissolved salt concentration, and imports of freshwater (lighter than sea water), the waters of the mixed layer have a variable specific mass (Chapter 1), but it is most often less than that of colder, deep waters, which are more saline and thus denser. Except at very high latitudes and even in winter, the mixed surface layer waters are effectively warmer than deep waters whose temperature is about 4°C.

The transition between this superficial mixed layer and the deep calm waters is rarely gradual: on the contrary, a very clear separation is noted (Fig. 2.2). There is a density gradient or pycnocline, which is less than or equal to a metre thick, and which separates waters of different densities (temperatures and/or densities differ). When this discontinuity separates waters of different temperatures, it is called a thermocline.

It is possible to have several very distinct strata superimposed on each other, in relation to their density, but coastal waters are characterised by having two very distinct layers (Fig. 2.2): the mixed layer which extends from the surface to a

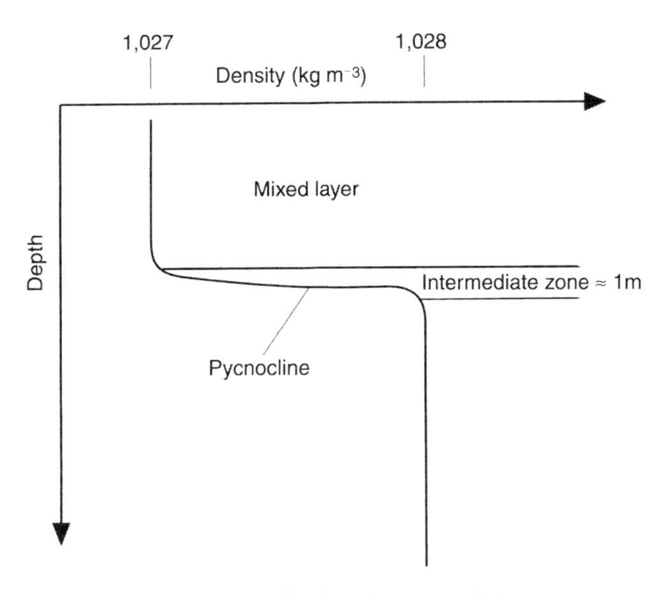

Fig. 2.2. Vertical density profile through two stratified water masses.

(variable) depth of several tens of metres, and the deep layer. These layers of water of different densities do not mix in practical terms when there is no turbulence; there is only a little mixing over a very thin layer, in relation to the scale of the aquatic environment.

Winds and waves, tidal or other currents can vary the thickness of the mixed layer, and the deep layer is only present when depth allows; it is absent in waters closest to the coast.

An input of energy from turbulent movements (waves, currents) is required to induce mixing at the level of the pycnocline, which constitutes a barrier to vertical transport. This well-mixed superficial zone ensures the transfer and equilibrium of dissolved gases between water and atmosphere and it is also at this level that the ocean exchanges heat energy with the atmosphere (often in the form of mists/heat hazes), hence its role in the climate (see preceding chapter). In biological terms, we shall see that the interactions are even more numerous (Chapter 4).

2.4 WATER MOVEMENT IN THE COASTAL ZONE

2.4.1 Swell and waves

Winds are at the origin of wave formation. It is important to distinguish at this point between forced and free waves. Waves caused by wind are called forced waves, and are characterised by powerful transport of water masses in the direction of the wind; when the waves leave the windy area in which they are formed, they are termed free waves, or swell.

Waves are oscillations at the sea surface, with short periodicity (the time that elapses between the passage of two consecutive waves). They entail rotational movement of water molecules, the speed of which reaches 2 m/s for 2 m high waves (approx.) with a period of 4 s. Such a swell creates orbital speeds or displacement corresponding to currents of 4 knots (7.4 km/h), reversing at each half-period. Waves 2 m in height are common in the open sea and thus it is these waves which induce the most rapid movement of water particles. Figure 2.3 (Bompais, 1991) demonstrates the movements, forces and speeds of movement of water created by waves in the shallow zone and their rapid attenuation with increasing depth.

Air molecules rubbing the surface of a wave transmit their kinetic energy to the water through contact; they are thus slowed down. The molecules above continue onwards. This creates compression (increase in pressure) behind the wave and a drop in pressure in front, and thus the height of the wave increases. When the wind speed exceeds around 25 km/h, the crest of the wave, which is moving more quickly than the rest of it, collapses forwards; it breaks and creates "white horses". The theoretical limit to the camber is 14%.

Taking into account the inertia of all submerged bodies, or simply the seabed, the constraints to be considered can be understood. When currents and waves are associated, their speeds combine dramatically when their direction coincides, increasing displacement, force and speed in the same proportion on all submerged bodies. Waves thus constitute one of the principal "forces of the sea", which all developments must take into account.

The height of waves created by wind depends on the speed and time for which it has blown, and also on the distance over which it acts (the "fetch"). A violent wind acting on a water surface several kilometres long cannot raise enormous waves; it requires 500 nautical miles to obtain the huge waves which are characteristic of the

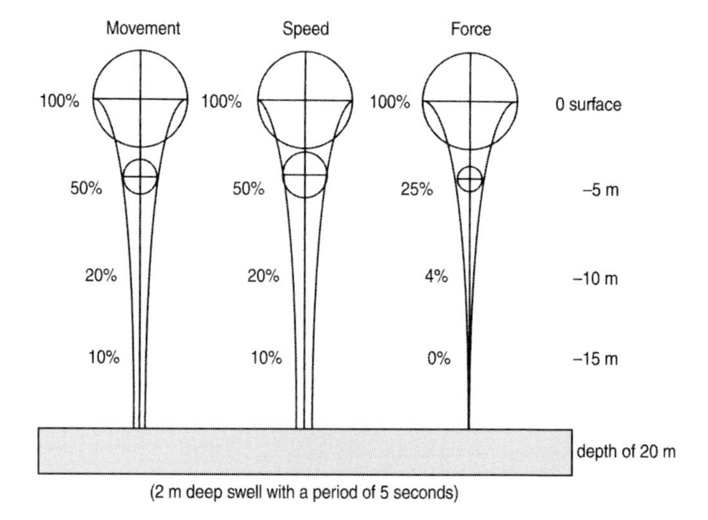

Fig. 2.3. Attenuation of movement and force of swell with depth (Bompais, 1991).

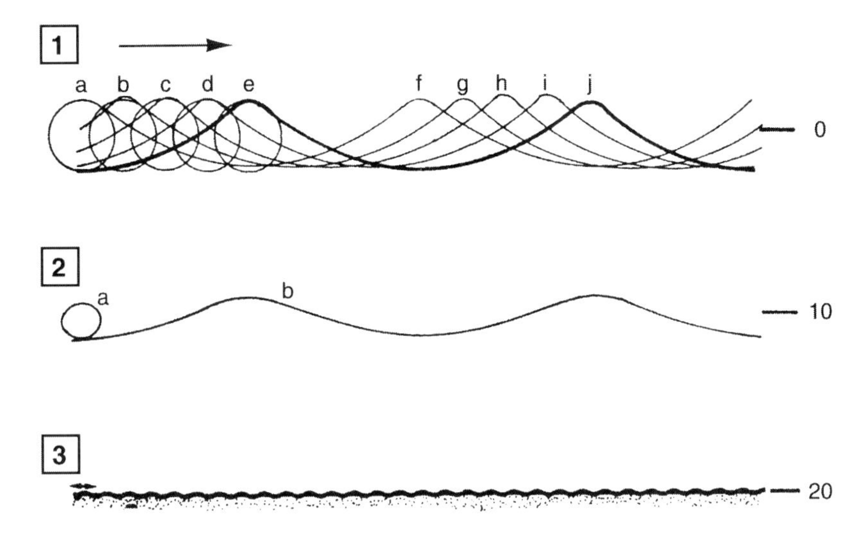

Fig. 2.4. Simplified representation of swell above a 20 m deep seabed.

oceans. In the Mediterranean, waves do not exceed 8 m in height because of the dimensions of this sea. In the North Sea, their height can reach 30 m for a wind blowing at 108 km/h for more than six hours; their period is 15 seconds and their wavelength 350 m (Schmied, 1994). The wavelength can reach 1.5 km, but it is often much less. The largest oceanic wave, measured by triangulation by C. Margraff (using the ship's mast as a reference point) on board the American Marine ship "Ramapo" measured 35 m in height.

The amplitude of waves or swell decreases more progressively than that of currents with decreasing depth in coastal areas (Fig. 2.4). This decrease is due to the friction of the waves on the seabed, which absorbs energy. On very flat beaches, the reduction can reach 30%. This decrease in amplitude is accompanied by a "pounding" and a flattening of movement which results in a to-and-fro motion of water particles on the bottom (beyond 30–40 m depth, such movements become very weak).

On approaching such beaches, when the relationship between water depth and wave height is around 1.3, the profile of the wave, slowed by the seabed, becomes extremely asymmetrical and the wave breaks. This breaking action carries the water to the coast and is compensated for by a return current underneath (undertow) which can be responsible for the erosion of beaches (Fig. 2.5). These currents are not always obvious; in some cases, they are very localised (15–30 m wide), leaving the beach and sweeping out to the open sea (rip currents), and are very dangerous to swimmers as a result of their speeds which, with strong winds, can reach 5.5 m/s (Popov, cited by Guilcher, 1979). When waves hit the shore obliquely, they carry littoral drift containing sediments, particularly sand from beaches, as a result of induced currents (Fig. 2.6). A detailed account of these phenomena can be found in the works of Brown and MacLachlan (1990) and Carter (1988).

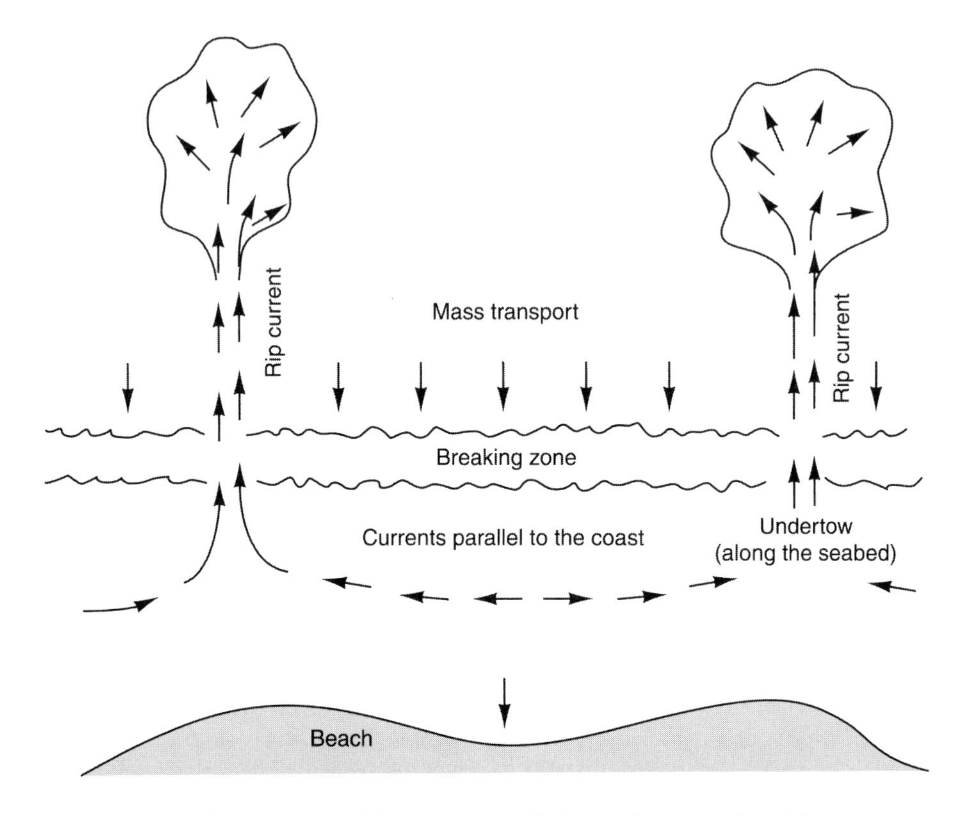

Fig. 2.5. Diagram of currents created by waves perpendicular to the coast (adapted from Brown and McLachlan, 1990).

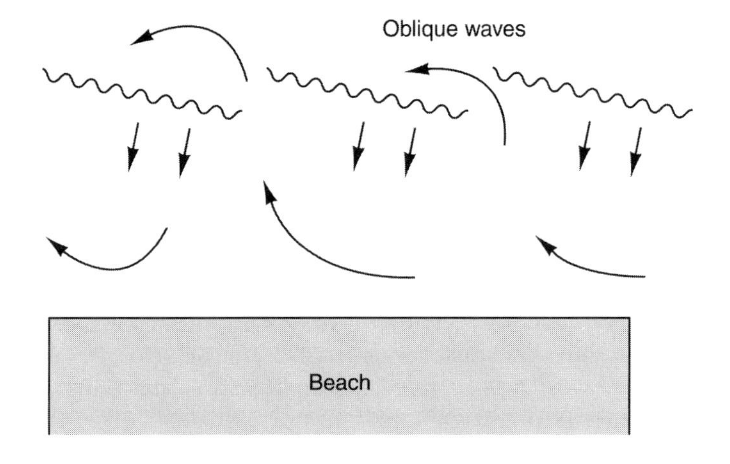

Fig. 2.6. Diagram of currents created by oblique waves (adapted from Brown and McLachlan, 1990).

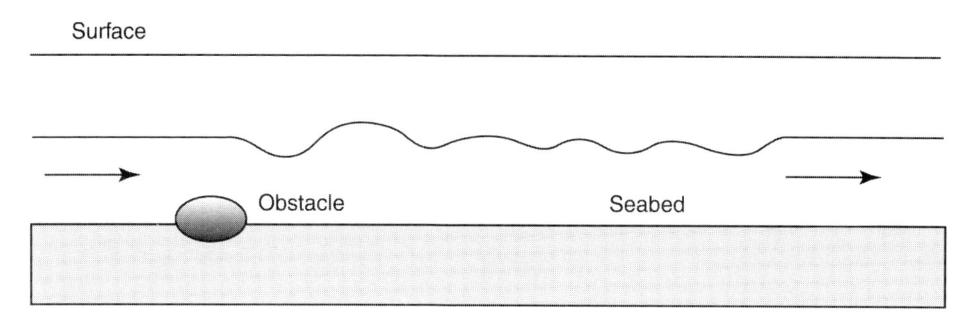

Fig. 2.7. Diagrammatic representation of a layer of water moved by a current above an obstacle and the internal wave which is created.

Surface waves are not the only ones to be observed at sea; when waters stratify into layers of different densities, internal waves can be found at the separation point (pycnocline); these have a much larger wavelength than that of the surface waves; they propagate at the level of the pycnocline. One of the most usual mechanisms producing internal waves at the pycnocline is a stable current meeting an obstacle: this can happen at a rock, a reef (see Figure 2.7) and also at the edge of continental slopes.

Particularly long and strong internal waves (50–70 m high, 30 km wavelength) can be created when strong tidal currents from the open sea, moving towards the coast, encounter a break in the slope of the continental shelf or a shallow plateau; there is acceleration of water speed and creation of an internal wave.

2.4.2 Currents

The energy imparted to the ocean by the sea, winds and tides and also large-scale movements creates not only waves but also currents*, i.e. movements of water characterised by one direction, speed and a precise flow. Various categories of current exist, but they all have turbulent flow. We restrict ourselves here to the study of currents in the coastal zone by excluding the large oceanic currents *sensu stricto*.

2.4.2.1 Tides and tidal currents

The forces resulting from tides are the strongest of all those exerted on the seas and an enormous amount of energy is bound up in the tides. A tide is a periodic oscillation of marine waters, linked to the attractive effect of the moon and sun on the fluid masses of the oceans; it creates very significant currents in coastal zones.

Although the sun has a mass 27 million times greater than that of the moon, it has less influence than the earth's satellite, as the gravitational forces between two bodies are proportional to the product of their mass divided by the square of their distance apart. The variable positions of the earth, the moon and the sun create many periodicities in the tides: the earth turns on its axis in 24 hours in relation to the sun, but the moon circles the earth in 24 h 50 min. The distance and inclination of the moon in relation to the earth and of the earth in relation to the sun also varies throughout the year. The periodicities of the wavelengths of the tides thus vary according to various rhythms: semi-diurnal, diurnal, bi-monthly, seasonal and annual. In simple terms, we say that the action of the moon adds to or masks that of the sun.

Because the action of the moon rotating around the earth is dominant, there are two periods of rising water (ending in high tide) and two periods of lowering water (ending in low tide) per day. From one day to the next, there is delay of about 50 minutes for each low or high tide, since each half period of the moon around the earth takes 12 hours 25 minutes (lunar day = 24 h 50 mins).

When the moon and sun are aligned (about every 14 days), the tidal force is maximal: a spring tide results (Fig. 2.8). When the sun and moon are at right angles, there is a neap tide (Fig. 2.8).

To this astronomical complexity must be added the roles of coastal topography and the seabed, which either amplify or dampen the actions of the heavenly bodies. The response of waters to the attraction of heavenly bodies is not instantaneous owing to inertia and friction of water on the seabed. Depending on the site, there can be a delay of several hours between cause and effect. One can thus associate a strong tide with a lunar phase when in fact it has been created by preceding phases. This occurs in the upstream end of estuaries or deep bays where the tide can manifest itself after a delay of several hours.

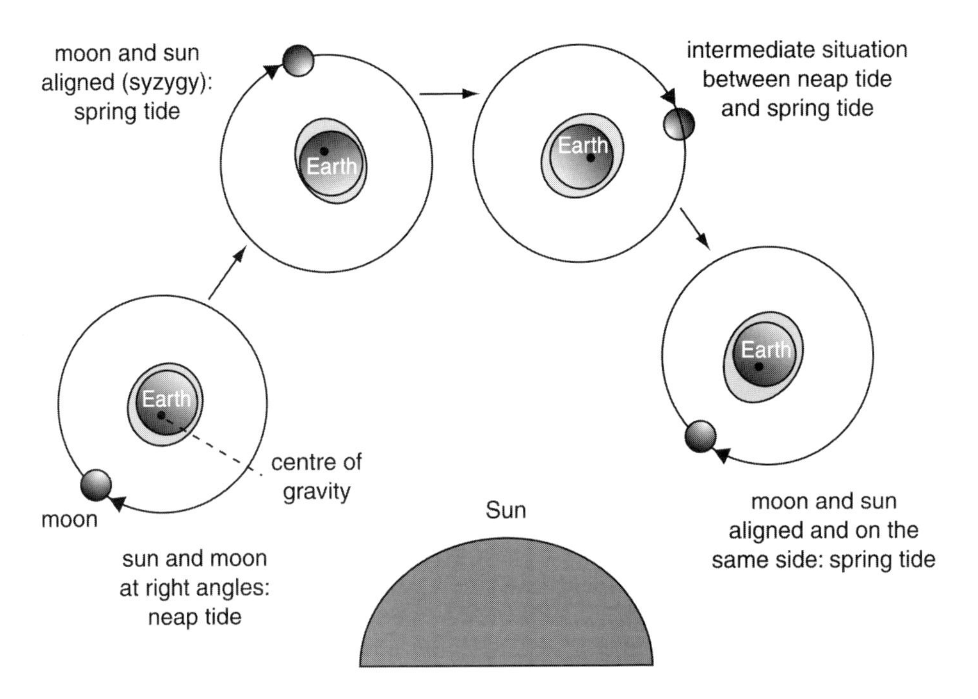

Fig. 2.8. Moon, sun and tides.

In shallow waters, the tide is associated with a tidal wave which can be either reflected or dissipated when it encounters a barrier to its progress. When the slope of the seabed is gentle, the wave energy dissipates there. The geometry of coasts (presence of bays, headlands, etc.) thus modifies wave propagation for tidal waves just as it does for other waves.

When a high tide enters a river and encounters obstacles slowing its progress at the mouth, the wave thus formed is called a "bore" (tidal). This wave can reach 2–3 m in height and exceed a speed of 20 km/h. The bore ascending the Amazon for 320 km and that of the Tsientang in China create a wave nearly 8 m high at high tides.

The vertical structure of coastal waters is dominated by the tides, as they cause mixing of waters from the surface to the bottom on many continental shelves: the mixed zone of the waters thus extends throughout their depth, while the movement created entails a horizontal displacement of water. In these conditions, tidal currents obviously create whirlpools and turbulences. They are likely to resuspend sediments and regrade them as a function of their granulometry (particle size). A more detailed description of these processes can be found in Guilcher (1979), Bonnefille (1980) or Carter (1988).

The coast constitutes a barrier for tidal currents. There is modification of the tide, when a tidal current encounters different depths, e.g. when it passes from a deep to a shallow zone such as on a bank, the size of the tide and the speed of the floodtide are increased. In these conditions, tidal currents create eddies and turbulences. Whirlpools created by tides in straits between islands, for example, are sometimes referred to as "maelstroms".

In most regions where aquaculture is practised, tidal movement is the driving force for activities; it renews the water bringing food and oxygen to the reared animals and allows temporary access to rearing sites, e.g. shellfish beds. When the morphology of the shore does not allow drift currents, tidal movement often shifts relatively confined water masses to and fro: in Saint-Brieuc bay, for example, a hydrodynamic study established by calculations that one particular particle returns to its starting point at the end of a tidal cycle.

The heights of tides vary depending on the site and their dynamics are complex. They create considerable movement of water, which influences its productivity. Tides lead to the existence of an intertidal zone, the foreshore*, marked by a very noticeable zonation of flora and fauna. The force of currents varies with the diurnal tidal cycle, and also with the alternation of spring and neap tides.

The tidal coefficient expresses the variations in the level of the tide with time. It can be applied to locations where the tide is barely noticeable as well as in others where it exceeds several metres. This coefficient varies from 20 to 120. From this coefficient, each geographical location is allocated a height coefficient (marked on marine charts) which is the half-way point of the highest tide for a period of mean spring tides (Guilcher, 1979). Where U is the unit of height and C the coefficient, the height above the mean level would be $U \times C$ and the tidal height* $2U \times C$.

With a coefficient of 120, the tidal height reaches 19.6 m in the Bay of Fundy in Canada, 16.1 m at Granville and exceeds 4 m on French Atlantic coasts. On a global scale, mean tidal height is lower, 1–2 m (see Guilcher, 1979 for more details). Tides create strong currents, reaching 8–9 knots in a channel close to Ushant in Brittany, but currents reaching 15 knots at a site in the Molucca Sea have been recorded.

The Baltic Sea and the Mediterranean Sea (except the Adriatic), separated from the large oceans, have reduced tides which renders them unusual environments.

Tidal currents lead surface waters to the seabed in most coastal zones. Current force varies with the diurnal tidal cycle and also with the alternation of spring and neap waters. In zones where the water is too deep for complete mixing from surface to bottom, water stratification develops in summer. Fronts between mixed waters on the shelf and stratified waters in the open sea are areas of high productivity. There is convergence of surface waters and "downwelling", i.e. the descending of surface waters to the depths; all organisms with sufficient buoyancy to allow them to avoid sinking will accumulate there.

The amplitude of tides is the difference between high tide and low tide. This amplitude allows three types of coast to be distinguished:

- microtidal coasts: tide is less than 2 m. These are open oceanic coasts and island coasts;
- mesotidal coasts: tide between 2–4 m;
- macrotidal coasts: tide > 4 m. These are the tides found in estuaries, in zones where bottom slope is shallow and in semi-closed bays or lagoons. Carter (1988) describes the hydrodynamic phenomena accompanying tides and these environments.

In general terms, the tide dominates in areas where wave energy is relatively weak. Tides move not only water but also fine sediments, in particular mud and clay, with specific densities of 1.03–1.3. At lower densities, these muds behave as a fluid: there is no longer a distinction between water and sediment. This type of transport is common in semi-closed intertidal zones such as bays, lagoons and certain estuaries. The dynamics of currents and deposits become extremely complex. Mont Saint-Michel Bay in France is a characteristic example of a tidal bay; the interplay between tides and sediments is the subject of physical modelling with a view to development.

A more detailed description of these processes can be found in Guilcher (1979), Bonnefille (1980) or Carter (1988).

2.4.2.2 *Superficial drift currents*

Winds, triggered by atmospheric movements, create drift currents oriented in the direction of the wind from which they originate. Winds move the superficial water molecules of the mixed layer and create many oceanic currents. Marine waters are often stratified, the current speed not changing between the surface and the thermocline* throughout the entire depth of the mixed layer.

The effect of winds varies depending on their intensity and their fetch (see above), but it is noticed that in the coastal zone or a lagoon, their influence is preponderant owing to the shallow depth. In addition to the currents created, the winds play a major role in the transfers between ocean and atmosphere over all ocean surfaces.

Current speeds at the surface are estimated to be one or two percent of that of the wind, which is not insignificant; several authors have provided numerical relationships:

- Guilcher (1979), the relationship between wind speed and superficial current speed (with V = current speed and W = wind speed) is:

$$V = W/100 \times 1.5$$

- This formula increases to $V = W/100 \times 5$ according to other authors.
- Crépon (1993) provided much more detailed and complex formulae relating winds and currents, as it must be realised that, in relation to mixed layer thickness, temperature and salinity (determining its density), the formulae would not be quite as simple as shown above.

Coastal currents often run parallel to the shore, which constitutes a barrier. Thus, in the English Channel a mean flux of water of 120,000 m^3/s passes Britain into the Baltic Sea; this represents two-thirds of the flow of the Amazon along the coasts (but here the littoral population is about 100 million inhabitants).

Figure 2.9 shows the major role played by this phenomenon which creates significant upwellings. The deviation of Rhone waters towards the Golfe du Lion,

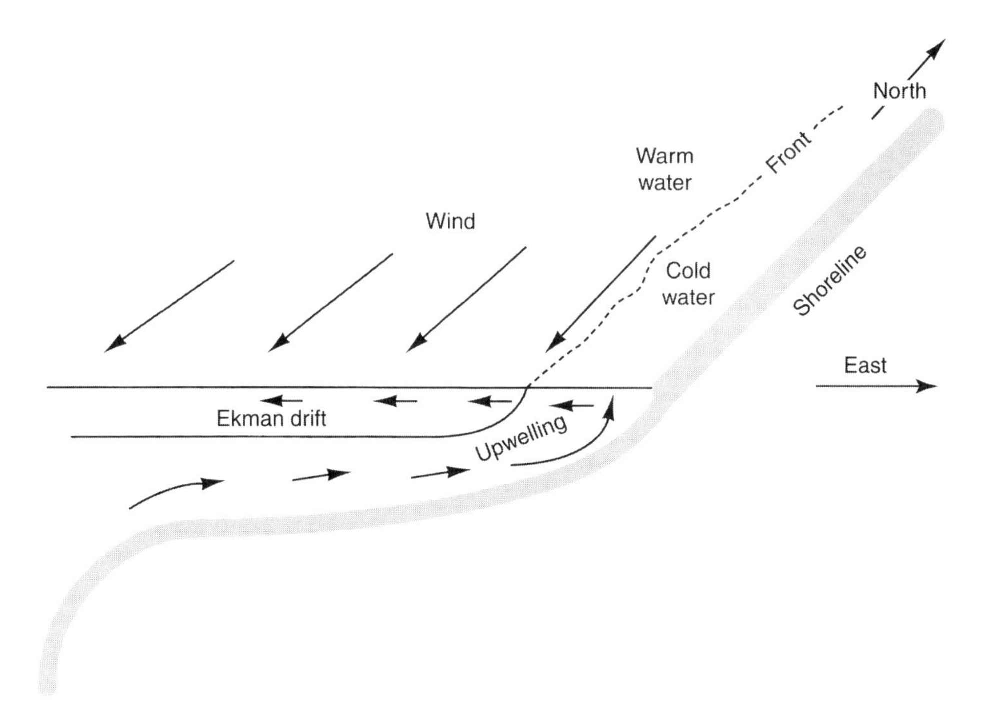

Fig. 2.9. Diagram of an upwelling region. The drift of surface waters created by the wind combines with Coriolis' force to create Ekman drift which carries warm surface coastal waters towards the open sea. They are replaced with cold, nutrient-rich waters from deep waters. The cold water remains separated from the warm water by a front. (Adapted from Mann and Lazier, 1991.)

where it fertilises shallow waters, constitutes another example, demonstrating Ekman drift (although other features would be likely to intervene). This type of current operates above the thermocline.

2.4.2.3 Compensation currents

As well as differences in density, differences in level can constitute another cause of currents. The penetration of Atlantic water into the Mediterranean, where evaporation is not compensated for by inputs from rivers, constitutes a compensation current.

2.4.2.4 Coriolis force, Ekman drift and coastal upwelling

The rotation of the earth on its own axis (1600 km/h at the equator), subjects moving fluid masses to the Coriolis effect. This turns the winds, and also the currents, towards the right in the northern hemisphere and towards the left in the southern hemisphere. This deviation of currents in relation to the wind which creates them is known as Ekman transport or drift. Thus the trade winds in the Atlantic and Pacific create north-equatorial currents.

Ekman drift results from viscosity forces and Coriolis forces and Ekman transport Me is described by the formula:

$$\mathrm{Me} = -\mathrm{T}/\mathrm{F}$$

where T = wind force at water surface, and F = Coriolis factor.

A force of $0.1\ \mathrm{N/m^2}$ and a Coriolis factor at a latitude of 45° of $\approx 10^{-4}$/s gives Me = 1000 kg/m/s, i.e. the transport of one tonne of water per second flowing at 90° to the right of the wind, for each metre, along a line parallel to the wind (Mann and Lazier, 1991).

If, as a result of wind direction and Ekman transport, the water moves away from the coasts, this leads to its replacement by deeper water. Superficial drift of water, due to wind, thus displaces enormous masses of liquid and forms the basis of upwellings: equatorial upwellings, for example, are approximately 100 m in thickness and 200 km wide and reach speeds of ≈ 1 m/s.

Upwellings constitute the best-known type of vertical current, but many more exist and play defining roles (see Bakun, in Troadec, 1989, pp. 186–187).

2.4.2.5 Oceanic fronts

When two currents with different directions meet, they create a "convergence" (or a divergence, in the opposite case). As these water masses with different characteristics do not mix, their border constitutes a "front". These fronts are often indicated by the presence of floating objects carried by the current which passively concentrate at these borders: visible accumulation of plankton can also be observed. Such zones are sought out by fishermen, as their richness in plankton also attracts the fish.

Islands, reefs and headlands create very complex, three-dimensional currents which in turn create physical and biological fronts. They have an influence on pelagic organisms, as secondary currents, with which they are associated, create convergent zones (see Fig. 4.4).

2.4.2.6 Taylor's stratified columns

In a rotating fluid, the liquid situated above a raised area tends to stay trapped there, which would explain, for example, the richness of banks such as the Rockall or Faeroes in the north Atlantic (pelagic larval stages are trapped in a plankton-rich column of water, thus ensuring good recruitment of exploited populations). These effects require large-scale banks, so that the Coriolis force can dominate the forces of friction (see Bakun, in Troadec, 1989, p. 171).

2.4.3 Contribution of freshwater in the coastal zone

When freshwater flowing down rivers reaches the sea, it is lighter and therefore lies above the denser sea water (Fig. 2.10). This difference in level leads to mixing, caused by the horizontal turbulences created by the friction between fresh and salt waters. However, as a result of the weakness of the turbulences and the high density of the salt water, their action has a very localised effect. The differences between the two layers thus blur very gradually.

The profile of the contact zone between fresh and salt water can vary depending on factors such as estuarine topography and the size of tides.

- In the absence of tides, the presence of a saline wedge is often noted (contact zone between fresh and salt water in the form of a gradient); in estuaries, salt water can ascend several kilometres up from the mouth of the river, beneath the freshwater (see colour plate section, plate 1). Mixing is limited to small turbulences created by the friction phenomena described above. Reaching the

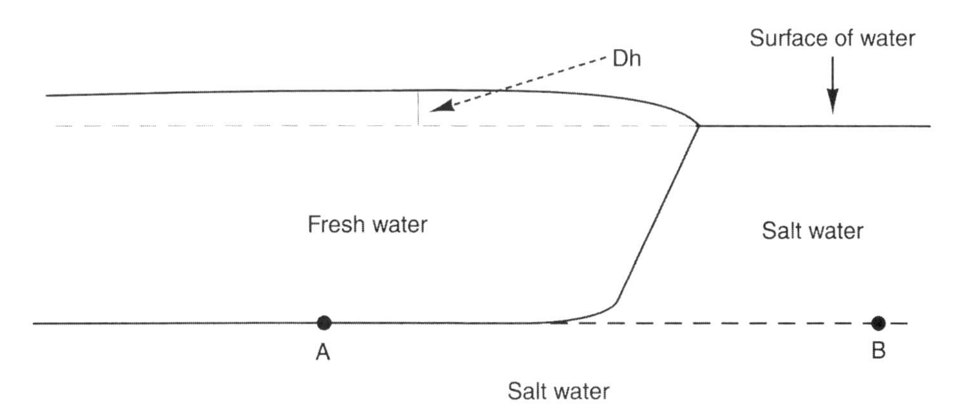

Fig. 2.10. Contact zone between the freshwater layer (on the left) and salt water (on right and below). If the pressures at A and B are equal, the height of the surface above A must be greater than the height at B. This difference in level, Dh, is a function of the thickness of the freshwater layer, usually a few centimetres thick. (Adapted from Mann and Lazier, 1991.)

sea, this freshwater can extend out above the sea water, forming a "plume": the Connecticut has a 2 m-deep plume of freshwater which extends for more than 10 km from the coast. The edge of the plume at the open sea is marked by a front.

- In the presence of tides, water mixing due to tidal currents and to the progression of freshwater towards the sea results in the estuary having well-mixed waters; there is a progression from freshwater to brackish to salt water.

At the mouth of a river or estuary, the water will be subjected to Coriolis forces and move to the right (northern hemisphere); the profile of the coast determines the course of this flow, which results in a coastal current parallel to the coast. This is the case for the Rhone current driven towards the Golfe du Lion, through which it travels until reaching Cap de Creus.

In the coastal zone, it is thus noted that the flow of freshwater can contribute to stratification, while the currents created by winds and tides are mechanisms which tend to mix. Horizontal movements of water are significant, in contrast to the open sea where vertical movements dominate (Mann and Lazier, 1991).

2.5 DYNAMICS OF MARINE SEDIMENTS

2.5.1 General

The coastal areas frequented by holidaymakers are mainly beaches. They are made up of an accumulation of sand, which is the best-known marine sediment. It delineates the "battlefield" between land and sea.

Sand is either the final product of direct erosion of rocks by the sea, from cliffs or submerged rocky zones, or is produced by alluvial transport from rivers. It can also have a biological origin: coral reefs or empty shells crushed by the sea are the best-known main constituents of sand. Sand composition varies and characterises its origin. While sands can be regraded by winter storms, beaches constitute one of the most stable coastlines as a result of their capacity for energy absorption (Brown and McLachlan, 1990).

The diameter of sand grains is between 0.1–2 mm, their density varies according to their composition between 2.66 (quartz) and 2.95 (carbonate). Granulometric analysis using a sieve allows a sand to be characterised by its granulometric frequency curve, which gives the mode, Do, i.e. the most frequent diameters of its grains (Bonnefille, 1980). Gravels (2–26 mm) and pebbles (2–50 mm) are more-or-less round pieces of rock. Average-sized sand particles (0.25–0.50 mm) are displaced by currents greater than 1 m/s.

Sand sticks together as a solid if its water content does not exceed a critical value. An excess of water transforms it into a liquid form again. On sandy bottoms there is thus a layer of fluid sand, of variable thickness, lying above the compact sand.

When the particles measure less than 60 μm and there is no organic matter, the term used is silt, while muds contain some organic matter or are linked by colloids (Bonnefille, 1980). The falling speeds of particles of fine mud (0.1–30 μm) are very

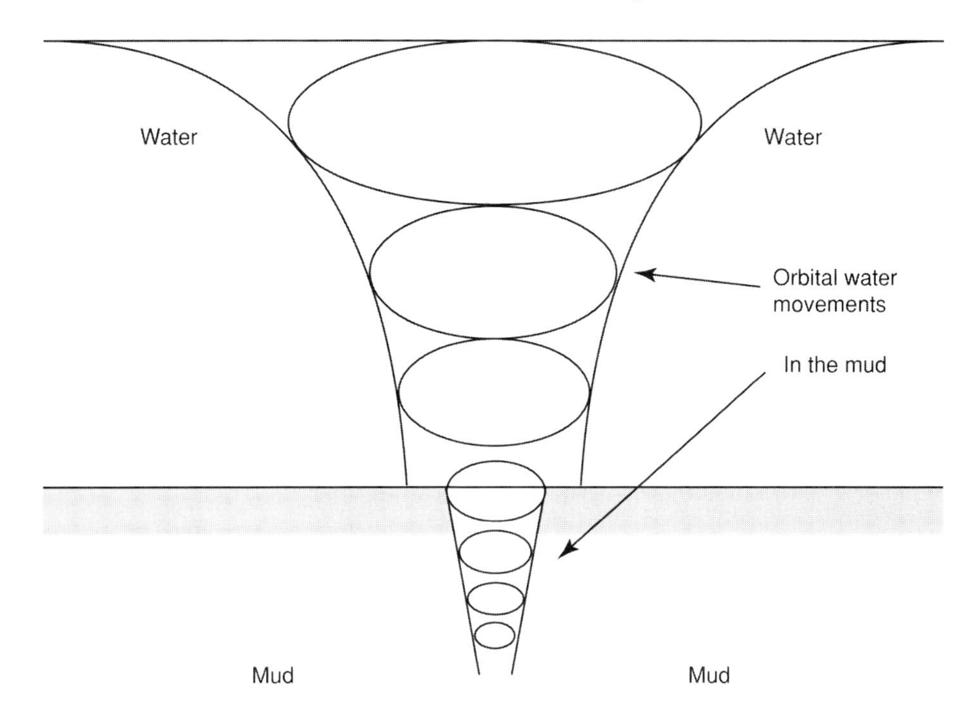

Fig. 2.11. Effects of swell on muddy seabeds.

slow (1 mm in 10–10^6 seconds). In sea water, characterised by dissolved salts, there is flocculation of particles, creating "floccules" of 0.5–1.5 mm, whose sinking speed is 10–100,000 times greater. A swell destroys these "floccules", which slows the sinking speed by a factor of 5–10.

The deposited floccules of mud constitute a flocculated suspension, which is still fluid and whose concentration is about 100 g (dry weight) per litre. Thus on muddy bottoms, there is a continuum between water and sediment. The water is eliminated slowly over time (concentration doubles in three months) and thus the layer is no longer fluid.

Very thick layers of mud still retain a certain fluidity. Under the influence of a swell, there is orbital movement in the mud, analogous to that found in the water, but weaker in amplitude, as the higher viscosity mud absorbs more energy (Fig. 2.11); there is also transfer of mud in the direction of swell propagation; this can result in resuspension of the mud.

2.5.2 Transport of sediments to the seabed

On a homogeneous bottom, an increasing current speed does not cause any movement until a critical speed is reached. Above this speed, some grains move, then asymmetrical ripples form with eddies, which modify flow.

The movement combines transport by carriage along the bottom, with suspension, when the hydrodynamic forces are capable of lifting particles off the bottom. A concentration gradient is therefore established immediately above the seabed, the heaviest particles being the most abundant close to the bottom and the finer ones lying well above.

At higher speeds, the ripples disappear and give way to dunes which progress in the direction of the flow. If the current increases again, anti-dunes, which go against the current, appear. Bonnefille (1980) details the calculations used to characterise these displacements, especially in the presence of swell which, being an alternative kind of flow, complicates the phenomenon.

2.5.3 Littoral transport by swell

Movements of the sea reshape and reclassify seabed materials, at least in the coastal zone. Beaches which have a variable slope of 1–5° are smoothest when the swell is deep and the sand is fine.

Waves transport sands along shores mainly when they meet the shore obliquely. The effect is maximal for angles of 50–65°. This is visible along the length of beaches, each wave taking sand from the bottom and carrying it high up the beach again; this is significant where waves do not break (camber <2.5°).

Waves with a greater camber break and create a bar. Between this bar and the coast, the wave which has broken carries large quantities of sand in a "littoral swell" parallel to the shore. Such currents can be modified by natural obstacles (rocky shores) or artificial ones; it is known that in the course of littoral swell, spits or breakwaters perpendicular to the shore are eroded on the downwind side and produce an accumulation of sand on the windward side.

Swell reshapes the seabed to about 30 m when its period is six seconds and to about 100 m for a period of 12 seconds. It is the deepest waves which transport the most sediment.

On the bottom, shallower sediments (coastal lagoons) reshaped by the swell or currents are oxygenated by contact with turbulences; nutrient salts, organic substances (but also pollutants) can be diluted and displaced again with the attendant consequences for marine life. Frontier and Viale (1991) compared these reshapings to "ploughing".

In contrast, below the thin superficial layer which is a few millimetres thick, the sediments are anoxic*.

The decomposition (by bacterial activity) of carcasses and various wastes from zooplankton (moulted exoskeletons, secretions, faeces) to their reutilisable mineral constituents is slow, and the waters rich in these nutrients* are thus waters of average depth: the productive layer is limited to the euphotic zone (where photosynthesis occurs). According to the classic diagram produced by Margalef, there are thus two superimposed systems which interconnect: the upper one exports its excess to the second, deeper one. This mechanism conditions the production of living organisms in marine waters and we shall return to it.

2.6 CONCLUSIONS

The physical phenomena which exist at sea are very varied and result in the setting in motion of waters on variable time and size scales. There is, however, a level of transition. As a result of the effect of the viscosity of water, described by the Reynolds number, living organisms smaller than 1 mm have difficulties in moving; these difficulties become progressively greater as their size decreases.

Waves, tides and currents create turbulences which result in the homogenisation of ambient conditions within coastal waters but, despite this, they retain a marked polarity: sunlight comes from above and, as a result of physical phenomena, it is the deep waters which provide the nutrients.

Movements of waters, whatever their origin, contribute largely to interactions between living organisms and their environments. Nutritive elements and dissolved gases are dispersed and can thus make contact with users which are not mobile (filter feeders, plankton). Eddies result in the homogenisation of ambient conditions within water masses, bringing together materials which interact (Frontier and Viale, 1991).

Physical phenomena are thus most often the cause of biological phenomena, while the opposite is rare.

2.7 BIBLIOGRAPHY

Bompais X. *Les filières pour l'élevage des moules*. Ifremer Éd., Plouzané, 1991.

Bonnefille R. *Cours d'hydraulique maritime*. Masson Éd., Paris, 1980.

Brown A.C. and McLachlan A. *Ecology of sandy shores*. Elsevier, Amsterdam, 1990.

Carter R.W.G. *Coastal environments*. Academic Press, London, 1991.

Crépon M. Initiation à la dynamique de l'océan. *Océanis* 1993; 19(2): 110.

Frontier S. and Pichod-Viale D. *Écosystèmes: structure, fonctionnement, évolution*. Masson Éd., Paris, 1991.

Guilcher A. *Précis d'hydrologie marine et continentale*. Masson Éd., Paris, 1979.

Mann K.L. and Lazier J.R.N. *Dynamics of marine ecosystems*. Blackwell Scientific, Boston, 1991.

Revaud d'Allonnes M. Turbulence océanique, l'effet coquilles Saint-Jacques. *Science et Avenir* 1994; 98: 24–29.

Thorpe S.A. and Hall J.A. Bubble clouds and temperature anomalies in the upper ocean. *Nature* 1987; 328(2): 28–51.

Troadec J.P. (Ed). *L'homme et les ressources halieutiques*. Ifremer Éd, Plouzané, 1989.

Vogel S. *Life in moving fluids: the physical biology of flow*. Willard Grant Press, Boston, 1981.

Schmied K. La mécanique des vagues. *Science et Avenir* 1994; 98: 40–46.

Weller R.A. Mixing in the upper ocean. *Nature* 1987; 328: 13–14.

3

Marine life and production of living matter

3.1 INTRODUCTION

It has been shown in preceding chapters how astronomic, atmospheric, climatic and hydrodynamic phenomena act together in a complex fashion to ultimately determine the conditions for aquatic living organisms. These are the conditions, either favourable or unfavourable, which will in their turn determine the abundance of populations of marine organisms, according to their true biological characteristics which must therefore be examined.

3.2 FEATURES OF MARINE ORGANISMS

Among the features of aquatic ecosystems, water is never a limiting factor as it is on land, and other limiting factors tend to be temporary. There are no physical barriers and no shelter in the three-dimensional aquatic domain, neither are there any spatial terms of reference: Hamner (1988) stated that there is no equivalent to a forest in the sea, in which phenomena can be studied on a fixed spatial base; life is small and dispersed. Variations in temperature are no greater than 30°C, while they can exceed 70°C on land. As a result, there are major differences between terrestrial and aquatic ecosystems.

All aquatic organisms possess several common features which differentiate aquaculture from terrestrial culture and rearing:

- On land, primary production created by the process of photosynthesis is carried out by plants of significant size, while primary production in the aquatic environment is due mainly to very small species (tens or hundreds of μm in size), which man is unable to harvest or which are difficult to identify.
- Aquatic animals live in an environment of a density close to their own, which allows a reduction in supporting structures (skeleton) and sometimes the disappearance of locomotory apparatus (sessile molluscs).

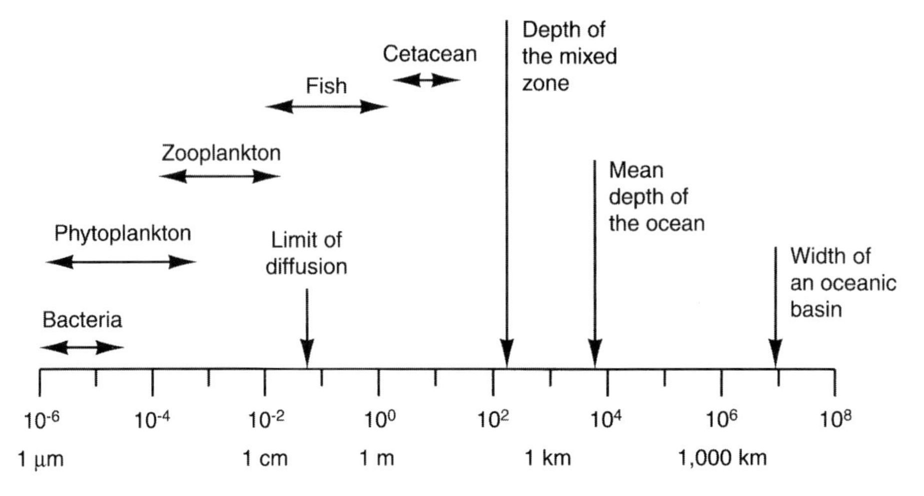

Fig. 3.1. Scale of sizes, from 1 µm to 100,000 km, showing characteristic organism sizes and the scale of physical dimensions (adapted from Mann and Lazier, 1991).

- Metabolism is based on proteins and not on carbohydrates.
- The size of the living organisms is the determining factor in the structure of food chains (Fig. 3.1).
- The absence of thermal regulation (poikilothermic animals) considerably reduces metabolic expenditure and allows better conversion of ingested food into flesh.
- The majority of aquatic species have a very high reproductive potential: a prawn or a fish spawns several hundreds of thousands eggs per year, a mussel or an oyster, several million eggs and the number of spores produced by algae is even greater.
- Reproductive methods are often different from those of farmed terrestrial species: that of the algae is very far removed from cultivated plants and those of molluscs, crustaceans and fish are equally far removed from those of birds or mammals which are the main species reared on land. Eggs of aquatic species hatch into very tiny animals, whose morphology and mode of life are very different to that of the adults. These forms, called larvae, very rarely receive parental care and are passively dispersed throughout huge volumes of water in the seas and oceans.

  The rearing of these planktonic stages is poorly managed as much of the basic biology is poorly understood: nevertheless, it is vital for the aquaculturist, since this stage provides the individuals which are the starting point of the industry.
- Filter feeding animals, such as bivalves (including oysters and mussels), have no terrestrial equivalent: they continuously pump sea water in order to extract phytoplankton or particles which occur there naturally and on which they feed.

Filter feeders are found as frequently in the coastal environment as in the open sea (planktonivorous fishes), both on hard substrates (mussels, oysters, worms, barnacles, etc.) and on soft substrates (scallops, oysters), or buried in the sediment (clams, worms, etc.). These filter feeders are capable of using primary planktonic production, and also the particulate organic matter which is either in suspension or deposited on the bottom. Their feeding spectrum extends beyond primary production *sensu stricto*, as some are detritivores; this is important from the point of view of their management.

- The duration of the life cycle of very many of the smaller species which constitute the first meshes of marine trophic networks is of the order of several days to several weeks, often less for the smallest forms. The fluctuation of physicochemical characteristics in an environment with no physical barriers is very quickly and intensely reflected in the biomasses of these populations, as a result of the brevity of their generation time.
- Lastly, the relative abundance of living organisms in the aquatic environment decreases rapidly as their size increases.

Figure 3.1 shows the relative sizes of marine organisms and the environments which they populate.

3.3 MAJOR CATEGORIES OF AQUATIC ORGANISMS

Organisms populating these environments belong to two categories:

- Those which live on the bottom (settled, attached or buried) constitute the benthos and are called benthic organisms. They exist in relation to a substrate and there is a distinction made between fixed benthos (oysters, mussels, algae or attached vegetation) and those which are mobile (scallops). Fish which live in contact with the bottom are designated benthic; demersal fish live close to the bottom, above the seabed.
- Pelagic organisms populate the open sea in large numbers. They move quite freely in this space: nekton designates species which can move over large distances (fish) as opposed to plankton. Tunas which frequent the mixed surface zone are well-known epipelagic species, but the fish constituting the most significant biomass in the ocean are the mesopelagic fishes less than 10 cm in length which live at depths of between 200 m and 1000 m.

We shall not be detailing the biology of these diverse groups or the ecosystems in which they live: they have been the subject of many zoological or ecological studies (see for example: Pérès and Devèze, 1963; Meadows and Campbell, 1993; Collignon, 1991; Barnabé, 1994) or in the magnificent treatise on biology by Purves *et al.* (1994).

The term "plankton" describes a vast array of organisms from bacteria to larval fish, mainly small in size, which are found in pelagic waters (Hamner, 1988). Most of these are capable of active swimming, or of movement by passive transport using changes in their buoyancy systems. We will adopt this definition, which is closer to reality than the old definition: "an assembly of living organisms incapable of horizontal movement, but capable of vertical movement in a mass of water".

Plankton is subdivided into size groups:

- megaplankton: can reach several metres (medusae, tunicates)
- macroplankton: >5 mm (krill, fish larvae, *Sagitta*)
- mesoplankton: 200–5000 μm (large diatoms, diatom chains, rotifers, copepods, mollusc and worm larvae, fish eggs, etc.)
- microplankton: 20–200 μm (single-celled algae, ciliates, nauplii and copepodite larvae of copepods, mollusc larvae)
- nanoplankton: 2–20 μm, partially made up of ultraplankton (2–10 μm)—unicellular algae, flagellates and small ciliates
- picoplankton: 0.2–2 μm, algae and bacteria, making up the "bacterioplankton"; picoplankton are caught in filters with a porosity of 0.2 μm.

The ultraplankton includes living organisms less than 10 μm in size. Other terms refer to different biological characteristics:

- holoplankton corresponds to permanent plankton; meroplankton to temporary plankton (crustacean, molluscan and fish larvae, for example). The tripton is made up of inert particulate material (non-living).

3.4 PHOTOSYNTHESIS AND FOOD WEBS

The dominant influence on marine life is the penetration of light into the water. Under the action of solar light, inorganic materials in the environment (carbon dioxide and nitrogenous and phosphoric compounds) are transformed by plants into organic matter (constituting living cells); this process is called photosynthesis. It occurs within water masses and is carried out by autotrophic organisms which belong either to the plankton or the benthos. This type of production is called primary production since it is the starting point of living matter both on land and in water.

Only a small part of incident light is utilised, as a result of the phenomena occurring at the water's surface, and also because photosynthesis is not a very efficient process. Only 20% of the radiant solar energy penetrates into the water and only 1% of this portion will be used for photosynthesis, which therefore is not limited by available solar energy.

Photosynthesis constitutes a wide array of physical, chemical and biological interactions which affect the primary producers and their environment; in physicochemical terms, it suffices to know that light energy (photons) causes electron transport within chemical compounds (chlorophyll pigments [chlorophyll a, b and c], carotene and its derivatives, and phycocyanin and phycoerythin). This energy is then transferred by specific transporters (ATP) and used in the construction of organic molecules unique to living organisms (carbohydrates, then proteins and lipids). The surrounding water is the source of hydrogen and oxygen as well as carbon in the form of carbon dioxide. It is the availability of nitrogen and phosphorus (constituting 8% and 2.5% of the biomass respectively) which is ultimately limiting, as we shall discover.

Photosynthesis occurs within pigments of plants and certain bacteria in shallow waters, penetrated by sunlight (euphotic zone). In simple terms, photosynthesis carries out the conversion of inorganic carbon (see Chapter 1) into organic carbon. A detailed, accessible and illustrated review of the process can be found in a comprehensive biology textbook.

Photosynthesis in the aquatic environment is subject to unique conditions relating to penetrating light and availability of certain essential mineral materials (see Chapter 1): the main elements required are dissolved at various concentrations in the water surrounding the living cells, allowing direct exchanges across cell walls. We shall see (Section 3.5) that other organic substances, present in infinitesimal quantities in the water, also play a determining role. The abundance of chlorophyll (linked to the number and size of the photosynthetic organisms) dictates the intensity of the primary production, but the rate of the process is dependent on temperature and many other factors, such as buoyancy of phytoplankton and movement of water masses.

This living plant matter becomes food for animals known as herbivores, which are incapable of synthesising the basic constituents of living material (carbohydrates, proteins or lipids). They constitute the second link of what is called the trophic (or food) chain and, as a result, are called secondary producers. These herbivores are in turn the prey of carnivorous animals, which make up the third link of the food chain, and which can be consumed by other carnivores, representing the 4th trophic level, and so on. The synthesis of living matter from organic matter by the phytoplankton characterises autotrophy and is unique to plants and certain bacteria. All other species are called heterotrophs and thus depend on autotrophs for their nutrition.

This transfer of living matter across successive levels in a simple food chain is not the only transfer occurring in water. Recent fundamental research has demonstrated other processes, both of primary production and interactions in the microscopic domain, which have been unnoticed until now: thus we shall use the term aquatic trophic network, in which the relationships between "eater" and "eaten" are numerous and complex.

3.5 PELAGIC PRIMARY PRODUCTION

The processes regulating this primary production are the same in principle, whether in fresh or salt water. We shall separate autotrophic organisms on the basis of size.

Phytoplankton communities dominate aquatic populations, since their habitat is the pelagic zone (the largest on the planet), but benthic or floating algae (sargassum, filamentous algae, diatoms, etc.) and certain types of attached bacteria are also aquatic primary producers. Macrophytic algae are of little importance in terms of oceanic production compared to planktonic species (they represent only 0.2 billion tonnes, in contrast to several tens of billions of tonnes of phytoplankton), but they retain an important position in terms of aquatic living resources; their direct consumption by man and their role in other forms of rearing/farming have made them a special type of producer.

Primary production is usually measured as the number of grams of carbon fixed per m^2 of sea surface per year (sometimes per day). Primary production must not be confused with standing crop, which is the amount of phytoplankton material present at any given instant in time ($g\,C/m^2$). The phytoplankton standing crop is measured by direct microscopic counts of sedimented water, or by extraction of chlorophyll from sea water using acetone.

Primary production is measured by estimating oxygen release or carbon dioxide uptake during photosynthesis:

$$CO_2 + H_2O \rightarrow (H_2CO)_6 + O_2 \qquad \text{(Eq. 3.1)}$$

However, the phytoplankton respires and consumes oxygen while producing carbon dioxide during respiration, according to the following inverse equation:

$$O_2 + (H_2CO)_6 \rightarrow CO_2 + H_2O \qquad \text{(Eq. 3.2)}$$

In order to measure photosynthesis, the intensity of respiration must thus be known. This is done by comparing the consumption of CO_2 in the light and in the dark. Two bottles, one opaque and one clear, are filled with sea water and suspended in the sea or placed in conditions of controlled light and temperature. Both photosynthesis and respiration occur in the transparent bottle, but only respiration occurs in the opaque bottle. The incubation lasts between 3 and 12 hours. The difference between the amounts of CO_2 absorbed by the two bottles allows the measurement of photosynthesis. At the start of the incubation, an isotope of natural carbon (^{12}C) is injected into the bottles in the form of ^{14}C as $^{14}CO_2$. This radioisotope can be detected by specially calibrated counters.

The estimation of the weight of organic carbon produced by global primary production of the oceans varies, depending on author, from 13 to 155 billion tonnes, or even 500 billion, with a mean estimate of between 20 and 50 billion tonnes. In comparison, terrestrial plant production is estimated at 25 billion tonnes, but the biomass of terrestrial plants is 500–800 times higher. Thus the production capacity per unit weight is much higher in the oceans than on land.

Pelagic primary production is often expressed as the carbon content of living organisms (assuming that 1 g of dry organic matter corresponds to 0.5 g of carbon). It varies from zone to zone; it is estimated at about 500 kg of organic carbon/ha/year for the open sea and 1000 kg of organic carbon/ha/year for coastal waters, which are much richer (see Chapter 1).

This natural production is low, as the concentration of essential inorganic salts in the surface waters is too low (0.001–0.08 mg N per litre, 0.002–0.003 mg P per litre) for good primary production (which requires 0.05 mg N and 0.005 mg P per litre). Much higher levels are found in deep waters (0.2–0.5 mg/l N and 0.05–0.09 mg/l P); also upwelling zones, such as the coastal waters off Peru, produce around 15 million tonnes of anchovy per year from an area of about 60,000 km^2, i.e. 2.5 t/ha/year (wet weight). Upwelling zones, which represent 0.1% of the oceans' surface, thus produce 50% of the exploited fish.

The transfer of living matter in food chains occurs with a yield of around 10% at each level of consumption. The classic plan of 1000 g of phytoplankton transformed into 100 g of sardine, to 10 g of mackerel and finally to 1 g of tuna is still accepted.

Primary production, in the strictest sense, constituting "initial" oceanic production (expressed in dry weight) is more than 700 times greater than that of fisheries resources expressed in wet weight (as a fish consists of around 80% water, the relationship in dry weight becomes 1/3500). Fisheries only correspond to 1/1000 of the basic production of the oceans: the main portion of catches is represented by carnivorous species (at the 3rd to 6th level in the food chain) and not by phytoplankton. From another point of view, not all oceanic food chains lead to species which are exploitable by man; many species are rejected.

It is outside our remit to describe the phytoplankton except in summary form, as it is an entire world in itself: the reader can refer, for example, to the works of Pérès cited in the preceding chapter and to the *Atlas of Phytoplankton* (volume 1: Sournia, 1986; volume 2: Ricard, 1987; volume 3 in press); *Manual of Mediterranean Planktonology* (Trégouboff and Rose), etc.

3.5.1 Autotrophic picoplankton and nanoplankton (0.2–20 μm)

This includes phytoplankton and bacterioplankton less than 20 μm in size. The phytoplankton is made up of monocellular algae, the type found in the oceans being *Synechococcus*, a little oval-shaped cell of 1–5 μm. In littoral waters and closed environments, species between 2–7 μm (*Chlorella nana* and *Nannochloris* sp. etc.) can proliferate and are sometimes cultivated for aquaculture purposes. The Coccolithophorids are autotrophic flagellates, 4–5 μm in size, with a calcareous skeleton. They are characteristic of warm and nutrient-poor waters. In the Mediterranean, they constitute 30–98% of the unicellular plankton organisms, with the genus *Coccolithus* being dominant (Pérès and Devèze, 1963).

The bacterioplankton is represented by photosynthetic cyanobacteria of 0.4–2 μm. These organisms can attain densities of tens of millions per litre in aquatic environments. Despite this abundance, for a long time their importance went unnoticed.

These species (with the exception of the Coccilithophorids which are easily identified by their calcareous skeleton) have been identified by fluorescent microscopy and electron microscopy. Their description, at the end of the 1970s, both in marine and freshwater ecosystems, relaunched the debate about the actual photosynthetic productivity of aquatic environments. In fact, the majority of primary pelagic production is carried out by very small cells. They demonstrate optimal photosynthesis and growth at very low light levels, at a level below the marine photic zone where only blue and ultraviolet radiation penetrate (blue-violet radiation of 455 nm). Thus it is noted, for example in the central zone of the Atlantic, that the maximum chlorophyll levels are at a depth of 90 m, at the bottom of the euphotic zone, where the light intensity is only 1% of that at the surface (Glover *et al.*, 1986). These plankton survive even in the dark and have been found at depths of

1000 m. They utilise nitrate, ammonia or urea and are responsible for 50–80% of the total nitrogenous consumption by all phytoplankton.

Primary production by picoplankton ranges between 0.01 mg of carbon per m^3 per hour (mg of $C/m^3/h$) to 31 mg $C/m^3/h$. Its biomass is expressed in chlorophyll and represents between 1 and 90% of the biomass of total phytoplankton chlorophyll. The contribution of the picoplankton to primary production is higher in the open sea, in oligotrophic waters (50–80%) than in coastal zones which are sometimes eutrophic (2–25%). In the Barentz Sea and in other northern areas, this plankton constitutes up to 90% of the biomass of chlorophyll in water (0.05–1 mg Chl/m^3). In the Mediterranean, it dominates in winter and in spring.

Freshwater picoplankton contributes to 16–70% of carbon production and 6–43% of chlorophyll biomass. This contribution increases with depth, reflecting the efficiency of their photosynthetic pigments in using blue light.

Phytoplankton biomass tends to be greater (1 order of magnitude) in the temperate than in the tropical zone (Sargasso Sea), which leads to tropical seas being considered as deserts. However, these differences are less in terms of production: the renewal rate of the biomass is in fact 3 times per day (Sheldon, 1984) and not 0.2 as was believed until then; this author utilised the data from nine series of different methods to achieve his demonstration.

This was a major discovery; it signified that, in the absence of predators and mortality, a biomass of 1 g of phytoplankton creates 3 g of living matter per day. In addition, these data agree with results for the zooplankton present, whose biomass is not particularly low (see below). The growth pattern of tropical oceanic phytoplankton is thus very rapid and the daily average generation time is estimated at 6.5 h, assuming a 12 h-long night (i.e. a generation time of 3 h during the day). With a generation time of 6.5 h and equilibrium between production and predation, 12 h growth results in a production of 23.4 mg of C/l/day, i.e. 1.2 mg C/m^2/day. As stated by Sheldon (1984), the oligotrophic regions of the oceans have been compared to deserts, but they only resemble them in terms of their low biomass; in their other characteristics (high growth rates and rapid recycling), they more closely resemble tropical rainforests.

While it is now well established that the pico and ultraplankton (less than 10 µm) are essential components of planktonic communities (Stockner, 1988), their abundance is limited by predation by flagellates, ciliates and rotifers (in freshwater), but not by copepods which catch slightly larger prey. They are also consumed by other planktonic organisms and by benthic macrophages (see Section 3.9). Such results constitute a new and determining element concerning the production of aquatic ecosystems, as well as for their development.

3.5.2 Autotrophic microplankton (20–200 µm)

Diatoms are often the main constituents of the larger phytoplankton, especially in temperate and cold waters, both in the open sea and in coastal areas. They are capable of concentrating close to the surface, as many are positively buoyant. Despite this fact, they mainly use blue-green radiation from the solar spectrum (red light which penetrates the surface waters has an inhibitory effect) and their maximal photosynthetic depth is between 20 and 30 m. Individual size varies between about 20 and 200 μm, but the frequent presence of protrusions and the fact that they are often assembled in chains result in colonies which can be several millimetres long. The genera *Chaetoceros, Skeletonema, Phaeodactylum, Thalassiosira,* and *Nitzschia* dominate the phytoplankton populations in coastal waters or lagoons. They proliferate in eutrophic waters (ponds, lagoons and lakes) and the first three species are often cultivated for aquaculture purposes. The blue colour of certain reared oysters is due to the presence of blue vessels containing marennine (a blue pigment) in the plankton of the claires*.

Peridinians (dinoflagellates and dinophyceae) are generally less common in temperate waters, but they are dominant in oligotrophic, tropical waters (in this size class) and in the summer in temperate waters. Species with no carapace (naked dinoflagellates) have been reared (*Gymnodinium*) to feed anchovy larvae in captivity. The armoured peridinians have a theca of cellulose plates. They provide food for numerous species of planktonivorous fish (anchovy and sardine) and benthic filter feeders. Their proliferation results in the phenomena of coloured waters. Even at low densities (200/l), some species produce toxins which are accumulated by the molluscan filter feeders that feed on them and are therefore toxic to consumers of this shellfish (cf. review by Alzieu, 1989). This poses problems for shellfish culture in the sea throughout most of the world.

Pelagic production reflects fluctuations connected to hydrology which in turn depends on climatic conditions; thus, in the Mediterranean in winter and at the start of spring, the genus *Chaetoceros* and the ultraplankton constitute the majority of the phytoplankton. At the end of spring, there is a proliferation of *Rhizosolenia* and *Nitzschia* in the surface waters. In summer, dinoflagellates are the predominant group, but diatoms are found in deeper water, at the maximum photosynthetic level (30 m). A second plankton proliferation in the autumn is similarly due to diatoms (Margalef, 1969).

Many other plankton belonging to other groups contribute in a limited way to primary aquatic production. The abundance of diverse categories of phytoplankton is linked to the nutrient richness of the water: diatoms prefer rich waters; peridinians, oligotrophic ones; and the ultra- and nanoplankton "ultra" oligotrophic waters.

From a theoretical point of view, Martinez *et al.* (1983) demonstrated the possibility of nitrogen fixation in the pelagic environment by masses of diatoms with intracellular, symbiotic bacteria, constituting a new source of nitrogen in oligotrophic waters. Recalling the demonstration of input of carbon dioxide into superficial oceanic layers (Thorpe and Hall, 1987), note that carbon and inorganic nitrogen are thus available in these layers and do not always constitute limiting factors for production.

3.6 THE "MICROBIAL LOOP"

3.6.1 Inert particulate material and viruses

The existence of inert particulate material (dead phytoplankton, protozoan tests, copepod exoskeletons, various carcasses, faeces and blobs of mucus, etc.), in parallel to primary production and the food chain described above, has been demonstrated. This is colonised by micro-organisms (protozoa and bacteria) which feed on them at least partially; these "flocks" are capable of growing, either by the absorption of organic substances in solution in the water (it is estimated that 90% of organic matter is in "inert" form, the living form only representing 10% of the total organic matter present in the ocean), or through the activity of the constituent organisms. This inert organic matter is characterised by its carbon content; the organic carbon forming the basis of molecules making up living matter (proteins, lipids and glucides) must be distinguished from the inorganic carbon of the carbon dioxide–carbonates system.

According to most authors, the particulate organic matter which sinks (Fig. 3.2) is an important route for energy and the flow of materials in the epipelagic zone, as it represents a support for the development of microbial activity (bacteria and protozoan detritivores). Between 10% and 20% of bacteria are attached to particles; the remainder are free in open water and constitute the heterotrophic bacterioplankton whose size varies between 0.3 and 1–2 μm. We shall see that this particulate material can also be consumed directly by the zooplankton (Section 3.7.1).

The existence of this inert particulate material, colonised by micro-organisms (protozoa and bacteria), provides the basis for real production "parallel" to primary production *sensu stricto*. This is termed "paraprimary" or microbial production and it constitutes one of the major problems of modern biological oceanography as this material in suspension is mainly utilised by filter feeding animals, which distorts production estimates. According to Newel (1984), this production is comparable to that of herbivores and exceeds that of first order carnivores. This suspended particulate material always forms a large part of the feeding bolus of molluscan filter feeders. Figure 3.3 shows the interactions within a microscopic aggregation of organic matter in the open sea.

The progress of scientific instrumentation allowed Koike *et al.* (1990) to show that some matter, considered to be dissolved, is in fact made up of very small inert organic particles (between 0.38 and 1 μm), which can be distinguished from bacteria and whose abundance is of the order of 10 million/ml. They can be deformed and pass through filters with 0.25 μm pores. Especially abundant at the surface, the confirmation of their existence allows a new approach to the recycling of organic matter in the sea.

In a similar way, Peduzzi and Weinbauer (1993) showed that bioactive substances, very similar to viruses, influence the dynamics of planktonic blooms* in relation to their concentration. The density of bacteriophagic viruses can reach 2.5×10^8/ml in unpolluted waters (Bergh *et al.*, 1989), but these are not viruses which are implicated in human diseases.

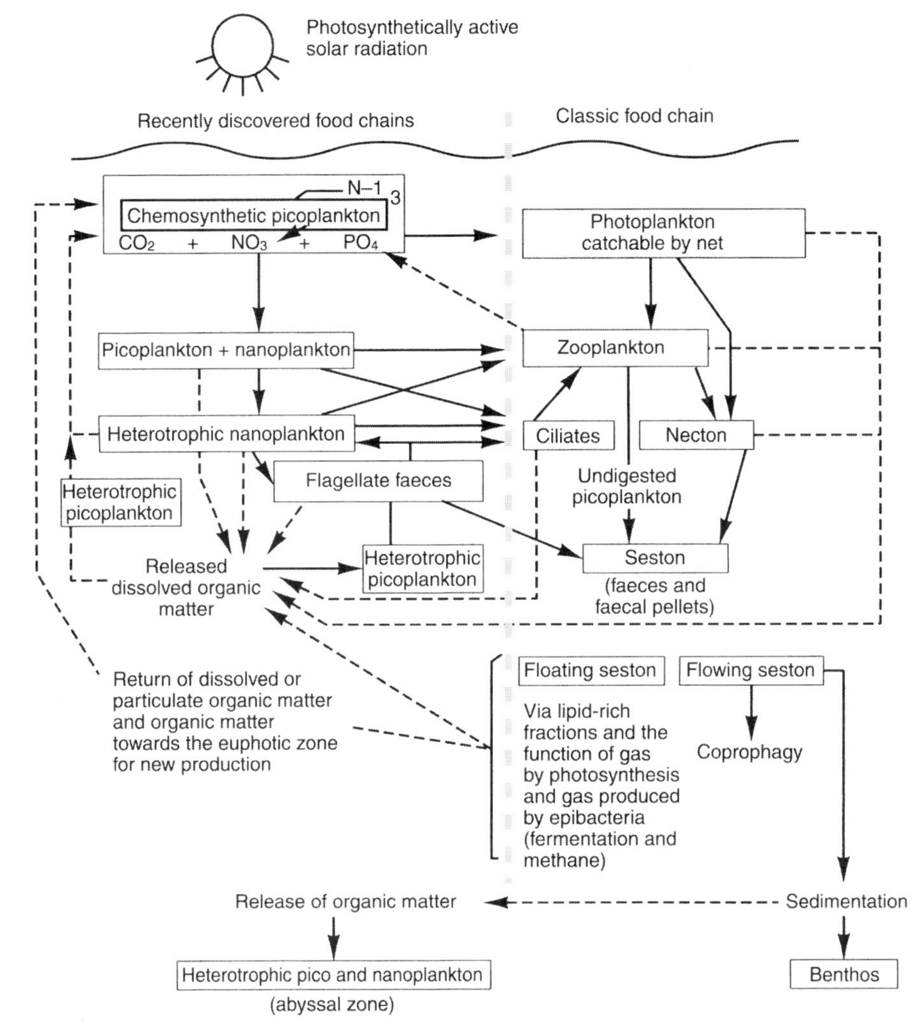

Fig. 3.2. Diagram showing the complexity of aquatic food webs. After a figure in Sieburth and Davis (1982), cited in Stockner (1988).

3.6.2 Dissolved organic matter and chemical mediators

This dissolved organic matter represents excretory products (faeces, urine, excreta and various substances) or the products of decomposition of living organisms after their death. Close to the coasts, terrestrial inputs from rivers or the atmosphere provide a small supplement. Dissolved organic matter is found in the form of simple molecules (e.g. amino acids) and also as more complex molecules ("marine humus") which have only been identified in the last decade using modern methods; polysaccharide molecules have also been isolated. The concentration of dissolved organic carbon in marine surface waters varies from 1 mg/l to several mg/l and the world's

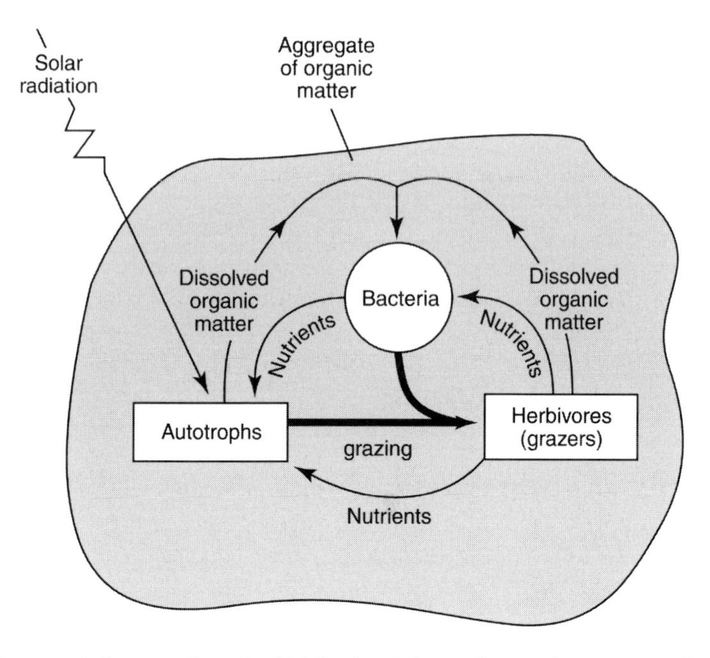

Fig. 3.3. Conceptual diagram of a microbial food web in a microscopic aggregate of organic matter (adapted from Mann and Lazier, 1991).

oceans constitute the largest reserve of organic carbon (estimated at 665 billion tonnes).

Particulate organic matter only represents about 9% of oceanic organic carbon; the majority (89%) is present in a dissolved state with the remaining 2% contained in aquatic living organisms, according to Fiala-Médioni and Pavillon (1988). Zooplankton corresponds to no more than 0.2% and fish to 0.002%. It should also be noted that coastal zones, which represent 8% of the oceans' surface, produce 24% of the organic carbon with a marine origin and 90% of the exploited marine resources.

Azam *et al.* (1983) estimated that 5–50% of the carbon fixed by primary production (transformed into phytoplankton biomass) is released into the environment in the form of dissolved organic matter. These organic molecules dispersed in the environment play several important roles, despite their low concentrations:

- Bacteria can extract them from water at very low concentrations and use them to increase their numbers (Section 3.6.3); this is also true for certain planktonic algae as well as for other small plankton such as the flagellates and ciliates.
- It has been proven that, at very low concentrations, substances such as biotine, thiamine and vitamin B12 can promote primary production (cf. review by Aubert, 1988).
- Other substances, whose nature is not always known (polysaccharides, lipids or proteins, for example) are secreted by certain categories of phytoplankton

and constitute inhibitors to other phytoplankton. Thus, for example, diatoms emit "telemediators" which control the development of dinoflagellates and vice versa, and certain dinoflagellates inhibit the development of other dinoflagellate species.

Chemical communication, based on the detection of substances at infinitesimal concentrations (chemoreception), has been demonstrated in other species and plays a role in many behaviours, including feeding (appetence/appeal of prey), reproduction (attraction between sexes at fertilisation). While it is certain that these mediators ensure the self-regulation of abundance of planktonic species, many other interactions remain unknown, as they are difficult to demonstrate.

Water therefore contains substances capable of acting, at very low concentrations, on the living organisms which exist there. In certain cases, this information has been applied in aquaculture (for example, some substances are known to be capable of instigating the fixation of planktonic molluscan larvae onto a substrate, and others which are appetite stimulants). These data should not be ignored and emphasise the continuing importance of research. Thus the specific composition of algal cultures, whose development in the wild is unpredictable, or the proliferation of dinoflagellates, whose toxins are accumulated by edible molluscs, could be controlled.

3.6.3 Bacteria

Dissolved organic matter is utilised primarily by bacteria. Their small size and large surface area allows the absorption of nutrients at very low concentrations. There is, therefore, a commensalism in the production of dissolved organic matter by phytoplankton and its utilisation by bacteria, and the production of dissolved organic matter is influenced by the availability of inorganic nutrients.

Bacteria concentrations in marine waters are very variable, but relatively low, as shown by the data reported by Azam *et al.* (1983):

Environment	Number ($\times 10^8$/l)	Biomass (μg of C/litre)
Estuaries	50	200
Coastal waters	10–50	5–200
Open sea	0.5–10	1–5
Deep waters	0.1	

These bacteria (0.3 to 1–2μm) are 70–80% free and move within the water column, the remainder being associated with particulate matter; they appear to be attracted by kinesis to plant cells, but stay about 10 μm away from living algae owing to antibiotic emission, while attaching to moribund algae.

Newel (1982) reported experimental data on free-living bacteria in the open sea which are found close to the residues of macrophytes or dead phytoplankton cells, utilising the carbon released by this detritus. Pulverised dried grass (at a concentration of about 30 mg/litre) resulted in bacterial proliferation reaching 10^7 cells/ml in

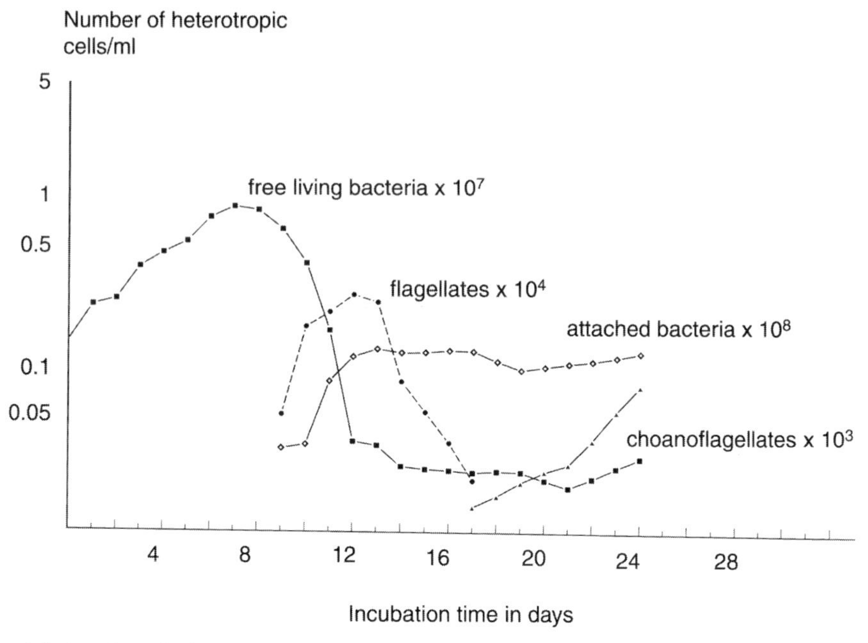

Fig. 3.4. Number of microheterotrophic organisms colonising dead, pulverised aquatic plant leaves at 10°C (14.5 mg of carbon/litre of water). Data from Newel 1982.

one week (Fig. 3.4). Such bacterial populations can provide food for numerous species. The bacterial density is thus returned to around 10×10^6/ml.

3.6.4 Protozoa

Protozoa are the main consumers of bacteria, both in the pelagic environment (Figs 3.2–3.4) and in other aquatic environments such as sewage settlement ponds (Curds, 1975).

In the example given by Newel (1982), the bacterial population declines rapidly as it is consumed by bacterivorous flagellates (*Monas, Oikomonas, Bodo*, etc.) and ciliates. Ciliates such as *Tetrahymena* can consume bacteria at a rate of about 500–600 per hour. The ciliate *Uronema marinum* (whose volume is about $1000\,\mu m^3$), can consume six times its volume per day, but small-volume flagellates ($12\,\mu m^3$) are capable of ingesting 10–30 times their own weight in bacteria per day. Thus these protozoan "browsers" have a major impact on the bacterial biomass in the natural environment. Zooflagellates (heterotrophic flagellates) consume particles between 0.5 and 2 µm in size (Sheldon *et al.*, 1986) and can attain densities of 3×10^3/ml. Ciliates of 12–15 µm consume particles between 1–3 µm in size. Oligotrich ciliates and metazoan larvae of 20 to 600 µm can capture particles of 3 to 15–20 µm. Bacterial density is limited (to about 10×10^6/ml) by heterotrophic flagellates of 3–10 µm. Fenchel's data (in Newel, 1984) also provided a numerical estimation of consumption rate of bacteria by microflagellates in the pelagic environment: on average, these flagellates consume 17 times their weight in bacteria, of which 27% is converted into flagellates. This yield is much higher than the 10% characterising average transfer between levels on a "classic" food web.

Estimates of population sizes of zooflagellates and ciliates have been made in the wild. Ciliates create a flux (measured by capture in traps submerged at different depths) of 0.32 million to 2 million cells/m^2/day and a biomass representing 5–158 times that of bacteria. The flux of zooflagellates varies between 25 to 955 million/m^2/day. This biomass, characteristic of the euphotic zone, decreases with increasing depth.

These predators of bacteria are capable of filtering 12–67% of the water mass per day (they consume bacteria and not the larger phytoplankton).

It is interesting to note that the various links in the "microbial loop" utilise food items which differ from their own size by an order of magnitude: size is the primary determining factor in the structure of food chains. Transfer of biomass in pelagic food webs towards larger organisms is partly effected by direct transfer by predation, implying a size difference between the consumer and the consumed (Azam *et al.* 1983; Borgman, 1983; Sheldon *et al.*, 1986). This is the basis for the basic oceanographic model.

It must, however, be noted that aquatic species play a role not only as consumer but also that of producer for another species; it is in this sense that the term "microbial loop" is used. Thus the excretion of faeces, emission of sexual products or resultant juveniles often constitute a possible form of food for other life forms (Figs 3.2 and 3.3). In this way, reproduction slows the transfer of biomass towards the higher levels of the food chain as it disperses biomass in the form of small gametes, thus returning biomass to the lower end (Borgman, 1983). This is significant since, in oyster culture for example, Héral (1990) estimated that 78% of the total energy absorbed is used for gamete production.

3.7 ZOOPLANKTON (SECONDARY PELAGIC PRODUCTION)

3.7.1 Microzooplankton

Flagellates and other protozoa, as well as the phytoplankton, in turn provide food for microzooplankton between 20 and 200 µm in size (Sherr *et al.*, 1986). This consists mainly of larval and adult stages of copepods, ciliates and larvae of various species, which are especially abundant in the euphotic zone. Transfer of matter between flagellate and copepod occurs with a yield of 30–40% (Sheldon *et al.*, 1986). In the Sargasso Sea, for example, the mean density of microzooplankton varies depending on area from 1900–5400 individuals/m^3 (Böttger, 1982). Copepods represent 90% of this total with molluscs and ascidians (Appendicularia) dominating the remainder. Nauplii larvae represent 51–68% and copepodites 28–42% of the copepodian fraction. Microplankton (specimens of less than 200 µm) constitute 52–68% of the total plankton.

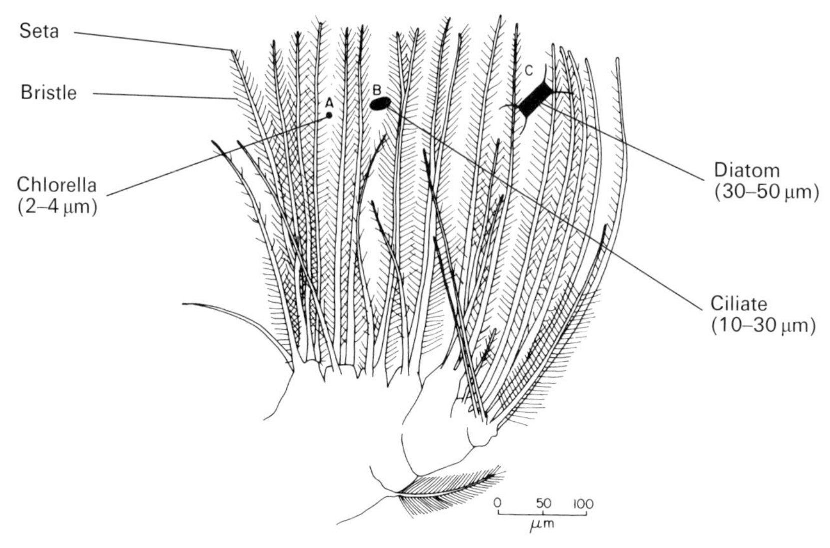

Fig. 3.5. Maxilla of a copepod (*Calanus helgolandicus*) and prey organisms (after Conover, 1976).

Given the enormous input by faecal bacteria produced by copepods or other marine species, the resultant bacterial floccules play an important role in terms of direct food for the zooplankton, and not only in terms of decomposing matter in the ocean. Filter feeders on zooplankton, such as copepods, are effectively capable of browsing on these particles which occur in the form of bacteria-rich floccules. Ogawa (1977) thus fed a range of copepods with faecal bacteria in suspension, produced by various marine species. This phenomenon has also been demonstrated in krill (bacteria–choanoflagellate–krill–vertebrate chain, coexisting with the diatom–krill–vertebrate chain) (Tanoué and Hara, 1986). Other types of detritus are utilised (dead planktonic cells, various carcasses), especially when living prey is rare, but assimilation is less good (9% for detritus, 24% for bacteria and 54% for phytoplankton).

Copepods filter water by creating a current using their appendages, which can beat at up to 600 beats/minute. The current passes over a filter trap (Fig. 3.5) carried on the jaws (maxillae) and particles are carried to the mouth. One *Calanus* can filter 1.5–3 l of water per day and it has been estimated that the zooplankton filters 3–6 million km^3 of water per day in the oceans, resulting in the extraction of 220–440 billion tonnes of suspended matter per year (Savenko, 1988); this emphasises the repetition of the process of resuspending particulate matter.

The significance of the plankton for fisheries has been the subject of a study in Louisiana (Mulkana, 1970). The fishery production of this state consists of 83% filter feeding fish.

3.7.2 Other zooplankton

Other zooplankton include all individuals greater than 200 μm in size (mega, macro and meso plankton). The adult forms of most copepods belong to either the micro or mesoplankton. *Calanus* is 3–5 mm long and is among the largest of the copepods, but many of the small copepods and all their larval forms, rotifers and many other species, escape traditional nets whose mesh are too large. It has been demonstrated that nets with 70 μm mesh size are required in order to capture nauplius stages (Bernhard *et al.*, 1973). Thus Mulkana (1970) showed that in terms of dry weight the biomass of zooplankton represents 113 times the biomass of plankton caught by traditional methods in a net with a mesh size of 330 μm. The majority of historical studies have been carried out using identical nets and it has been on this basis that production data have been calculated: the microzooplankton and larger zooplankton constitute a similar community which was poorly understood by historical planktonologists as a result of this mesh size problem in plankton nets.

The microplankton and larger zooplankton can be collected to feed the larvae of reared species (Barnabé, 1980, 1990); utilisation of prey is related to the development of the mouth size of the species being fed; thus the size of the prey increases as the feeders grow.

Many zooplanktonic species, greater than about 20 mm in size, are fished industrially (Omori, 1980): these are the Scyphomedusae in China, Japan and Korea, krill (Euphausids and Mysids) in the Antarctic, small decapods in various countries and copepods (Canada, Japan and Norway).

These species are used as bait or for rearing fish fry (Norway). It is therefore possible to have a complete symbiosis between fisheries and aquaculture, in terms of the collection and utilisation of natural plankton.

3.8 THE PLANKTONOPHAGIC NEKTON

Plankton constitute the food of molluscan, crustacean and fish larvae, and also for certain adults of these groups, including reared species. Fish larvae belong to the plankton initially (reduced mobility) and then to the nekton where they reach a length of several centimetres (and move actively), demonstrating that all stages of transition can exist between these two groups.

Although they are associated with water masses, plankton are capable of small horizontal movements and some (copepods) carry out vertical diurnal migrations of several hundreds of metres. The relationships between this distribution on a small scale and the possibilities for predation by fish larvae were studied at sea by Jenkins (1988) for two species of flounder: flounder larvae encountered greater variations in prey density by moving several metres vertically than by moving horizontally by about 10 metres. On the horizontal plane, densities vary rapidly. The author noted the presence of multispecies swarms showing an identical positive response to the same factors in the environment, but also negative responses leading to different distributions of other species. The abundance of microplankton at the site studied is well below densities used in the rearing of larval fishes, while flounder larvae show the same growth rate as those reared in the laboratory at prey densities 10 times higher; not only is the hypothesis according to which the fish larvae feed in swarms not verified, but the rearing procedures overestimate the abundance of prey required for reared larvae. In all cases there is a very wide divergence between results observed in the wild and in rearing conditions (Barnabé, 1991).

Contradictory data have been provided by Owen (1981), who studied the grouping of plankton in the environment of anchovy larvae in southern California. He concluded that both mobile and non-mobile species are affected by clustering phenomena on a small scale but found a greater amplitude of distribution on the vertical plane than on the horizontal plane (swarms having a lens-like appearance); this variation is greater in amplitude at the pycnocline.

Many fish remain planktonophages all their lives: this is most often true for small, pelagic species (sardine, anchovy, herring, etc.) but with some exceptions (basking shark). The food of herring has often been studied and the work of Last (1989) collated this knowledge: while the species reaches 30 cm in size, its food is composed mainly of copepods, then euphausids ("krill"), sand-eels, various fish eggs (15% in number, but 1.8% in weight) and sprat larvae. The food items vary depending on the size of the fish. This diet illustrates the way in which reproduction, through production of gametes and larvae, contributes to the recycling of organic matter in pelagic food chains. As we have seen (Fig. 3.2), the food chains leading to the nekton are very complex.

Predation by planktonophagic species on larvae is high and constitutes the main cause of mortality in the natural environment (Hunter, 1984).

3.9 BENTHIC PREDATION

3.9.1 Primary production

Higher aquatic plants and seaweeds produce around 0.2 billion tonnes annually and their role is marginal in global oceanic production. In contrast, these benthic algae are cultivated in south-east Asia for direct human consumption (a Japanese consumes 50 g/day) or for agro-food industries, forming part of manufactured feeds.

Their production is localised in the zones where light can reach the seabed and thus includes lagoon and estuarine ecosystems and also shallow coastal zones and continental waters (ponds, dams, lakes, rivers, etc.). These sites are particularly suitable for development.

The thickness of the water layer in which photosynthesis occurs is less here than in the deep oceanic environment and it might be supposed that, as a result, photosynthesis is less active. In these areas, macrophytes (forming fields of algae such as the laminaria or marine flowering plants such as posidonia or zostera) can attach to the seabed resulting in a total production (phytoplankton + macrophytes) similar to the high pelagic production resulting only from phytoplankton encountered in upwelling zones (Newel, 1982). Total production is close to 1.5% of the incidental light energy in both cases. This author showed that the most productive ecosystems in the pelagic domain (upwellings and coastal zones and lagoons) attain the maximal level of primary production (0.3 to 0.5×10^5 kJ/m^2/year depending on latitude).

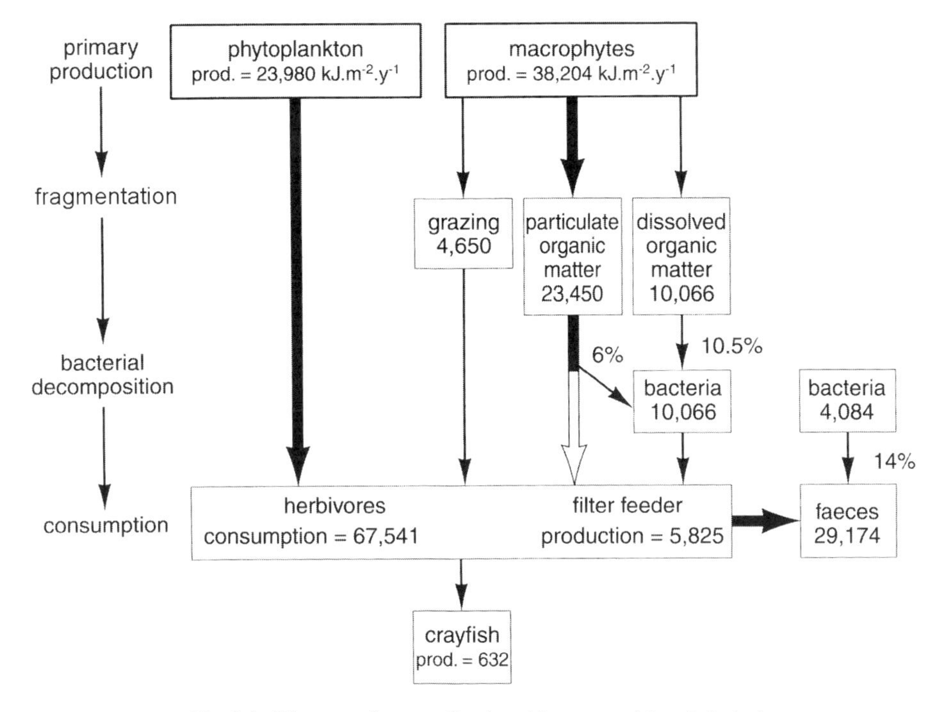

Fig. 3.6. Diagram of energy flux in mid-water and in a kelp bed.

Analysis also shows that primary production of plant material exceeds its consumption by herbivores. Detritivores associated with this layer of macrophytes utilise plant debris and achieve a secondary production estimated to be close to 10% of primary production. This figure demonstrates that plant debris is used directly (without decomposition or remineralisation). Living phytoplankton particles and particles of detritus thus constitute a major trophic resource for consumers in lagoons, estuaries and coastal waters.

The data provided by Newel (1984) also allow the comparison of a chain using phytoplankton waste with a chain utilising waste from macrophytic algae (Fig. 3.6).

One particular type of primary producer merits special mention on account of its role in aquaculture: symbiotic dinoflagellates (zooxanthellae) participate in the nutrition of Tridacnia or giant clams (bivalve molluscs) by fixing carbon by photosynthesis, which is then used by the mollusc. This contribution has been studied, for example, by Fitt *et al.* (1986). The survival and growth of juveniles are greater in the presence of symbionts, but they are not obligatory. In contrast, adults extract a significant part of the carbon necessary for their growth and metabolism from their zooxanthellae. In tropical waters where, as we have seen, primary production is carried out by small-scale organisms and remains "dilute"—and therefore largely inaccessible to man—the symbiotic dinoflagellate–tridacnia production constitutes a short food chain, and the shellfish is directly accessible to man (the reproduction of these molluscs is controlled, at least experimentally). These edible bivalves are in great demand and their mode of nutrition allows a promising aquaculture development which is being pursued in the Pacific and the Caribbean.

Mud covered in water possesses a surface skin which is very rich in life when exposed to sunlight: the biogenic layer of diatoms, cyanophaceae, protozoa, ciliates, flagellates and bacteria feeds very many living species in the mud.

Benthic primary production is very variable. Collignon (1991) reported the following data (in grams of carbon per centimetre squared per year (g C/cm^2/year)):

- Aberdeen area (Scotland), sandy bottom, 10 m deep: 4–9 g C/cm^2/year
- Estuarine mud, Aberdeen, foreshore: 31 g C/cm^2/year
- Sandy mud on the continental shelf, USA: average of 30 g C/cm^2/year
- Unstable sand zone, Marseilles, 1–12 m deep: 41–71 g C/cm^2/year
- Ébrié lagoon, 0–5 m deep: 192 g C/cm^2/year
- Coral reefs with associated cyanophyceae: 730 g C/cm^2/year.

Coral reefs constitute a world which it is not possible to study in detail, but their production is considerable and is estimated at between 2000 and 5000 g of C per metre squared per year. The biomass of fish there is of the order of 500 kg/hectare and production is 100–300 kg/hectare.

3.9.2 Bacterial activity

Bacterial activity on the seabed is particularly important on the continental shelf and in lagoon areas since 83% of mineralisation is carried out there (see Caumette (1989) for more details).

3.9.3 Benthic filter feeders

In the benthos, fixed animals ingest particles in suspension in the open water by filtration; this happens in fouling organism communities. Among the fixed species, molluscan filter feeders incontestably constitute one of the main products of aquaculture. In general terms, filter feeders preferentially retain particles in suspension of between 5 and 25 μm, but when the density of particulate material is low, smaller elements (1 μm) and larger ones up to 40 μm are consumed (Mook, 1981). Lubet (1991) gave details of the biological basis for rearing these species including their food consumption. Note that the same particles can be recycled by several species before falling to the seabed.

The consumers of pelagic protozoa (predators of bacteria) can therefore be cultivated species. In the mussel, the estimated transformation of primary production energy into mussel (62,190 kJ/m^2/year) occurs with a yield of 9.3% (5825 kJ/m^2/year). These mussels can in turn provide food for crayfish (transfer rate of 10.8%); the energy equivalent is 632 kJ/m^2/year. The best illustration of how the energy losses increase the further one moves away from primary production, at different levels of a food web, is shown in Fig. 3.4, using data from Newel (1984).

Despite the complexity of these food webs, it has been calculated that the total "ecological yield" is about 10% for large carnivores, which constitute the highest level of this web. The yield in aquatic food webs is thus comparable to that for terrestrial ecosystems.

3.9.4 Benthic biomasses

The order of magnitude of biomass in the littoral zone is about 1000 g/m^2. This biomass is of the order of 50 g/m^2 at a depth of 200 m and less than 1 g/m^2 in the deep sea; however, these figures can vary enormously depending on site (up to 100 kg of mussels/m^2 on certain artificial reefs).

Mann (1976) estimated that 80% of the 6–7 billion tonnes of living matter representing the benthic fauna of the world's oceans are localised on the continental shelves where the shallow waters allow active photosynthesis. The majority of benthic species have planktonic larvae, which facilitates the recolonisation of new substrates and gene mixing. In long-lived species, such as lobster, the ratio of production* to biomass* (P/B) is low, less than 1, while short-lived animals, such as mussels, have a (P/B) ratio which can reach 10. As a result, the annual production is of the order of double the biomass, that is around 13 billion tonnes.

The economic value of the benthos to man is largely in the form of fish from the seabed exploited by fisheries, but crustaceans and molluscs are often more valuable than fish, despite their lower tonnage.

3.10 THE IMPORTANCE OF THE INTERTIDAL ZONE

We have frequently emphasised the importance of the "mixed layer" in coastal waters; the intertidal zone, the exclusive point of contact between the two biological worlds of land and sea, constitutes their upper limit.

It is at this interface that water turbulence has its most violent effects; however, these movements also create a great nutrient richness, which is easily distributed by this turbulence. As the waters are shallow, gaseous exchange with the atmosphere is equally facilitated, as is the penetration of sunlight.

Specific ecosystems occupy these environments:

- *Coral reefs:* These are constructed by corals and also by algae and some worms and constitute oases of productivity (1000–2000 g of C/m^2/year) in environments where such levels are generally low. They are characteristic of tropical waters (temperature greater than 18–20°C) and grow in well-lit waters at a maximal depth of 20–30 m. Coelenterate corals such as *Lophelia pertusa* are also found in the cooler waters of the North Atlantic. Zooxanthellae, symbiotic algae, live in the surface tissues of corals. These algae are the primary producers of the reef, but the polyps forming the coral can also filter and ingest particulate material, bacteria or zooplankton which are present in the water. The process of calcification conducted by innumerable coral polyps results in gigantic constructions such as the Great Barrier Reef of Australia which extends for over 2000 km.

These reef systems are very sensitive to the effects of storms (wind and waves, but also freshwater) and, once damaged, only recover very slowly, taking 25 years for complete recovery. They are also subjected to invasion by predators such as the starfish *Acanthaster planci* which destroyed large areas of reef in the South Pacific in the 1970s. It is not known to what extent this phenomenon occurs naturally, or whether it is induced by man.

- *Algal ecosystems*: Macroalgae (seaweeds) grow in circulating coastal waters with temperatures below 20°C. They require rocky or gravelly seabeds, to which they can attach and resist the forces of the sea. They occupy both the zone uncovered by the tides and also part of the infralittoral zone (Chapter 1), but there is zonation: the Laminaria, such as *Laminaria* or *Macrocystis*, the kelps, have colonised immense coastal areas up to depths of 20–40 m. Their thalli contain flotation devices and they can reach 50 m in length and extend to the surface. They constitute veritable forests which are very dense, and their presence breaks the waves, contributing to the protection of the coast. They also slow currents (drag coefficient: see Chapter 2). The production of laminaria zones can reach 1500 g of C/m^2/year or more than 3 kg of dry matter/m^2/year.

  These forests shelter not only abundant populations of sea urchins (which threaten biodiversity) but also abalone and crustaceans. The presence of sea otters, which consume sea urchins and thus reduce their grazing, re-establishes the biodiversity (Carter, 1988).

  In shallow zones or along the shore, the fucus type seaweeds replace the Laminarians only on rocky substrates, as shores of sand or too-small pebbles do not suit them. The fucus type seaweeds withstand being left above the water and maximal density is reached in zones with moderate water movement. Rocks which are too severely pounded by the waves remain bare, but in temperate or warm waters calcareous encrusting algae can grow there.

  The higher plants require a root system in soft but stable seabeds. In the Mediterranean, the Zostera and especially the Posidonia constitute vast fields in the infralittoral zone and similar plant aggregations are widespread throughout the world. In tropical lagoons another species (*Thalassia*), occupies these shallow waters.

- *Rocky shore ecosystems:* These are very diverse and we shall not repeat details of their plant cover. Bivalve molluscs live attached to the rock while gastropods use the interstices or cavities; crustaceans, fish, worms and many other animal groups are also found here. Mussels and limpets colonise the highest levels and are able to resist the power of the waves. Mussels often constitute the dominant populations of rocky shores until the depth exceeds 10 m. Their biomass in the wild can reach 100 kg/m^2 on "mussel beds" situated either along the shore or in a deeper area. We shall see (Part C) that shellfish culture is one of the main forms of development of the intertidal zone. Because it offers a solid substrate for attachment or shelter, rocky shores are very often completely covered by fauna or flora and there is no space for newcomers. Reef seabeds are very similar to rocky shores, especially in their dead parts.

- *Beach ecosystems:* The beach substrate extends up to a depth of several metres towards the open sea and is characterised by low biological productivity and physical instability. The micro and meiofauna which occupy the interstices between the grains of sand are numerically abundant but, because of their small size, their biomasses are low. Only burying species can survive.

  In contrast, the shallow waters above these seabeds are rich: abundant populations of Mysids capable of hiding themselves in the loose seabed exploit the planktonic production, and juvenile fish which assemble in these shallow waters towards the end of their larval lives, feed on these crustaceans or on other fish. These zones constitute littoral nurseries and their significance in the recruitment of exploited species must be emphasised.
- *Estuaries:* Since they consist of zones of contact between fresh and salt water, we have seen (Chapter 2) that these water masses often remain very individual with saline wedges penetrating below freshwaters. This does not detract from the fact that estuaries constitute complex ecosystems with, for example, a peculiar plankton assemblage dominated by species tolerating wide ranges of salinity. Differences in salinity and the presence of fresh or marine waters, often full of terrestrial inputs, make estuaries rich and productive, but also turbid, areas, where sedimentation is significant. The accumulation of these products can lead to the formation of deltas.

3.11 VARIATIONS IN PRODUCTION

3.11.1 Periodic variations

Since primary production depends on light, its cycle is therefore diurnal. The study of daily variations in production shows that the maximum is always observed between 11:00 h and 14:00 h (maximal sunlight). Very significant variations in oxygen concentration, pH and many other factors affect shallow environments and aquaculture in these areas is dependent on them. As a result, rearing ponds are often equipped with devices used to counteract such variations (aerators, for example). Taking into account the high specific heat of water (Chapter 1), this diurnal periodicity only alters the temperature by a few degrees (often less than 3°C, between night and day).

Seasonal variations modify the temperature and day length; temperature regulates the growth rate of poikilotherms*, but these periodic variations also intervene in the triggering of various phenomena (sexual maturation, migration, diapause). In the North Sea, primary production changes from 10 mg C/m^2/day in December to 700 mg C/m^2/day in August to 70 mg in November. In the Tunis lake this production goes from 500 mg C/m^2/day in spring to 10 mg in summer and 2000 mg in autumn. In warm regions, seasonal variations are blurred.

3.11.2 Irregular variations

Briefly, the production of the majority of fisheries and that of cultivated species utilising natural resources (shellfish culture, for example) shows variations whose periodicity can be between a few years and several decades; this has been noted everywhere and the chronological series of data for nearly a century of oyster production in France demonstrates this very well (Héral, 1990).

These variations also affect fisheries, but while the oceanography of the fisheries has sometimes shown evidence of relationships between hydrological phenomena and fish production, many other fluctuations remain inexplicable; we shall return to these in Part B.

3.12 POLLUTION AND FOOD WEBS

Pollution* introduces organic carbon, inorganic salts or new molecules to aquatic ecosystems. We have seen the role played by these substances in the production process, even at low concentrations, but the pollutants introduced by man into the aquatic environment (pesticides, detergents, hydrocarbons, heavy metals) also have an effect on the mediator substances not in the diet, whose activity can be greatly modified, and on the metabolism of living organisms. For example, very low doses of a herbicide such as pyridazinone are capable of stimulating chlorella in culture, increasing their diameter by 30% and stimulating the production of fatty acids (Herczeg *et al.*, 1980); while other man-made substances are known to be toxic, the effects of the majority are unknown.

While considering that chemical pollution is low in the sea (according to the National Network for Observation of the Marine Environment), it must be accepted that certain pollutants can interfere with mediator substances at very low concentrations. For example, they can change the equilibria in phytoplankton communities, which can result in the proliferation of toxic dinoflagellates, found all around the world and which may be limiting to aquaculture and fisheries.

The waters in shellfish ponds are regularly analysed for bacteria, to determine the density of those which are pathogenic to man, but the effect of pollution by a bacterial route can be mitigated, or reversed, by depuration. Such pollution can nevertheless have catastrophic economic consequences for aquaculture as has been demonstrated by the prohibition of sale of shellfish from the Étang de Thau in 1989–1990 due to the presence of salmonella bacteria, detected by IFREMER but strongly contested (Piétrasanta, 1993).

3.13 CONCLUSIONS

Initial production of living matter in the aquatic environment through the activity of numerous very small organisms spreads throughout a three-dimensional environment. In comparison to the terrestrial ecosystem, there is a large amount of contact between the surfaces of the tiny organisms and the surrounding environment and there is not the massive concentration of organic matter found in the support structures of terrestrial plants (trunks and stems).

Some of these producers have only been known for 20 years, and their detection requires sophisticated methods. Because of their small size, these organisms can easily exchange nutrients or other substances with the environment in which they are immersed. The exchanges can be very rapid, creating a daily production which can represent several times their biomass. The producers bathe in an "organic soup" made up of very small particles (less than 1 μm) whose density in surface waters is about 10 million per millilitre (Koike *et al.*, 1990).

The great majority of aquatic environments are already modified by human activity: synthetic molecules, which do not exist in nature, are released in great quantities into the aquatic environment and are active at low concentrations (nitrates, trace elements, etc.). The commonness of black tides and pollution of beaches by hydrocarbons or other obvious waste, only constitutes the visible part of a state which is in fact much more serious but less visible: widespread chemical pollution of most coastal zones, due largely to agriculture and industry. According to the German Association for the Protection of the Environment, the North Sea may be dead from a depth of 9 m as a result of pollution from Poland and the CIS (800,000 tonnes of fertiliser discharged every year by agriculture).

The sensitivity of aquatic organisms to various substances, at infinitesimally low concentrations, can modify the subtle equilibria which regulate biological production processes. Their concentrations can build up along the food chain, sometimes leading to toxic alterations in the natural composition of living organisms used for human foodstuffs (e.g. *Dinophysis* with dangerous toxins and Minamata disease caused by the accumulation of mercury) or to total mortalities in rearing systems (e.g. proliferation of microalgae in Norwegian or French bays where salmon are reared).

New data on planktonology and production dynamics have overturned traditional oceanography, but the exploitation of living aquatic resources, depending by definition on the production of living matter from the aquatic environment, is one of the areas of application of these disciplines, although it may appear that there is little contact between them today. Nutrition of molluscs, fish and crustaceans in extensive aquaculture and larval rearing of most reared species make demands on one or other component of the plankton.

For future development, an approach should be found which can be supported by international and national organisations which study and exploit aquatic environments.

3.14 BIBLIOGRAPHY

Alzieu C. L'eau, milieu de culture. In: G. Barnabé (ed.) *Aquaculture*. Tech & Doc Lavoisier, Paris, 1989; 17–45.

Aubert M. Eutrophie et dystrophie en milieu marin: origine et évolution. *Rev. Int. Océanogr. Med.*, 1988; LXXXXI: 3–16.

Azam F., Fenchel T., Field J.G., Gray J.S., Meyer-Reil L.A., and Thingstad E. The ecological role of water column microbes in the sea. *Mar. Ecol. Prog. Ser.* 1983; 10: 257–283.

Barnabé G. (1980) Système de collecte du zooplancton à l'aide de dispositifs autonomes et stationnaires. In R. Billard (ed.) *La pisciculture en étang*. INRA Publications, Paris; 215–220.

Barnabé G. Harvesting zooplankton. In: G. Barnabé (ed.) *Aquaculture, Vol 1*. Ellis Horwood, Chichester, 1990; 264–271.

Barnabé G. L'élevage des premiers stades en écloserie. In: G. Barnabé (ed.) *Bases biologiques et écologiques de l'aquaculture*. Tec & Doc Lavoisier, Paris, 1991; 361–401.

Barnabé G. (ed.) *Aquaculture: Biology and ecology of cultured species*. Ellis Horwood, Chichester, 1994.

Bergh J.C., Borsheim K.Y., Bratbak G., and Heldal M. High abundances in viruses found in aquatic environments. *Nature* 1989; 340: 467–468.

Bernhard M., Nassogne A., Zattera A., and Moller F. Influence of pore size of plankton nets and towing speed on the sampling performance of two high-speed samplers (Delfino I and II) and its consequences for the assessment of plankton populations. *Marine Biology* 1973; 20: 109–136.

Borgman U. Effect of somatic growth and reproduction on biomass transfer up pelagic food webs as calculated from particle-size conversion efficiency. *Can. J. Fish. Aquat.* Sc. 1983; 40(11); 2010–2018.

Böttger R. Studies on the small invertebrate plankton of the Sargasso Sea. *Helgolander Meeresunters* 1982; 35: 369–383.

Caumette P. Les lagunes et marais maritimes. In: M Bianchi, Marty D., Bertrand J.C., Caumette P., Gauthier M. (eds) *Micro-organismes dans les écosystèmes océaniques*. Masson, Paris, 1989; 249–282.

Collignon J. *Écologie et biologie marines*. Masson, Paris, 1991.

Curds C.R. Protozoa. In: C.R. Curds and Hawkes (eds) *Ecological aspect of used-water treatment. Vol 1*. Academic Press, London, 1975; 203–268.

Fiala-Médioni A., and Pavillon J.F. La matière organique dissoute. *Synthèse, Océanis*, 1988; 14(2): 295–303.

Fitt W.W., Fisher C.R., and Trench R.K. Contribution of the symbiotic Dinoflagellate *Symbiodinium microadriaticum* to the nutrition, growth and survival of larval and juvenile Tridacnid Clams. *Aquaculture* 1986; 55: 5–22.

Glover H.E., Keller M.D., and Guillard R.R.L. Light quality and oceanic ultraphytoplankters. *Nature* 1986; 319(6049): 142–143.

Hamner W.M. Behavior of plankton and patch formation in pelagic ecosystems. *Bull. Mar. Sci.* 1988; 43(3): 752–757.

Héral, M. Traditional oyster culture in France. In: G. Barnabé (ed) *Aquaculture, Vol 1*. Ellis Horwood, Chichester, 1990; 342–380.

Herczeg T., Lehoczki E., Rojik I., Vass I.S., Farkad T., and Szalay L. Stimulatory effects of pyridazinone herbicides on Chorella. *Plant Science Letters* 1980; 285–294.

Hunter J.R. Inference regarding predation on the early life stage of cod. In: E. Dahl, Danielsen D.S., Moksnes et Solemdal, P. (eds) *The*

Propagation of Cod Gadus Morhua L. Flodevingen Rapport. Series 1, 1984; 533–562.

Jenkins J.P. Micro- and fine-scale distribution of microplankton in the feeding environment of larval flounder. *Mar. Ecol. Prog. Ser.* 1988; 43: 233–244.

Koike, I., Shigemitsu, H., Kazuki, T., and Kazuhiro, K. Role of sub-micrometre particles in the ocean. *Nature*, 1990; 345: 242–244.

Last J.M. The food of herring *Clupea harengus* in the North Sea (1983–1986). *J. Fish Biol.* 1989; 34: 489–501.

Lubet P. Bases biologiques de la culture des mollusques. In: G. Barnabé (ed) *Bases biologiques et écologiques de l'aquaculture.* Tec & Doc Lavoisier, Paris, 1991; 99–210.

Mann K.H. Production of the bottom of the sea. In: D.H. Cushing and Walsh, J.J. (eds) *The ecology of the seas.* Blackwell Scientific Publications, Oxford, 1976; 225–250.

Margalef R. Composicion especifica del fitoplancton de la costa catalano-levantina (Méditerraneo occidental) en 1962–1967. *Inv. Pesq.* 1969; 33(1): 345–380.

Martinez L., Silver M.W., King J.M., and Aldridge A.L. Nitrogen fixation by floating Diatoms mats: A source of new nitrogen to oligotrophic ocean water. *Science* 1983; 221(4606): (not paginated).

Meadows, P.S., and Campbell, J.I. *An introduction to marine science.* Blackie Academic, London, 1993.

Mook D.H. Removal of suspended particles by fouling communities. *Mar. Ecol.*, Prog. Ser. 1981; 5: 279–281.

Mulkana M.S. Significance of nanoplankton to commercially important finfish and shellfish in Barataria Bay, Louisiana. *Proceedings of Louisiana Academy of Science, XXXIII*, 1970; 38–43.

Newel R.C. The energetics of detritus utilisation in coastal lagoons and near shore waters. *Oceanol. Acta* 1982: 347–355.

Ogawa K. The role of bacterial floc as food for zooplankton in the sea. *Bull. Jap. Soc. Scient. Fish* 1977; 43(3): 395–407.

Omori M. *Étude du zooplancton appliqué aux pêches dans le monde.* Enseign. Océanogr. Biol., Univ. Paris, 1980.

Owen R.W. Microscale plankton patchiness in the larval anchovy environment. *Rapp., Reun, Cons. Int. Explor. Mer*, 1981; 178: 364–368.

Peduzzi P., and Weinbauer M.G. The submicron size fraction of sea water containing high number of virus particles as bioactive agents in unicellular plankton community successions. *J. Plankton Res.* 1993; 15(20): 1375–1386.

Pérès J.M., and Devèze L. *Océanographie biologique et biologie marine.* P.U.F. Éd, Paris, 1963.

Piétrasanta Y. *L'écharpe verte. Combats pour une nouvelle écologie.* Albin Michel Éd, Paris, 1993.

Purves W.K., Orians G., and Heller H.C. *Le monde du vivant: Traité de biologie.* Flammarion Éd, Paris, 1994.

Richard M. *Atlas du phytoplancton marin. Vol. 2.* Éditions du CNRS, Paris, 1987.

Sheldon R.W. Phytoplankton growth rates in the tropical ocean. *Limnol. Oceanogr.* 1984; 29(6): 1342–1346.
Sheldon R.W., Nival P., and Rassoulzadegan F. An experimental investigation of a flagellate-ciliate-copepod food chain with some observations relevant to the linear biomass hypothesis. *Limnol. Oceanogr.* 1986; 31(1): 184–188.
Sherr E., Sherr B., and Paffenhdffer G.A. Phagotrophic protozoa as food for metazoans: a missing trophic link in marine pelagic foodweb systems. *Marine Microbial Food Webs* 1986; 1(2): 61–80.
Sournia A. *Atlas du phytoplancton marin. Vol 1.* Éditions du CNRS, Paris, 1986.
Stockner J.G. Phototrophic picoplankton: An overview from marine and freshwater picosystems. *Limnol. Oceanogr.* 1988; 33(2): 765–775.
Svenko VS. The role of biosedimentation. In: *The formation of bottom deposits.* Vodn. Resur 1988; 4: 12–129 (in Russian).
Tanoué E., and Hara S. Ecological implications of fecal pellet produced by Antarctic krill *Euphausia superba* in the Antarctic Ocean. *Mar. Biol.* 1986; 91(3): 359–369.
Thorpe S.A., and Hall A.J. Bubble clouds and temperature anomalies in the upper ocean. *Nature* 1987; 328(2): 48–51.
Tregouboff G., and Rose M. *Manuel de planctonologie méditerranéenne. Vols 1 and 2.* Éditions du CNRS, Paris, 1957.

4

Physical factors, water movements and biological production

4.1 WATER MOVEMENTS IN RELATION TO AQUATIC ORGANISMS

In Chapters 1 and 2, we have seen which physical factors affect coastal waters, while Chapter 3 describes the characteristics of marine life. In this chapter we shall try to integrate these interactions between the environment and the living organisms inhabiting them.

Routes of biological production will not be examined: numerous studies have described these processes (see for example, Frontier and Pichod-Viale, 1991; Collignon, 1991 for a more global approach). We shall limit ourselves to examining the relationships between hydrology and living organisms that condition biological production.

Numerous studies have effectively established that much of the variability in marine productivity is the result of water movements (Sherman and Alexander, 1986; Troadec, 1989; Mann and Lazier, 1991); we shall return to these when examining problems with fisheries (Chapter 7). We shall refer most often to Mann and Lazier's excellent review.

4.1.1 Water and phytoplankton movements

Initial production by the ocean is mainly the result of phytoplankton consisting of microalgae, most often smaller than 100 μm in size. We have seen (Chapter 2) that organisms greater than 1 mm in size are hardly affected by water viscosity, while for smaller organisms the apparent viscosity increases as size decreases and thus the Reynolds number becomes smaller; viscosity dominates the lives of those species between 1 mm and 1 μm in size; turbulences in the form of shear forces can also affect organisms between 1 mm and 100 μm (Fig. 4.1).

Below 100 μm, molecular diffusion becomes more rapid than water movements in transport across the boundary layer. The situation has been compared by Mann and

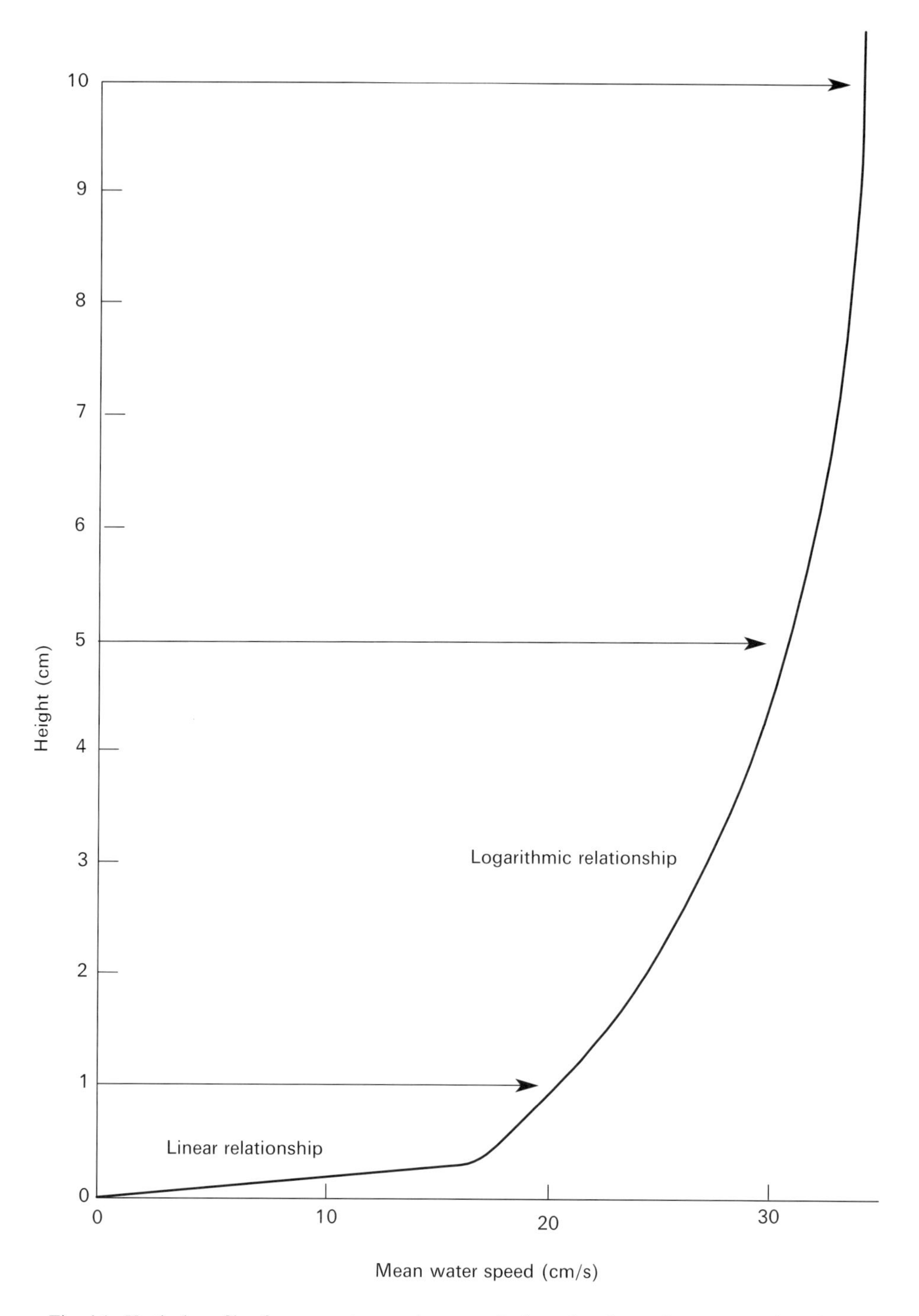

Fig. 4.1. Vertical profile of mean water speeds across the boundary layer above a smooth surface.

Lazier (1991) to that of a person swimming in honey; the turbulent movements which circulate water around these small organisms have little effect. These minute beings depend on molecular diffusion for the transfer of nutrients and waste, while for slightly larger organisms nutrients and waste are carried by turbulent diffusion which is thousands of times more efficient.

Phytoplankton cells without movement in the water use up the nutrients in the surrounding environment and their intake of nutrients is then limited by the speed at which these are replenished by the diffusion around them. It was traditionally considered that plant plankton is passive in the water; however, to obtain its nutrients, it must overcome its limitations of diffusion and for this it uses two displacement mechanisms.

4.1.2 Phytoplankton translocation

Translocation can be passive or active.

While a small part of the phytoplankton and the particulate waste which it creates are positively buoyant (by trapping air bubbles or high lipid concentrations), the majority is doomed to sedimentation. Most phytoplankton is denser than water ($d = 1.021$–1.028, the density of the cytoplasm $= 1.03$–1.1). The theca of diatoms has a density of 2.6, that of *Coccolithophoridae* plates reaching 2.7–2.9. Their sedimentation rate (size $<500\,\mu m$) approximately follows Stokes Law calculated for spheres, but shape and size have an effect on the drag coefficient (Chapter 2). Thus there is continuous elimination of phytoplankton by sedimentation. As far as its physiology permits, the phytoplankton compensates for these losses through its growth rate; it remains in the superficial layers by regulating osmotic pressure, movements of appendages and flagellae, together with lipid synthesis. If these do not occur, it sediments at a speed of several metres per day, since its mean density is around 1.10, while that of sea water is about 1.025. In the Baltic Sea, it has been calculated that 8–96% of diatoms sediment daily (Kononen, 1994).

Temperature modifies density and viscosity; thus, between 0 and 25°C, a particle of $20\,\mu m$ will double its sedimentation rate. Note that old cells sediment more rapidly, owing to variations in concentrations in lipids and NH_4^+.

In mixed waters, wave movement counteracts the tendency to sink; in upwelling zones, ascending water opposes sedimentation, which explains the high concentrations of phytoplankton encountered. Without turbulence, diatoms, for example, would end up in deep water where there is insufficient light for photosynthesis.

The other way in which to surmount this limitation is to move in relation to the water, i.e. to swim.

This type of movement is the solution adopted by the flagellates, but it is far from simple as they have to overcome the problems of increased viscosity on a small scale, and the phenomena due to the boundary layer. At this scale the most efficient mechanisms are flagellae and cilia. Many flagellates greater than $10\,\mu m$ in size are capable of swimming at more than 10 times their body length, allowing them to ingest nutrients while swimming. Swimming or sedimentation cannot avoid the limits imposed by diffusion, but make differences which affect the issue of interspecific competition. Phytoplankton do not move to improve their nutrient exchange, but to find zones which are richer in nutrients.

During the course of his study of phytoplankton in the Baltic Sea, Kononen (1994) showed that stratification of the water column had no effect on the dinoflagellates, which control their position by vertical migration, in contrast to diatoms. Dead dinoflagellates disintegrate in the water before sedimenting in the form of detritus particles. When nutrients become scarce, a section of the dominant dinoflagellate population (*Peridinium hangoei*) forms resistant cysts which sediment to the seabed and constitute the starting point for new generations. This characteristic is unique to the dinoflagellates, including toxic species, and explains the rapidity with which they can colonise the water. In general, motile dinoflagellates tend to replace diatoms in stable waters characterised by low nutrient concentrations.

Small phytoplankton have Reynolds numbers and the forces of viscosity dominate; Mann and Lazier (1991) established that a picoplankton of 1 μm had a movement comparable to that of a human in treacle, in which they could only move at 1 cm/min! Yamazaki and Kamykrowsy (1991) showed that the swimming capacities of different species influence their vertical distribution, even in a layer mixed by strong winds: at the base of this layer, light can be limiting for the diatoms which are trapped there, while the flagellates can maintain their position in the well-lit layer. Studies on turbulence, carried out in a submarine, show a dampening of movements with depth, which slightly exceeds calculated values.

Water movement is not the only physical factor affecting phytoplankton and its sedimentation. In the Red Sea, Bijma *et al.* (1994) showed that the sedimentation of Globigerinoides is related to the lunar cycle as the periodicity of reproduction is in phase with the lunar cycle. This results in a flux of the same periodicity in sediment traps; for the dominant species, the carbonate flux was estimated at 1 g/m^2/year.

This doubtless sounds very little, but 99.5% of the CO_2 produced during geological times has been transported to the bottom of the oceans by marine organisms in the form of calcareous skeletons.

4.1.3 Zooplankton: predation in a viscous environment

4.1.3.1 Capture mechanisms

Like phytoplankton, zooplankton live in a viscous environment. This is true for the copepods which are the most abundant example of the zooplankton; forces of viscosity dominate around the copepod's appendages. These animals create a pulsed current across their ventral surface, using their secondary antennae, mandibular palps, first maxillae and maxillipeds. When a particle appears, these appendages, which are covered in bristles (*setae*), beat asymmetrically, bringing the prey towards the medio-ventral axis. A projection and then retraction of the secondary maxillipeds brings the prey between them, in a small volume of water. On a large scale, this is similar (according to Mann and Lazier, 1991) to using one fork to collect breadcrumbs on another fork, with both submerged in honey!

These movements, which are costly in terms of energy, are only shown when there is prey, as its capture occurs where the Reynolds number is low; copepods show detection and capture behaviour and do not act as passive filters. They are capable of detecting their prey using chemosensors of smell and taste and can have mechanoreceptors. Detection range extends to 1 mm. Tiselius and Jonsson (1990) provide details of swimming behaviours and feeding strategies of several species and have made hydrodynamic analyses.

Copepods can use three different capture mechanisms. The first is filtration, the particle being retained in a space narrower than its width. The second is direct interception by a bristle: once in contact, taking into account the small Reynolds number, the particle sticks. The last is the random contact of a moving particle, carried by the current, with a bristle. A similar catch could be achieved by gravity (sinking particle).

Planktonic crustaceans are sensitive to small scale turbulences; their metabolism, which can be assessed from their cardiac rhythm, varies in relation to whether they are in calm or moving water. This rhythm increases by 14% in daphnia between calm water and a turbulence of $5\,mm^2\,s^{-3}$; in a marine copepod, a similar change leads to a 93% increase in the rhythm (Alcaraz *et al.*, 1994).

4.1.3.2 Zooplankton production

Since zooplankton search for their prey, it could be expected that they would be most abundant at the level of maximum chlorophyll, i.e. of phytoplankton, but this might result from nutritional or physical factors. There is little information on the zooplankton that can migrate to deep water and about these migrations.

It is assumed that large zooplankton consume healthy phytoplankton, that the microzooplankton (flagellates) live on the senescent phase and that its best feeding will be when biomass is at its maximum. Turbulence doubles the contact between zooplankton and phytoplankton and increases the contact between bacteria and phytoplankton by 3–4 orders of magnitude.

Note that the phytoplankton, the biological cycle of which only lasts a few days, is capable of responding quickly to favourable conditions: zooplankton, with its longer biological cycle (several weeks or months) responds more slowly; as a result, the production of phytoplankton normally exceeds its consumption by zooplankton and is therefore rarely limiting.

Zooplankton growth is relatively slow even in upwellings, since mixed waters warm slowly; zooplankton production is delayed and the peak is reached in midsummer. Thus the phytoplankton bloom escapes predation and the majority sediment at the end of their lives. Thus, for example, in the North Sea off Belgium, competition between primary production and its consumption by copepods only exists in June and July.

The influence of physical factors on the spatial distribution of zooplankton has been studied in the Saint Lawrence estuary by Laprise and Dodson (1994); stratification and salinity regulate the longitudinal distribution of stable groups of species (in fresh, estuarine and marine euryhaline waters) whose spatial distribution is a function of salinity and vertical stratification; turbidity and temperature play a lesser role. Lowest abundances are found in the most stratified waters.

4.1.4 Fish larvae

Correlations between the abundance of fish larvae and oceanographic or climatic phenomena have long been in evidence and we shall see that they contribute most of the variability to the recruitment of exploited populations (Chapter 7).

There are many examples of relationships, but recent studies have provided ever-more precise information (ICES, 1994):

- cod eggs and larvae transported by currents onto the Georges Bank have a high chance of remaining in this zone which is favourable to their survival;
- the level of light intensity regulates both the spatial distribution of cod larvae in their endotrophic* phase and the longevity of this phase (Skifesvik, 1994);
- the abundance of 0+ age class—that is larvae and juveniles of the year—is related to water temperature; a simple multiple regression model between temperature and recruitment of cod in the north-east Arctic explains 46% of the variability in recruitment between 1946 and 1988;
- other studies show that the feeding rhythm of cod larvae increases by a factor of 7 when surface wind speed increases from 2–10 m/s, i.e. in conditions with a large amount of water movement. This result contradicts the hypothesis put forward previously, notably by Mann and Lazier (1991), according to which calm, well-stratified waters favour larval survival (as a result of a high density of phytoplankton with maximum numbers of dinoflagellates), while strong winds and well-mixed waters would not be so good for them. Optimal conditions can change with species and it is during periods of least wind that conditions are best for first feeding of anchovy (Peterman and Bradford, 1987).

The presence of "abiotic structures" (fronts, for example) or biotic structures (presence of aggregates or marine snow, or gelatinous plankton such as jellyfish) also play a role as larvae and juveniles assemble close to these structures (Kingsford, 1993).

Data from experimental aquaculture in ponds have confirmed the significance of water in slow convection to larval survival for all reared species (Barnabé, 1991). We have also observed, on dives, that banks of larvae assemble around artificial reefs, but sheltered from the current, "downstream" of the structure; physical shelter cannot be the primary motivation as turbulences can also facilitate the capture of zooplankton.

To return to the cod, for many authors (ICES, 1994), the key variable affecting stocks would certainly be biological production and parameters such as temperature or wind intensity; they constitute a multivariate mixture of factors which guide this productivity. Other factors, such as salinity, buoyancy of eggs and the timing of reproduction in relation to temperature, have proven effects.

In the Bahamas, Thorrold *et al.* (1994) showed that the abundance of three of the six families of the most common fish larvae was associated to lunar periodicity, while in other families their abundance was correlated with a certain type of wind. They concluded that the relationships between larval behaviour and the physical environment are complex.

In Delaware Bay (USA) Rowe and Epifano (1994a) established that the abundance of *Cynoscion regalis* larvae follows a periodicity of 12 hours, identical to the duration of diurnal half-tide. The same authors calculated (1994b) that the displacement speed of larvae is identical to that of the water, whatever the depth.

A study of the joint impact of hydrological and chemical variations on the ichthyofauna of the Danube (Bacalbasa-Dobrovici, 1994) showed that the influence of chemical substances (especially those produced by agriculture) is harmful, not only to migratory species, but also to riverine species (spawning, larval survival). Sturgeons are under threat.

4.1.5 The nekton

Because of its size, which is always greater than a centimetre, and its swimming capacity, which by definition allows it to escape the movements of water masses, the planktonophagic nekton is not directly affected by these movements. Thus, after the larval phase, juvenile fish whose swimming capacity increases are able to exploit turbulences.

Despite this fact, the planktonophagic nekton is indirectly very dependent on water movements throughout its search for planktonic prey. The distribution of these prey-species in fact dictates that of their predators, and zones rich in plankton (fronts, upwellings, mixed zones and coastal zones) are very rich in small pelagic species such as sardines or anchovies, themselves hunted by larger predators (mackerel and tuna). This type of food chain is structured in relation to size (see Fig. 3.1). The classic feeding behaviour all along this chain is the visual selection of the largest prey compatible with the mouth size of the predator.

4.1.6 Benthic plants

Because of their fixed position, the key to the benthic plants' productivity lies with the movements of the water with which they are in contact. This has been demonstrated in freshwater for several species: an increase in water movement markedly increases the uptake of P. Large brown algae constituting the kelp have their fronds close to the surface where they are subjected to orbital wave movements. For the fronds of *Macrocystis*, photosynthesis and the uptake of nutrients increase with the current up to a speed of 5 cm/s. In turbulent coastal zones where these species are found, the amount of turbulence is well above the optimum: this has been noted in experiments on seaweed in calm and choppy water (see Mann and Lazier, 1991, for more details).

4.1.7 Benthic animals

4.1.7.1 Filter feeders and detritivores in the benthic limiting layer

Most often these are quite large species whose Reynolds number is high. As is the case for seaweed, access to particulate food transported by waters to attached animals is regulated by currents or turbulences.

In very calm waters, there is an accumulation of dissolved organic matter on the seabed that is easily utilised by animals which directly ingest the sediment. At the other extreme, strong currents erode the sediments, leaving little organic matter and rendering filtration difficult by the absence of suspended matter. For intermediate stages, a graded abundance of filter feeders is noted.

The high benthic production of estuaries is thus attributed to the dominance of tidal currents. Currents of 0.1–1 m/s create a vertical turbulence coefficient with a component of 50–500 cm^2/s, indicating that living matter produced at the surface becomes accessible to benthic filter feeders, at several metres' depth, in less than an hour. In zones where the current above the bottom exceeds 30 cm/s, it has been shown that the fauna is stressed and impoverished.

Filter feeders decrease the concentration of particles close to the bottom and moderate currents generate enough turbulence in the limiting layer of the bottom to increase the availability of food by turbulent diffusion. Mann and Lazier (1991) reported the similar results of several authors, showing that, above a bivalve mollusc bed, the density of particulate organic matter is negatively correlated with current speed. When the mussels are placed 5 cm and 1 m above the bottom, those situated at 1 m have more food and grow faster. Food consumption by filter feeders is as much dependent on current speed in the benthic boundary layer as it is on other previously-studied and better known factors, such as food concentration or temperature. Currents also affect filtration speed in barnacles and bivalves.

It should be noted that, in the currents, the heaviest particles (often minerals) have their maximal flux close to the bottom, while lighter particles are more abundant several centimetres above the seabed; there is, therefore, a biological advantage for benthic animals in placing their feeding organs several centimetres above the seabed.

Detritivores are also very abundant on the bottom, whether they are species which burrow in the sediment or move around on it, or are fixed to a substrate. The distinction between filter feeders and detritivores is not easy; the mussel, for example, feeds on inert organic particles or bacteria. Small detritivores constitute a very varied world—the meiobenthos; a pertinent and illustrated insight is given by Adey and Loveland (1991).

4.1.7.2 The boundary layer and larval attachment

Many benthic species have planktonic larvae which seek to attach to the seabed at the end of their pelagic lives. These larvae utilise currents and turbulences which transport them passively during this dispersal phase, as their own speed of movement is very slow. Mann and Lazier (1991) compared each tidal current to cyclonic winds which blow twice a day, for larvae seeking attachment.

In the case of 300 μm polychaete larvae, precise measurements have shown that when the currents are of the order of 5–10 cm/s, the turbulences are so close to the bottom that there is no area in which the larva can move by itself; thus, larvae cannot settle during 40% of the tidal cycle.

The only place where the water is sufficiently calm is in the 100 μm thick boundary layer immediately above the seabed (Chapter 2). In this layer, the larvae hardly have space to manoeuvre; thus they do not seek their habitat by horizontal swimming and it appears that they only use their swimming capacity to descend and explore the substrate and come up again if it is not suitable. This is the "hot air balloon technique". Lefèvre (1990) described the problem of attachment in larval polychaetes as well as the role played by various factors in larval distribution. She showed that the actual dispersion is the result of the interaction between biological (behaviour) and physical factors.

Caceres-Martinez *et al.* (1994) determined the modes of larval attachment in the mussel *Mytilus galloprovincialis*. The attachment substrate can be very variable: hard substrate in a calm or exposed environment, seaweed stalks or byssus* threads, or rough shells of adult mussels. Differences in the numbers attaching depend on the currents. After this initial attachment, the young mussels can still move and form clusters. They only move when the water is calm, but in moving water attach to anything by their byssus.

The transport and subsequent attachment of larvae of two species of barnacle studied by Pineda (1994) in California illustrate these rules. The temporal variability in attachment noted by this author can be correlated both to the abundance of the cohort of larvae and the physical processes of larval transport, while the spatial variability of their distribution can be associated with behavioural responses and substrate availability.

Thus the importance of the properties of the boundary layers for larval attachment can be understood. Tidal currents above the bottom exceed the swimming speed of larvae, but they decrease logarithmically as the distance from the bottom decreases, since the current is slowed by water friction close to the bottom; there is minimal turbulence in the viscous limiting layer contiguous to the bottom (Fig. 4.1). This is the only place where pelagic organisms can rest on the bottom; above this, they are carried by the current.

4.2 SPATIAL HETEROGENEITY AND "SWARMS"

As we have just seen, physical factors control the main events which set the abundance and distribution of marine planktonic species. Light, nutrients and movement determine the specific composition of phytoplanktonic communities, but there are two biotic factors—competition and predation—which regulate the structure of planktonic populations (the abundance of one species or the scarcity of another, for example); the interactions between these diverse categories of factors are numerous and not all known. Consequently, the intensity of primary production is not uniform and depends on the hydrodynamics, concentration of dissolved substances and other factors.

Although phytoplankton cells (microalgae) are in passive suspension in water, they are not distributed homogeneously within the water masses even though the majority are incapable of moving themselves. They form patches or "swarms", the size of which varies depending on the dominant species (from 1–12 m according to McAlice (1970), but in fact they showed different densities when their samples were taken more than 10 cm apart). Other authors, sampling phytoplankton from stations tens of metres apart, counted 215 phytoplankton cells (microalgae) belonging to 20 species in passive suspension in the water then, successively, two cells of one species and none in the last sample. The extension of these swarms in the same direction as the wind has sometimes been noted, as has their concentration at interfaces between water layers of different densities or temperatures. Given the stratification of waters as a result of thermoclines or pycnoclines, the distribution of plankton communities has been well studied at this level: there is simultaneously a concentration of phytoplankton and an aggregation of zooplankton; the biomass of prey and predators makes the thermocline a place of predation and competition (Harris, 1987). Interfaces such as fronts also play a key role in phytoplankton production (Loder and Pratt, 1985).

Similar results have been reported by Matsushita *et al.* (1988), who concluded that the distribution of fish larvae swarms and their zooplankton prey is under the influence of oceanographic phenomena (upwellings and fronts). These two discontinuities are characterised by significant turbulence of the water. In the settlement of larvae, it has been established that zones of eddies and turbulence are favourable to attachment (Raimbault, personal communication). Therefore, oceanographic phenomena dictate the major fluctuations in stock variations in pelagic species. In the 1970s, for example, a significant drop in the anchovy production in Peru was noted, owing to the abnormal persistence of a warm current called "El Niño", but it resulted from a general phenomenon which has been widely studied (Sherman and Alexander, 1986).

The juvenile stages of copepods, which have little mobility, have their best chance of survival within these passive concentrations of phytoplankton caused by hydrodynamic phenomena. This is the group which constitutes the majority (between 80–90%) of the zooplankton, at least in marine environments (Dagg, 1977). The adults, with much more capacity for movement and which carry out vertical migrations, have a greater chance of survival, whatever the spatial distribution of the phytoplankton. This provides an explanation of mortalities affecting juveniles. While the hydrodynamics of the waters are responsible for the heterogeneous distribution of living organisms in natural environments, such discontinuities (e.g. pycnocline) have been demonstrated in the aquarium and are encountered in aquaculture ponds and especially in cages.

Islands and headlands cause such turbulences (Wolanski and Hamner, 1988) and, before setting up aquaculture units based on planktonic production, it is wise to determine the location of these turbulences and to choose sites in relation to these data. This type of detailed study shows how much production processes, whether fisheries or aquaculture, are linked to hydrological conditions.

Despite the predominant effects of hydrology on the spatial heterogeneity of the plankton, some plankton concentrations have another origin; certain interfaces create the presence of assemblages, as was observed by Tanaka *et al.* (1987a, b) in a Japanese bay. Swarms of *Acartia* in the form of a carpet (30 cm thick) or an ellipsoid (long axis of 50 cm), each with only a single species, were discovered at the seabed. The different shapes of the swarms showed that there was a biological effect caused by behavioural effects. The numbers present were 326 and 511 individuals/l respectively, that is 10 times more than elsewhere. Larvae of the Japanese sea bream hunt in these swarms close to the coast in less than 15 m of water. Similar phenomena have been reported by Jenkins (1988).

An extraordinary phenomenon of planktonic concentration has been described by Alldredge *et al.* (1984). From a single-seater submarine they directly observed concentrations of copepodites of the species *Calanus pacificus* at depths of 450 m, in a layer between 17 and 23 m thick, 100 m above the bottom, in the Santa Barbara basin, California. The mean density per m^3 of copepods in this layer was about 14×10^6 ($\pm 11 \times 10^6$), while it fell to $500/m^3$ above and below this. These concentrations are not associated with a discontinuity of physicochemical parameters and so, on the basis of biochemical analyses, the authors concluded that these populations were in a state of diapause during periods of low availability of food in the surface waters. This is another example of a behavioural phenomenon.

Significant differences between species also exist. It has been shown in the laboratory that *Acartia tonsa* and *Centropages typicus* require a constant density of food and are, without doubt, sensitive to the presence of swarms of prey (dinoflagellates). In contrast, *Pseudocalanus minutus* and *Calanus finmarchicus* are capable of surviving long periods of fasting. Within the swarms created by hydrodynamic effects, certain species could thus take advantage of others, but it is assumed that the swarms disperse sufficiently rapidly to avoid one species predominating (which is what we are led to believe from experiments carried out in large-volume ponds or tanks). The phenomenon is further complicated by the different developmental times of the various components of the plankton and the species which constitute the food web (Fig. 4.2).

The uniformity of pelagic life is thus only an apparent one and the elucidation of all these questions makes the plankton one of the most difficult, but also most interesting, communities to study.

4.3 OCEANIC STRATIFICATION AND BIOLOGY OF SURFACE WATERS

Coastal waters are also surface waters of the ocean, in continuity with waters of the open sea. While limited to the continental shelves, they represent around 8% of the ocean, equivalent to the surface area of Africa.

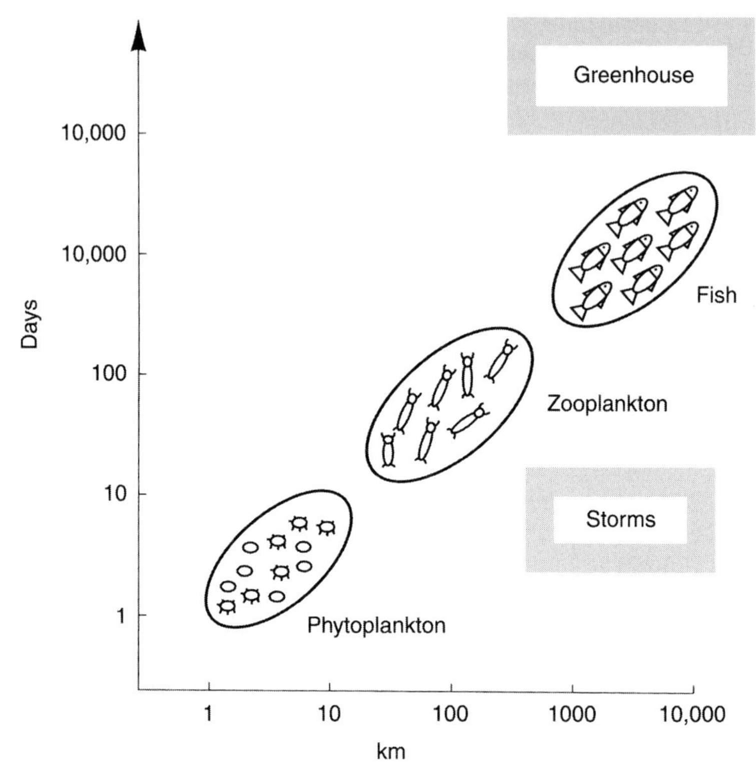

Fig. 4.2. Size scale of populations of marine species and time scale of their life cycles. Durations and scales of storms and the greenhouse effect are included for comparison (from Mann and Lazier, 1991).

4.3.1 Plankton and water masses

The association between water masses rich in nutrients and intensity of primary production has been amply demonstrated in the two preceding chapters. Note, however, that production is sometimes limited by the absence of a single substance; Martin *et al.* (1990) showed that Antarctic waters suffer from an iron deficiency, limiting primary production in water rich in other nutrients. Two zones comparable for their concentration of N and P, one coastal, rich in iron (7.4 nmol/kg), probably of terrestrial origin, had a primary production of 3 g of C/m^2/day; the other, situated in the open sea, deficient in iron (0.16 nmol/kg) and manganese, had a production of 0.1 g of C/m^2/day, as the phytoplankton is incapable of utilising more than 10% of the main nutrients present in the water. This is an example of Liebig's "Law of the Minimum".

4.3.2 Stratification of the tropical ocean

We have seen in the preceding chapter that one of the problems linked to the phytoplankton (primary production in the ocean) is that they require light which comes from above, while their source of nutrients is in the depths. In a water column with no turbulence, the euphotic zone becomes poor in nutrients following their utilisation by the phytoplankton. These data allow an understanding of the development of spring phytoplankton "blooms"*: they are caused by the confinement of organisms to the euphotic layer and by vertical stratification developed by the sunlight warming the waters. As a result of the confinement created by this stratification, mineral salts are rapidly used up and the phytoplankton dies.

As a result of the limited movement of surface waters by moderate winds (the trade winds), large portions of the oceans are permanently stratified, with a high concentration of nutrients below the pycnocline and a low concentration above, the pycnocline acting as a barrier to diffusion (see Chapter 2). This is true for most of the tropical oceans, representing half of the world's oceans. New production in the oligotrophic waters of the Pacific Ocean, for example, is very low and this is due to the barrier created by the pycnocline, which limits transfer of nitrates and other nutrients.

This does not completely inhibit photosynthesis, as phytoplankton browsers excrete NH_4 that can be utilised by the phytoplankton which remains at a certain level, but this situation is unstable and production low. Recycled nitrogen is in the form of NH_4, while nitrogen coming from deep areas is in the form of NO_3, which allows distinction between regenerated and new production. The constancy of conditions throughout the year suggests that the two processes (new and regenerated production) are in equilibrium. Thus in the vertical structure typical of a tropical ocean, there is great variation in primary production and the production of nitrates and chlorophyll in the surface layers.

Tropical oceanic environments (between 30°N and 30°S) are thus characterised by a thermocline, often situated between 130 and 200 m deep, which limits the ascent of deep waters in the absence of upwelling or other phenomena. The shortage of nutrients results in a production of about 70 g C/m^2/year. Recycling of living organic matter is very rapid at high temperatures, although the biomass present is very low (Sheldon, 1984). Consequently the water is very transparent; it is tropical "blue water".

Despite the limitations described above, it has been noted that photosynthesis can occur in waters where nitrogenous compounds are not detected and this phenomenon is known as the growth paradox of poor environments or "paradox of poor waters"; the phytoplankton is capable of acquiring nutrients, even if it only stays in nutrient-rich water for 4% of the time. Other studies show that the diffusion zones of zooplankton wastes are a significant source of nutrients for the phytoplankton. Another source would be the decomposition by bacteria of organic matter within the microbial loop (Chapter 3). Distributed throughout the sea are abundant aggregates called "marine snow", which are very amorphous and fragile and contain phytoplankton, bacteria and protozoa. It is assumed that, in these aggregates, matter is quickly recycled and generates a turnover of phytoplankton (see Fig. 3.3), but this hypothesis is difficult to test although photosynthesis and respiration can be measured using microelectrodes.

4.3.3 Stratification and mixing in temperate and polar waters

The situation is very different in the temperate zone where conditions change with the seasons; the tropical situation corresponds to the end of summer. The temperate ocean is dominated by turbulent movements created by strong winds and tides. The phytoplankton depends entirely on this turbulence which brings nutrients into the euphotic zone where they can be used for photosynthesis (Fig. 4.3).

In some places, the pycnocline may be poorly defined, but there are progressive variations in water density as the result of a movement which decreases gradually with depth. Nitrates thus spread at the base of the mixed layer and, when the light is sufficient, photosynthesis commences, consuming nitrates. This explains the maximum chlorophyll levels found in the depths, in a layer limited towards the bottom by light and towards the surface by nutrients (Fig. 4.3). Study of the diffusion of nitrate by eddies shows that this diffusion regulates primary production; in addition, turnover of production is a function of new production.

Moving towards the poles, there is a breaking up of the mixed layer as a result of the stronger winter winds: turbulences penetrate deeper and deeper into the rich

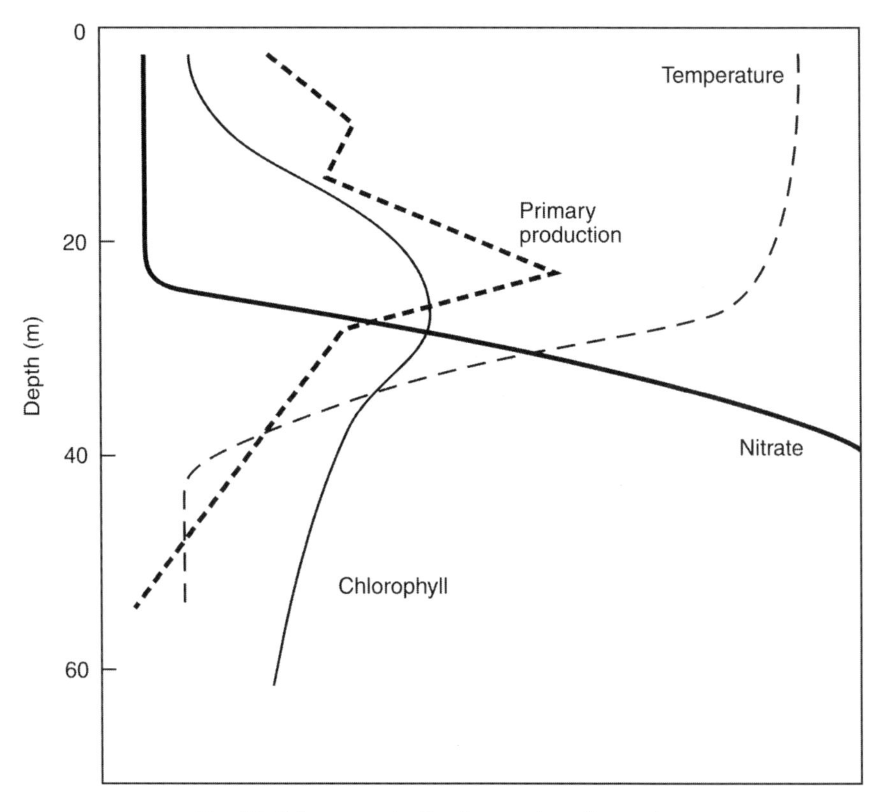

Fig. 4.3. Primary production in oceanic surface waters.

waters and bring nutrients back up into clear waters. In the spring, the phenomenon is reversed: the waters are warmed, the mixed layer becomes thinner and the phytoplankton spends more time in a well-lit zone—there is a great burst of phytoplankton, the spring bloom.

The spring bloom is directly dependent on the stability of the water column. The increase in primary production is created by a stabilisation of the water column after a period of strong mixing; these phenomena are also found in coastal and upwelling zones. The development of the phytoplankton cycle in temperate waters can be summarised as follows:

- ascent of nutrients by mixing of waters, caused by winter winds;
- start of primary production in surface layers;
- acceleration of production by increase in the photoperiod and light intensity (spring bloom of diatoms);
- during summer, stratification as in tropical waters: phytoplankton composed of flagellates (better surface/volume ratio) which can move to utilise nitrogenous compounds;
- autumn bloom with breakdown of the mixed layer;
- moderate winter production.

This explains why the maximal values for oceanic chlorophyll are situated in boreal latitudes. Although temperatures there are low, the richness in nutrients of the surface waters is adequate. The production (spring bloom) in temperate Atlantic waters coincides with the 12°C isotherm.

Diurnal and seasonal changes are significant in the mixed layer. For example, north of the Azores (41°N 27°W), the thermocline is at around 100 m in January, its depth decreasing to just 10 m in July.

The position of maximum chlorophyll (Fig. 4.3) in temperate waters depends on the sedimentation speed of the phytoplankton; biomass increases close to the surface, but it sediments most at maximum depths. Mann and Lazier (1991) discussed this problem in detail, and also the facts and the hypotheses relating to mechanisms of oceanic primary production in surface waters. It should be noted that the chlorophyll maximum can cause a slight warming of the water which increases vertical mixing and the deepening of the mixed layer.

The alternation of these physical conditions (tides, seasons, wind speed and direction) increases biological production. Variability is thus necessary. Periods of light winds and reduced upwelling are those of maximal chlorophyll production, but this production is the result of strong mixing, followed by stratification of the water column. The mixing agent is the driving of water particles by the wind and the stratification agent is the sun. This alternation is one of the most important sequences in marine ecology.

4.4 ASCENT OF DEEP WATER (RESURGENCES OR UPWELLINGS)

In order to obtain nutrients, the phytoplankton can utilise the mixing process which occurs in the surface layers as well as significant imports from atmospheric dust but, for real proliferation, it has to keep returning either continuously or periodically to the nutrient salts brought from the depths to the euphotic layer. The result is a zone of ascending deep water called resurgences or "upwellings"; vertical speeds of ascent in these areas vary between a few metres to about 30 metres per day. The enrichment by nutrient salts leads to fertilisation of the euphotic zone, but the ascending movement of waters also increases the mean staying time for a plant cell in the euphotic zone, the major parameter of productivity.

We have seen that upwellings are created by large oceanic currents or by regular winds (trade winds) inducing currents. These are very localised and constitute the best fishing grounds in the world (33% of the production from 1% of the oceans) and phytoplankton production can reach 10 tonnes of dry matter/ha/year. These fertile waters are most often characterised by their abundance of diatoms and nutrient-poor waters (oligotrophic) by that of flagellates. Sedimentation of diatoms is counterbalanced by the ascending current, which is why the primary production is exceptional.

Primary production can be modulated by one or several factors. Lee Chen and Yuh Ling (1994) described how, during the warm season, nitrate concentration alone regulated chlorophyll distribution at the surface of the Kiroshio upwelling; in contrast, in the cold season, both temperature and nitrate concentration played significant roles.

Minas *et al.* (1986) calculated the primary production of various resurgences from chemical and hydrographic data:

- in Peru, production is about 0.6 g C/m^2/day (219 g/C/m^2/year);
- south west of Africa: 1.1 g C/m^2/day (401 g/C/m^2/year);
- north west of Africa: 2.3 g C/m^2/year (839 g/C/m^2/year).

The coincidence of the physical process of turbulence and the biological process in natural ecosystems is thus the determinant in the success of the production of living matter within an ecosystem. Thus, in the Pacific close to the coast of Peru, the upwelling created by the trade winds allows the proliferation of phytoplankton and then of the fish which are widely exploited (15% of the world's fisheries).

The planktonophagic fish are the most abundant in the upwelling zones; sardine and anchovy, for example, consume zooplankton, but also phytoplankton ingested at the same time as zooplankton. It is not known if the ingested plant matter is digested. Larger fish such as the horse mackerel feed on large zooplankton.

The feeding problem is different for fish larvae, which require prey of a certain size in significant quantities: thus, Hunter (1976) estimated that it requires 1790 dinoflagellates/l for the survival of Californian anchovy larvae. It has been seen how this kind of plankton develops under certain specific hydrological conditions, demonstrating again that physical factors have a strong influence on fish abundance.

We have seen (Sections 1.3–1.5) that in certain years (1982, 1991–1993, 1997–1998) in the Pacific, the trade winds (considered the most regular winds in the world), weakened and that the upwelling was replaced by a warm current called "El Niño". With the phytoplankton production decreased in these nutrient-poor waters, the fishery decreased by 80%. An increase in rains in North America was noticed (with catastrophic floods in the Mid-West in 1993) and an abnormal intensity of the monsoon in south-east Asia (floods in Bangladesh); this was the "Southern Oscillation". This assembly (El Niño + Southern Oscillation) is known under the name of ENSO (see also Section 1.4). Merle (1995) gave an overview of these phenomena and their incidences.

A particular case of upwelling of interest to aquaculture in marine waters consists of the ascent of deep, rich waters in Galicia, a mussel-producing area (Hanson *et al.* 1986). Under the influence of winds from the north-east (which push surface waters from rias towards the open sea), deep waters coming from the central Atlantic rise. They ascend onto the continental shelf up to the rias of north-west Spain (the origin of these deep waters has been discovered from their salinity, temperature and nutrient content). These periodic enrichments are responsible for the nutrient-rich waters and, consequently, the high mussel productivity. This is a direct demonstration of the interactions which can occur, even at great distances, between the open sea and coastal waters in an environment with no physical barriers under the effect of climatic phenomena.

We shall now examine various oceanographic phenomena, differing from upwellings:

- Convergences resulting from the meeting of two opposing surface currents are visible at the surface and are marked by a front; floating material gathers at convergences, and plankton (and often larval stages of fish) are also associated with them. Concentrations 6–40 times greater than in the rest of the water at this level have been found. Convergences are considered to be nutrient pumps.
- Internal waves with large wavelengths create undulations at the surface, several kilometres apart, detectable from space by radar. Internal waves situated at the level of the pycnocline can be particularly significant (50–70 m in amplitude, 30 km wavelength). They are created at the break in the slope of the continental shelf when strong tidal currents advance from the open sea towards the coast. They are also found in the straits of Gibraltar where they are created by the entrance of Atlantic waters into the Mediterranean.

4.5 TURBULENCES AND PRODUCTION IN COASTAL WATERS

In coastal areas where the waters are mixed, dead biological material and detritus accumulate on the bottom and the freed nutrients can be re-utilised by the primary producers. Telluric inputs also contain vital nutrients and primary production can reach 0.4 g C/m^2/day with peaks of 2 g (Collignon, 1991). Coastal waters thus present characteristics of fertility similar to those of upwelling zones, although the mechanisms of enrichment are different (see Table 4.1).

Table 4.1. Summary of some average productivities (g C/m^2/year) according to area.

Zone	1° Production	Trophic levels	Fish production
Oceanic (temperate)	50–100	5.0	0.005
Coastal	>100	3.0	0.34
Upwelling	200–400	1.5	36

4.5.1 Mixing due to tides and production of living matter

Mann and Lazier (1991) stated in their work that the alternation of a period of vertical transport of nutrients with a period of stratification is the key to success for high primary production in a great variety of environments.

Nevertheless, the phenomena found in coastal waters lead to the adoption of a different conclusion. The same authors considered that the mixing created by the tides in shallow waters prevents stratification of the water column, but the potential harmful effect on plankton productivity is more than compensated for by the increased flux of nutrients into the water column from the sediment, resulting in primary production which is higher than the average in coastal waters. While phytoplankton is transported by turbulence through the water column and is exposed there to abundant nutrients, the intensity of the spring bloom depends mainly on the seasonal increase in light and not on the breakdown in stratification. The result is that the production cycle starts earlier in waters which are mixed by tides.

On the Georges Bank close to the Gulf of Maine, and on the Dogger Bank in the North Sea, primary production is much higher than on the continental shelves where the waters are stratified. The situation is similar on the Faeroes Bank.

Herring spawning grounds are often located in these areas with tidal movement. In the summer, these elevated seabeds are bordered by a front towards the side of the open sea (tidal front) which constitutes a natural barrier for the larvae: the length of time they remain there will limit the size of the stock produced. In addition, the herring lays its eggs on a bed of gravel of a certain size, which is found especially in areas where the tide mixes the waters well (the granulometry of the bottom reflects the importance of hydrodynamics; see Chapter 2). According to another hypothesis, it is the presence of a persistent physical characteristic which is important in defining the spawning grounds. It can be seen that the explanations of any single phenomenon can vary.

While large-scale climatic variations guide large ecosystems on long annual cycles, such phenomena exist on all time-scales. In the Étang de Thau, a 7500 ha lagoon in the northern Mediterranean where oyster culture is carried out, a study by Tournier and Deslous-Paoli (1993) produced a numerical relationship between the speed of the north-west winds over a 36-hour period and the oxygen deficit close to the bottom of the pond which was obviously prejudicial to the reared molluscs. The north-west winds alternate with periods of calm characteristic of anticyclones, which are unpredictable in the long term.

These examples demonstrate the problem of constancy in climatic phenomena (or at least regular alternation), but the climate remains unpredictable and so one can never be sure of having simultaneous optimal conditions within natural ecosystems.

4.5.2 Currents and turbulences caused by islands

The perturbation of currents caused by an island can lead to the upwelling of water from below the thermocline and an enrichment of surface waters. Good examples are those of the Scilly Isles (south-west of Great Britain) and St Kilda (off north-west Scotland): the temperature of surface waters there is 3°C colder than that of the stratified summer waters. Such phenomena explain the density of seabirds found on these islands; the tidal currents double in speed, carrying away surface waters and allowing the ascent of deep waters. A plume in which chlorophyll production exceeds 4 mg/m^3 extends for over 50 km, while concentration in surrounding waters does not exceed 0.5 mg/m^3. Around St Kilda, this phenomenon is less accentuated, but extends over an area of 5000 km^2.

On a smaller scale, Wolansky and Hammer (1988) noted large aggregations of zooplankton in the turbulent area downstream of a promontory when the tidal current was maximal (Fig. 4.4).

Such phenomena cannot be generalised, as they depend on current intensity; they are not found, for example, in the French islands in the Pacific although the thermocline is much deeper there (see Merle, 1995).

The increase in production of living matter around islands far from continents and above seamounts is evident, however, although the mechanisms remain unknown.

4.5.3 Biological effect of freshwaters in the coastal zone

Rivers reach the sea by means of an estuary. Lie (1983) gave the following definition of an estuary: a mass of semi-enclosed coastal water freely accessible to high tide and in which sea water is mixed in a measurable quantity with freshwater coming from continental drainage.

4.5.3.1 Direct effect of material carried by rivers

Modifications of rivers carried out by man affect the impact of floods: many rivers have been dammed for the production of electricity. Water is stored when flow is high and released when it is low. Boudouresque (1995) described how the Rhône now only transports 2–3 million tonnes (that is 5–10%) of what it used to transport in solid materials at the end of the 19th century (40 Mt). The situation is the same for the rivers Èbre and Danube and many others. In the Nile, almost all of the 120 Mt of sediments is now trapped in the Aswan dam. Flow in the Nile has fallen from 100 to 3 km^3/year as a result of irrigation.

In San Francisco, the Sacramento carried 34 km^3 of water into the sea each year. In fact, 40% is taken for local consumption and 24% channelled towards south and central California, leaving only 36% to flow into the estuary. This reduces primary

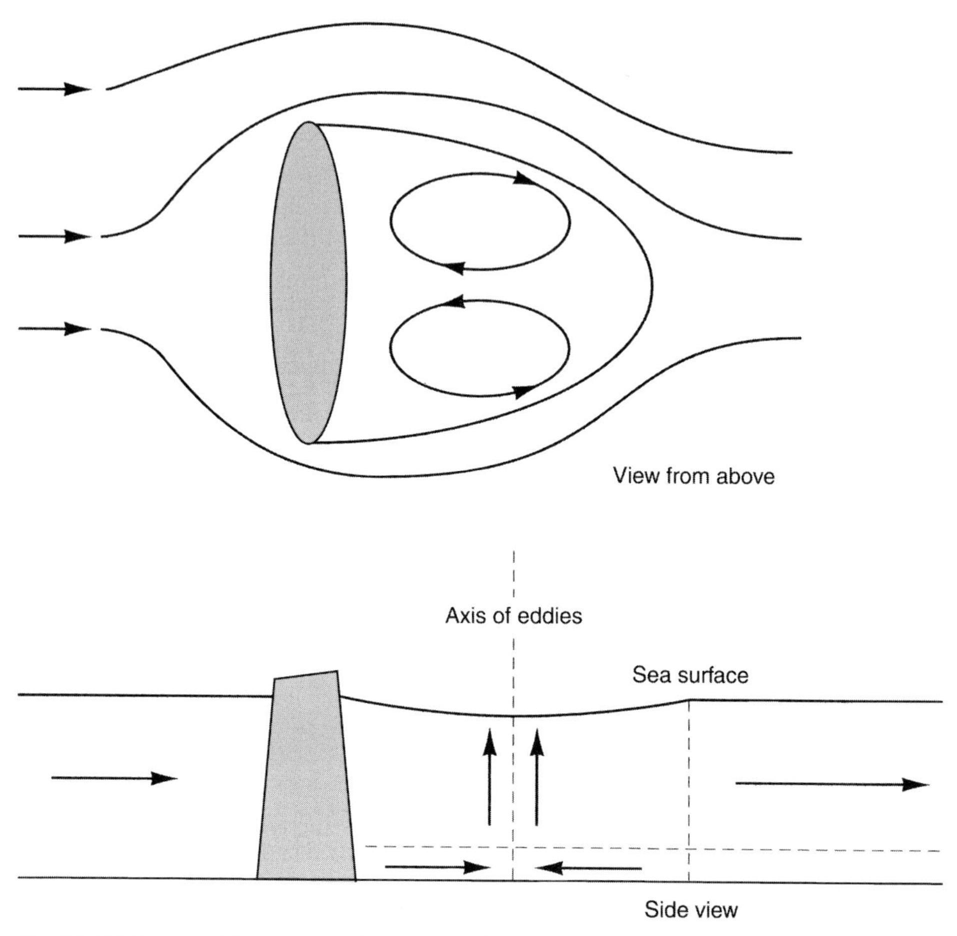

Fig. 4.4. Eddies caused by a current downstream of an island; the vertical depth scale (about 40 m) is exaggerated in relation to the horizontal scale (about 1 km) (adapted from Wolanski and Hammer, 1988).

production at the mouth, but sea water ascends higher up the estuary: in addition to the reduction in primary production, there is accelerated degradation of spawning beds owing to the ascent of salt, as well as pollution, as a result of the reduction in dilution. The production of striped bass has decreased by 75% and that of chinook salmon by 70%. In the past, fishery production was correlated to spates.

Floods have a marked effect on productivity and they are accompanied in the coastal zone by diatom blooms. Even with normal flows, large rivers affect production in the coastal zone. At the mouth of the Mississippi, a chlorophyll-rich area extends for 50 km to the east and south and 125 km to the west in the coastal current. In the Amazon, low salinities and an increase in planktonic production are noted as far away as Barbados. A correlation between catches of *Penaeus indicus* and floods in the Zambezi has been established, but it is not known which part of the life cycle is favoured.

In contrast, a decrease by a factor of 5 in fisheries has been noted in the Dniepr estuary in the Black Sea. The fertilisers used by agriculture and the nutrients carried in waste water increase phytoplankton production both in the estuary and in the estuarine plume. Uneaten plankton sediments and decomposes. In summer, when the waters are warm, discharges of freshwater are released from dams, increasing water stratification in the estuary; as a result, oxygen concentration in the lower layer is close to zero and mass fish mortalities result. These oxygen depletions existed before the construction of dams, but anoxic events have become more frequent, reducing the catches of turbot, plaice and crabs. Despite the increase in primary production, stratification and changes in salinity kill fish.

An identical situation is found in France in the Étang de Berre, to which waters from the Durance have been redirected in order to produce electricity. These are examples of development by man carried out without any ecological consideration, which have resulted in not only environmental, but also economic, disasters.

A similar phenomenon, that of stratification, can (depending on conditions) lead either to beneficial or deleterious effects on the environment. For example, in Iceland the stratifications that generate warm freshwater (from volcanic activities) are correlated with high phytoplankton production and early blooms.

In estuarine plumes, the available nutrients are often consumed in the estuary. If this load is too high, it creates turbidity which limits primary production through lack of light.

4.5.3.2 The effect of current-based transport

As a result of water circulation in estuaries, nutrients resulting from the decomposition of organic matter on the bottom are carried to the surface when salt water is pushed away by freshwater. It has been noted that, when estuaries discharge freshwater onto the continental shelf, the deeper waters are pulled towards the mouth of the estuary, often from far away: thus bottom tracers released 70 km out to sea from the Chesapeake estuary (USA) return towards the mouth.

These few examples show that floods are a factor in marine productivity, although humans are keen to stop them in order to preserve their terrestrial activities. They also show that there is often very little difference between systems which will later evolve in very different ways; this is the "scallop effect" reported by Revaud d'Allonnes (1994) where identical initial conditions lead to different situations (identical causes resulting in different effects). We shall see later that "forcing" these initial conditions is one path for development.

4.5.4 Vertical structure and movement of coastal waters

Mann and Lazier summarised the situation for coastal waters by saying that there are four factors influencing the vertical structure of the water column:

1. the warming and cooling resulting from changes in solar radiation;
2. the input of freshwater creating a lower density layer at the surface and intensifying vertical stratification;

3. tidal currents creating turbulent mixing and impeding the formation of a thermocline and a halocline;
4. wind-generated currents which can create coastal upwelling or which can impede stratification.

Given that there is no general rule of interaction for these factors, each geographical case constitutes a particular entity, but general trends can be identified:

- The coastal biological cycle is characterised by a spring bloom and by a relatively high biomass during the summer.
- The oceanic cycle is marked by a bloom resulting from a thinning of the mixed layer, caused by less mixing due to lighter winds.
- The biological cycle characteristic of the continental shelf results partly from the input of freshwater and partly from the shelter from dominant winds. The maintenance of relatively high spring production, similar to that found in the coastal cycle, is made possible by the mixing arising from tidal action, which renews the nutrients in the surface waters; it may also result from the circulation conditions of waters in stratified estuaries.

4.6 CONCLUSIONS

Phenomena linking water movements to production of living matter occur on scales which vary from molecular diffusion to an entire oceanic basin.

Around organisms, there is a boundary layer and viscosity governs exchanges over distances of less than a millimetre. At this level, transport of essential substances (nutrients, CO_2 and food particles) close to organisms depends on viscosity, but the last part of the voyage is inhibited by the boundary layer and the adherence of water molecules. This is the realm of molecular diffusion.

Beyond the millimetre scale these phenomena are dependent on water turbulence: turbulences reaching the boundary layer around organisms transport nutrients and waste products and influence the rhythm of encounters between predators and their prey. Pelagic organisms utilise currents and buoyancy for their migrations, while sessile* organisms attach where they are deposited by currents. Physical energy thus aids transfer of energy between trophic webs without the intervention of the organisms. This water turbulence is regulated by the tides, large marine currents and climatic phenomena.

What is important is the simultaneous variation in space and time of hydrological and biological factors, resulting in maximal efficiency of ecological processes. The synchrony of biological phases and hydrological events thus constitutes the key to the success of biological production. It is the organisms which determine the time scale. Populations of small organisms with short life-cycles can change most quickly.

In this sense, and as we have seen above, water movements constitute vital energy to the ecosystem. This circulation energy is termed as "auxiliary" or ecological (Frontier and Pichod-Viale, 1991).

Littoral ecosystems are characterised by terrestrial or marine imports, and also by a regeneration process linked to water movement. Lie (1983) stated that, in contrast to oceanic systems and upwelling of deep coastal waters which have a photosynthetic basis, littoral ecosystems are mainly based on detritus.

These data lead to the revision of much knowledge acquired from standing water in the laboratory. They also present great possibilities for the exploitation of marine living resources: because of physical factors, primary production in the oceanic environment is far from reaching its biological potential.

Water movements and their richness thus have an impact on fisheries production. A study of yields over a period of 10 years (Chaussade, 1994) showed that in warm seas with stratified waters, fishing yields were less than 800 kg/km^2/year in contrast to 3000–4000 kg/km^2/year in the north-east Atlantic and the north Pacific and 8000 kg/km^2/year in upwelling regions (reaching 56,000 kg/km^2/year). In warm waters, coral reefs are very productive zones (2000–5000 g of C/m^2/year) and for fish production there is around 10,000–30,000 kg/km^2/year for a biomass of 50,000 kg/km^2/year, i.e. 100–300 kg/ha/year.

4.7 BIBLIOGRAPHY

Adey W.H. and Loveland K. *Dynamic aquaria*. Academic Press, San Diego.

Alcaraz M., Salz E. and Calbet A. Small scale turbulence and zooplankton metabolism. *Limnol. Oceanogr.* 1994; 39(6): 1465–1470.

Alldredge A.L., Robinson B.H., Fleminger A., Torres J.J., King J.M. and Hamner W.M. Direct sampling and in situ observation of a persistent copepod aggregation in the mesopelagic zone of the Santa Barbara Basin. *Mar. Biol.* 1984; 80: 75–81.

Bacalbasa-Dobrovici N. Effect of changed hydrology and chemistry on the fish fauna. *Biologie der Donau* 1994; 2: 267–272.

Barnabé G. Hatchery rearing of the early stages. In: Barnabé G. (ed.) *Aquaculture: Biology and ecology of cultured species*. Ellis Horwood, Chichester, 1994.

Bijma J., Hemleben C. and Welnitz K. Lunar-influenced carbonate flux of the planktic foraminifer *Globigerinoides sacculifer* from the central Red Sea. *Deep Sea Res.* 1994; 412(3): 511–530.

Boudouresque C.F. *Impact de l'homme et conservation du milieu marin en Méditerranée*. Cours DEUG B. Univ. Méditerranée, Fac. Sc. Luminy; GUIS Posidonie Éd. 1995, Marseille.

Bougis P. *Écologie du plancton marin. Vol 1: Le phytoplancton; Vol 2: Le zooplancton*. Masson, Paris 1974.

Caceres-Martinez J., Robledo J. and Figueras A. Settlement and post larvae behavior of *Mytilus galloprovincialis*: Field and laboratory experiments. *Mar. Ecol. Prog. Ser.* 1994; 112(1–2): 107–117.

Carter R.WG. *Coastal environments*. Academic Press, London 1991.

Chaussade J. *La mer nourricière, enjeu du XXI siècle*. Institut de Géographie Publ., Nantes 1994.

Collignon J. *Écologie et biologie marines*. Masson Éd., Paris 1991.

Dagg M. Some effects of patchy food environments on copepods. *Limnology and Oceanology* 1977; 22(1): 99–107.

Frontier S. and Pichod-Viale D. *Ecosystèmes: Structure, fonctionnement, évolution*. Masson Éd., Paris 1991.

Hanson R.B., Alvarez-Ossorio M.T., Cal R., Campos M.J., Roman M., Santiago G., Varela M. and Yoder J.A. Plankton response following a spring upwelling event in the Ria de Arosa, Spain. *Mar. Ecol. Prog. Ser.* 1986; 32: 101–113.

Harris R.P. Spatial and temporal organization in marine plankton communities. In Fee J.H.R. and Giller P.S. (eds) *Organization of communities*. Blackwell Science, Edinburgh 1987; 327–346.

Hunter J.R. Culture and growth of northern anchovy *Engraulis mordax*. *Fish. Bull.* 1976; 74(1): 81–88.

ICES. *Cod and climate change*. Proceedings of ICES Mar. Sci. Symp. Vol. 198, Reykjavik, 23–27 August 1993.

Jenkins G.P. Micro- and fine-scale distribution of microplankton in the feeding environment of larval flounder. *Mar. Ecol. Prog. Ser.* 1988; 43: 233–244.

Kingsford M.J. Biotic and abiotic structure in the pelagic environment: Importance to small fishes. *Bull. Mar. Sci.* 1993; 53(2): 393–415.

Kononen K. Sedimentation of vernal and late summer phytoplankton communities in the coastal Baltic Sea. *Arch. Hydrobiol.* 1994; 131(2): 175–198.

Laprise R. and Dodson J.J. Environmental variability as a factor controlling spatial patterns in distribution and species diversity of zooplankton in the St Lawrence Estuary. *Mar. Ecol. Prog. Ser.* 1994; 107(1–2): 67–81.

Lee Chen, Yuh Ling. The importance of temperature and nitrate to the distribution of phytoplankton in the Kirushio induced upwelling northeast of Taiwan. *Proc. Natl Sci. Counc. Rep. China. Part B*; 18(1): 44–51.

Lefèvre M. Interférence sur le recrutement benthique entre hydrodynamisme des masses d'eau et comportement larvaire. *Océanis* 1990; 16(3): 135–148.

Lie U. Les écosystèmes marins: Recherche et gestion. *Impact, Science et Société* 1983; 3/4: 299–314.

Loder J.W. and Pratt T. Physical controls on phytoplankton production at tidal fronts. In: Gibbs P.E. (ed) *Proceedings of the 19th European Marine Biology Symposium*. Cambridge University Press 1985; 3–21.

Mann K.H. and Lazier J.R.N. *Dynamics of marine ecosystems*. Blackwell Scientific, Boston, 1991.

Martin J.H., Gordon R.M. and Fitzwater S.W. Iron in Antarctic waters. *Nature* 1990; 345: 156–158.

Matsushita K., Shimizu M. and Nose Y. Food density and rate of feeding larvae of anchovy and sardine in patchy distribution. *Nippon Suissan Gakkaishi* 1988; 54(3): 401–411.

McAlice B.J. Observations on the small-scale distribution of estuarine phytoplankton. *Mar. Biol.* 1970; 7: 100–111.

Merle J. Environnement climatique du Pacifique sud. In: C.R. Colloque: *Quelle recherche environnementale dans le Pacifique sud: Bilans thématiques*. Paris 28–31 March 1995. MESR Éd. Paris; 3–16.

Minas H.J., Minas M. and Packard T.T. Productivity in upwelling area deduced from hydrographic and chemical field. *Limnol. Oceanogr.* 1986; 31(6): 1182–1206.

Peterman R.M. and Bradford M.J. Wind speed and mortality rate of a marine fish, the northern anchovy (*Engraulis mordax*). *Science* 1987; 235: 354–356.

Pineda, J. Spatial and temporal patterns in barnacle settlement rate along a southern California rocky shore. *Mar. Ecol. Prog. Ser.* 1994; 107(1–2): 125–138.

Revaud d'Allonnes M. Turbulence océanique, l'effet coquilles Saint-Jacques. *Science et Avenir* 1994; Hors Séries 98: 24–29.

Rowe P.M. and Epifano C.E. Tidal stream transport of weakfish larvae in Delaware Bay (USA). *Mar. Ecol. Prog. Ser.* 1994a; 110(2–3): 105–114.

Rowe P.M. and Epifano C.E. Flux and transport of weakfish larvae in Delaware Bay (USA). *Mar. Ecol. Prog. Ser.* 1994b; 110(2–3): 114–120.

Sheldon R.W. Phytoplankton growth rates in the tropical ocean. *Limnol. Oceanogr.* 1984; 29(6): 1342–1346.

Sherman K. and Alexander L.M. (eds) *Variability and management of large marine ecosystems*. AAS Selected Symposium. Westview Press, Boulder, CO, 1986.

Skifesvik A.B. Impact of physical environment on the behavior of cod larvae. In: *Cod and climate change*. ICES Marine Science Symposium 1994: 646–653.

Tanaka M., Ueda H. and Azeta M. Near-bottom copepod aggregation around the nursery ground of the juvenile red sea bream in Shijiki Bay. *Nippon Suisan Gakkaishi* 1987a; 53(9): 1537–1544.

Tanaka M., Ueda H., Azeta M. and Sudo H. Significance of near-bottom copepod aggregation as food resources for the juvenile red sea bream in Shijiki Bay. *Nippon Suisan Gakkaishi* 1987b; 53(9): 1545–1552.

Thorrold S.R., Shenker J.M., Maddox E.D., Mojica R. and Wishinsky E. Larval supply of shorefishes to nursery habitat around Lee Stocking Island. *Mar. Biol.* 1994; 118(4): 567–578.

Tiselius P. and Jonsson P. Foraging behavior of six calanoid copepods: Observations and hydrodynamic analysis. *Mar. Ecol. Prog. Ser.* 1990; 66: 23–33.

Tournier H. and Deslous-Paoli J.M. Variation spatio-temporelle estivale de l'oxygène dans les secteurs conchylicoles de l'étang de Thau. *J. Rech. Oceanogr.* 1993; 18 (3 and 4): 71–73.

Troadec J.P. (ed.) *L'homme et les ressources halieutiques*. Ifremer Pub., 1989.

Wolanski E. and Hammer W.M. Topographically controlled fronts in the ocean and their biological influence. *Science* 1988; 241: 177–181.

Yamazaki H. and Kamykrowsky D. The vertical trajectories of motile phytoplankton in wind-mixed water column. *Deep-Sea Res.* 1991; 38(2); 219–241.

Part B

Man and coastal waters

... if current trends are maintained, civilisation as we know it will not be able to survive.

European Commission

Our sea is sick because of man, this superindustrialised and superabundant Homo sapiens

André Siegfried

5

Man and the shore

5.1 THE SHORELINE: BORDER AND INTERFACE

The shoreline which separates the terrestrial domain from that of marine waters constitutes a very marked physical barrier for living organisms that populate the "blue" planet. Life commenced in the waters and amphibious organisms frequent both these environments, while for others life is either aquatic or terrestrial.

Life is terrestrial for man and, for him, the shoreline constitutes both an attraction and also a physical limit, although some individuals do not hesitate to claim their aquatic origins, marked by the webbed nature of their hands, the presence of fat under the skin and the capacity for apnea (Morgan, 1988). We shall leave the debate about our aquatic origins to the evolutionists, but it must be recognised that they are useful in explaining our attraction to the liquid elements, water births and diving babies!

Confined to exposed land representing little more than a quarter of the planet's surface, man is concentrated in coastal zones. It has been estimated that 60% of the human population lives in a coastal band 50 km wide, and half of them in developing countries. The United Nations' estimate for the year 2000 is 75%. We shall see that this is not exclusively for material reasons. First, ships under sail, then steam and finally oil have generated fishing activity at its most active close to the continents and fishery products from this coastal zone represent 80% of total catches.

The sea is vast and the standard of living for people in western societies puts living space high on the agenda: correlations between blocks of flats and space occupied (surface area and number of residents, for example) have been demonstrated. If for man, space is thus associated with luxury, it must be remembered that, in ethnological terms, division of space and the idea of territory are essential for all species; man has already overpopulated dry land and this will probably only get worse. Thus, the shore is overpopulated, often overdeveloped (rich countries) or underdeveloped (poor countries): in 1991, the cyclone which devastated Bangladesh and killed 200,000 people was predicted and warnings given, but there were not sufficient shelters for the whole population.

In contrast to this pressure of human populations on the terrestrial coastal zone, contact with the waters (other than pollution) is reduced on the littoral fringes. Throughout history, various navies have allowed safe passage across seas and oceans—a bit like travel across a desert— from one place to another without any colonisation of oceanic spaces occurring, as man is not an aquatic species. With the rare exception of cross-ocean yacht races, industrial fishing boats or commercial ships, the high seas remain empty as far as humans are concerned.

A large part of the world's oceans is isolated from all visitations. When we are shown lists of temperature measurements (carried out by fishing, commercial or research vessels), we can see that areas almost as large as the whole of South America remain empty in the Pacific, for example; this is also true for the polar oceans, while coastal zones are covered superabundantly.

It is comforting, rather than a regret, that such immense areas of ocean remain virgin with regard to all human presence; it's like a reservoir for the future, even though the shoreline is no longer a frontier. This does not imply that these zones are untouched; as we have seen in preceding chapters, the functioning of the climate, which links ocean and atmosphere, does not allow events elsewhere to be ignored, since everything is mixed in this gigantic machinery on a global scale.

5.2. THE IRRATIONAL ATTRACTION OF THE SHORE

This is not the place for the consideration of the symbolism of the sea or Freudian theories, but the strong and often irrational motivation underlying human actions can no longer be ignored. In short, coastal waters exert their influence on human activities well beyond their geographical effects. The shore line—where land, sea and the skies meet—is an interface, a unique and almost "magical" place. The coast represents the idea of an air–land–sea interface.

Thus, for Tamburrino (1991), reflection became philosophical: "... the sea shore is one of the places where one is led spontaneously to reflect more on the meaning of life. This is, without doubt, because it provides a focal point for nature: here, it is more varied and more comprehensible through its elements of water, air, land and energy. Nature reminds us of its existence and invites us to speak in turn to it about ourselves. It thus creates an interactive process." Others see the sea shore as a place for relaxation which is there to be looked at.

The irresistible attraction of the sea for man indicated to Laubier (1990) that a person is a thalassic organism for whom the attraction for the sea is far from being rational; observing the sea is relaxing, constituting a pleasure, but it also has a role in human beings' physical and mental health. This author remarks that the coming together of work and leisure is found in the fact that technology centres are set up close to seaside tourist resorts (Silicon Valley in California and Sophia Antipolis in France for example; one can add Montpellier and Montpellier-Technopole and Port-Mariane inland, but linked to the sea).

Effectively, industries such as research, computing and electronics have little requirement for basic materials and can thus be set up with regard to the preferences of their personnel. In an indirect fashion, direct democracy, as described by Naisbitt (1982) is attained: "Decision-making processes, without affecting institutional powers in any way, but more in terms of reinforcing them, must be opened up to public participation in order to be as coherent and constructive as possible".

In France, the "Loi Littoral" of 3 January 1986 determined (article 1) that "the shore is a geographical entity which requires a specific policy for development, protection and exploitation". Protection constitutes the main worry for the legislator; development is considered in depth in relation to preceding development on the sea front. Expansion of activities is aimed mainly at research and innovation, protection of biological and ecological equilibria and preservation of sites and activities linked to the sea.

In order to observe coastal waters, various observation networks have been established in France: RNO (National Network for Observation of Quality in the Marine Environment); REMI (Network for Microbiological Monitoring of Shellfish Culture Zones) and REPHY (Network for Monitoring of Toxic Phytoplankton).

5.3 THE TOURIST TRADE; ADVANTAGES AND DISADVANTAGES

Parallel to the permanent inhabitants, another form of human activity is coastal tourism which utilises coastal waters as a resource, bringing more and more littoral areas into contact with roads, ports or industries.

5.3.1 Degradation linked to the tourist industry

UNESCO* clearly sees tourism as a "passport to development", but also indicates its possible deleterious effects on the environment, notably when concerning conservation of natural heritage and biodioversity. The effects of tourism on the environment are diverse and have been evaluated as follows:

Destruction of vegetation	17%
Increase in illegal practices	22%
Risks to wildlife	22%
Lack of respect for site	23%
Development of infrastructures	Not evaluated

In Phuket, a well-known tourist resort in Thailand considered to be "like paradise", 60% of the reefs are in a poor state; this change has occurred very suddenly over the past decade. It is due to sedimentation, pollution from terrestrial areas close to the coasts, to dynamiting in order to catch fish and also to the anchoring of boats, the discharge of wastes and the increased tourist traffic.

Another problem, linked to the degradation of the reefs, is that of ciguaterra, spread from the Antilles in the West Indies to the Pacific. Human poisoning occurs via a toxin, ciguatoxin, after eating tropical fish. These have browsed directly on micro-algae living on dead coral, amongst the red algae, following either natural phenomena such as cyclones or perturbation of the reefs by man. Fish which consume these organisms concentrate the toxin and become unfit for human consumption, thus creating a marketing problem for the fishermen. The micro-alga responsible, *Gambiediscus*, has been isolated, but no remedy has yet been found. In Mururoa (Polynesia), work linked to nuclear research and the tremors linked to explosions have created ciguaterra phenomena, while at Ranguiroa it has been caused by the construction of a marina and elsewhere by the wreck of a ship after grounding.

Another adverse aspect of tourism is that it utilises natural resources but pays little attention to the cultural aspect of the host country. We shall see (Chapter 8) that, often against the interests of conservation, 53% of the local population lives in sites classed as being important for natural heritage, with tourism as the primary economic activity.

5.3.2 Economic impact of tourism

The simple bather appreciates warm water, but also the sun and the sandy beaches while the diver explores the rich rocky shores and the sailboarder seeks out windy sites. This tourist presence creates very significant economic activity. For a convincing demonstration, compare the state of the Languedoc or Spanish beaches in the 1960s with that of today. The number of new bathing beaches can no longer be counted.

All authors agree, whatever the country concerned, in recognising the importance of tourism in the coastal zone.

Tourism employs 130 million people in the world and its capital investment in tourism projects exceeds €305 billion or more than the construction or oil industries. Between 1970 and 1990, world tourism increased by 300% and was predicted to increase by half again from then until now. The annual turnover represented 6% of the gross world product in the mid-1990s.

In OECD* countries, tourism constitutes 27% of service exports and up to 14% of manpower. It is a form of industry which is expanding greatly; the OECD (1993) stated that it represents 13% of the total gross national product of these countries with a turnover of €2,600 billion and 250 million employees. According to the World Tourist Organisation (1995), 50% growth could be expected in the following 10 years with the creation of 125 million jobs.

It has been estimated that tourism around the Mediterranean represents one-third of the world's 350 million tourists, that is around 100 million, with 4000 km of the coast devoted to them. This would mean 0.4 tourists per metre of coast! Two-thirds of tourist destinations are defined as littoral (but 80% in Tunisia, for example). Tourism represents 5–10% of the world's economy and increases at a rate of 5% per year (Goldberg, 1994). In the Caribbean, tourism represents 43% of the gross national product.

In Spain, 51 million tourists visit the coasts annually, producing a turnover of about €11 billion. In Canada, industry connected with the exploitation of the ocean provides a million jobs representing approximately €6 billion, according to the OECD. In France, two-thirds of the holiday locations and campsites and a third of hotels are in coastal regions (France received 60 million tourists in 1994).

The economic consequences are not limited to the property rental and real estate sectors and commercial activity in general; infrastructures are also required (roads, water distribution, waste water treatment stations, etc.). The marinas of these new towns shelter numerous tourist fishing boats and, in fine weather, several hundred can be counted in the fishing zones close to the coast. The pleasure fisherman pays a lot for his fish, but at the same time he is at leisure, which to him is priceless.

In Europe, coastal zones for leisure activities are utilised mainly in the summer, but air travel or the development of new equipment allow the perennial tasting of the delights of tropical waters or protection from the cold under the ice. Here, technology has no limits: sailboarding as a minority sport has already been superseded by the bodyboard or the jet-ski.

Thus it can be noted that the main economic activity linked to the sea is tourism, which by far exceeds fishing and aquaculture.

5.4 ECOLOGICAL ASPECTS OF ANTHROPOLOGICAL PRESSURE ON THE SHORE

Coastal seas are an area of economic activity (dwelling, work, transport, fishing) and also leisure and holiday (tourism) activity; they are often visited by multiple users and the diversity of uses to which it is put is very wide. We have borrowed Table 1 from the OECD (1991), which lists the different pressures applied by man on the marine littoral environment.

As a result of the growth in world population, in the year 2000, 6000 million people will populate a finite amount of land. The colonisation of the seas has already begun: the artificial islands in Japan are known but, closer to home, on all the shores space is being taken from the sea. This impact has been measured in Provence-Côte d'Azur (France), it represents 3059 ha of ports, marinas and various developments, much more than protected sites, as we shall see.

It is estimated that 10% of the entire population of France live around the coast, occupying 4% of the total land. Forty-three per cent of construction is being carried out in coastal regions.

Man's actions are adding to the slow rise in sea level (around 2 mm/year since the start of the century) and, worldwide, 70% of beaches are in retreat. The extraction of sediments from beaches is one cause, as well as dams built on rivers. The solid content (the muds transported by rivers) has been reduced to one-tenth of that in the 19th century (see also Chapter 4). Since the construction of a dam on a coastal river in Togo, the neighbouring coast road had to be reconstructed three times after being eroded by the sea. Only 10% of the world's beaches gain ground over the sea. In France, out of 5000 km of coast, 1850 km of coast are retreating at 0.5–1 m/year (Lacaze, 1993).

Living on or under the sea is no longer a utopic vision. In Japan, the creation of artificial islands constructed for these ends is very advanced (17 islands, creating 615 ha of surface, situated between 500 m and 5 km from the shore (Goldberg, 1994)). While this type of construction has not been developed in Europe,

Table 5.1. Antropological pressures on the marine environment (OECD, 1991).

Pressure	Substance or activity involved	Main anthropogenic sources	Possible effects
	Nutrients	Waste water, agriculture, aquaculture, industry	Eutrophication
Wastes entering the oceans	Pathogenic organisms	Waste water, agriculture	Diseases and infections, contamination of molluscs and crustaceans
	Oil	Industry, waste water, shipping, cars, urban sewage	Oiling of birds and other animals, pollution of sea-foods, pollution of beaches
	Synthetic organic compounds	Industry, waste water, agriculture, forestry exploitation	Metabolic problems
	Radioactive waste	Nuclear weapons testing, treatment of irradiated fuel, discharges into the sea, nuclear power station accidents	Metabolic problems
	Heavy metals	Industry, waste water, discharges into the sea, cars	Metabolic problems
	Plastic material and refuse	Unauthorised dumping, shipping refuse, fishing bycatch	Tangled up wildlife digestive problems
	Soil organic and inorganic waste	Waste water, discharges to the sea, industry	Deoxygenation, filling-in of habitats
Reorganisation of the environment	Coastal development	Dredging, industrial residential and tourist development	Reduced aesthetic value and dis-appearance of habitants, ecosystems modification
Exploitation of resources	Fishing for fish, molluscs and crustaceans	Fisheries activities	Reduction in stocks, modification of ecosystems
	Oil exploration	Drilling, accidents	Pollution by oil and chemicals
	Mining exploration	Dredging, dumping of waste, extraction	Alteration of water quality, coastal erosion
Incidence of climatic change	Carbon dioxide, CFC, other gases and the greenhouse effect	Energy production, transport, agriculture, industry	Estimation of sea level: coastal flooding, dis-appearance of marsh-lands, deterioration of infrastructures, habitat perturbation, decrease in oceanic production

degradation of the shoreline as a result of construction or port development is still significant (cf. Meinez *et al.*, 1991, for the French Mediterranean). Numerous prototype submarine houses have seen the light of day, but they seem to have more to do with the dreams of their inventors than with use in real life, owing to their lack of economic feasibility.

Sometimes, existing but threatened islands are the object of valid concern. Thus, Japan invested 28 thousand million yen to repair the base of a small island 1.7 km by 4.5 km situated 1700 km south of Tokyo. This allowed Japan, by virtue of exclusive zone rights, to reclaim an exclusive economic zone as big as the island of Hokkaido. (Source: *Flash Japan*, no. 58, May 1995.)

It is almost impossible to compile even a summary table of the impacts of human activity on the coast and coastal waters. For an idea one can consult the works of Mauvais (1991) about the impact of ports, those of Guillaud and Romana (1991) on urban waste, the publication of the articles of the COMAR colloquium (UNESCO, 1992) and the very comprehensive review by Aubert (1994) about the Mediterranean. The Equinox review has edited a special edition (no. 32) on the pollution of French shores in 1990 and indicates the players involved in the management of the littoral environment. Details of the retreat of the coastline, from a geomorphological viewpoint, can be found in the work of Paskoff (1993).

At an ecological level, the well-documented review of Lacaze (1993) is clear. It states that 77 marine species, to which must be added 151 species of the coastal margins, have disappeared from French shores. This is attributed to the destruction of habitats, to pollution and overexploitation by man. On an international scale, Goldberg (1994) attempts to demonstrate the inevitable conflicts arising from the use of coastal zones; like other authors, the management of coastal populations is a main problem. The work of Pirazzoli (1993) constitutes an interesting review in geographical terms.

A study of oyster production in Chesapeake Bay (Maryland, USA) by Roschild *et al.* (1994), shows that production has decreased by more than 50 times since the start of the century as a result of the mechanical destruction of habitats and overfishing.

In global terms, as stated by Hale (1992), in most of the world, the quality of coastal waters is declining, particularly in estuaries, with estimated stocks of species decreasing and vitally important habitats (wetlands, reefs and seaweed beds) being destroyed or degraded. Towns extend out over rich alluvial plains or on the coast; conflicts between users are exacerbated and the percentage of increasingly poor coastal inhabitants continues to grow. These trends towards degradation or overutilisation are not unique to coastal systems, but are particularly obvious along the coasts.

The combined action of climate and its changes linked to the greenhouse effect (Chapter 1) and the concentration of people on the coasts result in other consequences: the passage of cyclone "Andrew" over New York cost $50 billion in insurance. After "Andrew", premiums increased by 25% and reinsurance raised its premiums from 400% to 500%. Twenty-one of the biggest natural catastrophes in the history of American insurance (up to 1997), occurred in the preceding 10 years, 16 being caused by the combined action of wind and water. The increasing number and intensity of hurricanes, storms and floods over the whole planet have been attributed to the rapid increase in temperatures of the air and the sea, i.e. the greenhouse effect.

Europe is also under threat, since the area of low pressure long centred over the north Atlantic has moved towards Russia, thus installing a "fan" and creating violent winds over our continent.

5.5 HUMAN ACTIVITY AND THE DEGRADATION OF COASTAL WATERS

Fishing and aquaculture will not be examined here, as they form the subject of a separate chapter. The OECD (1993) reported that, in the USA, 50% of coastal wetlands have been destroyed since 1970 and that, in the same country, fishing for molluscs and crustaceans is prohibited over a 200,000 ha area on both east and west coasts.

In Italy, the Orbetello lagoon in the Tyrrhenian Sea has seen the composition of its vegetation change completely between 1975 and 1993 (Guerrieri *et al.*, 1993), with *Cladophora* replacing *Cydomoteum*. The wet weight of biomass represented by the former in 1993 was estimated at 100,000 tonnes. Phytoplankton density has been estimated at 1500–7000 times greater than it was in 1975–1976. Nitrogenous and phosphorus imports resulting from the riverine population (54 tonnes of N and 13 tonnes of P), intensive fish culture (37 tonnes of N and 1.2 tonnes of P), and tourism (65 tonnes of N and 13 tonnes of P), added to unmeasured quantities from agricultural runoff, are at the heart of this situation. Fishing, which exceeded 500 tonnes/year decreased to less than 200 in 1991. Two types of intervention have been proposed: water circulation and gathering of algae.

Other examples can be given. Herring and cod eggs are killed by waste water at concentrations of 0.1%. Concentrations of methylmercury of between 0.02 and 0.005 ppm* result in deformities in fish. The survival of sea bass *Dicentrarchus labrax* is affected by detergent concentrations in sea water of 0.0034 ml/l and the survival of eggs of the same species declines when the same water contains 5 mg/l of copper.

In an article entitled "Chronicle of the sea as a dustbin", the weekly *Le Marin* (12 August 1994, p.23) reported that statistical extrapolations give an estimate of 175 million pieces of rubbish each weighing about 100 g occurring over 90,000 km^2 of the Bay of Biscay, that is 194 items per ha. Plastic constitutes between 75% and 95% (3% being plastic supermarket bags). Off Marseilles, the density is 346 waste items per ha in the coastal sea. It is estimated that 5 million tonnes of non-organic rubbish are thrown into the sea every year.

Weber reported that, every hour, the world's fleet of 35,000 commercial ships release several thousand m^3 of water which have been taken on board at a distant port, thus introducing thousands of new species to another point on the globe. It is estimated that 600,000 tonnes of hydrocarbons are released into the sea each year, although oil pollution from ships decreased by 60% between 1981 and 1989.

In addition to these impacts, marine coastal waters receive freshwater from rivers (widely used for human activity) and urban waste, either treated or untreated. These inputs of water, whether or not they are affected by man's activities, often upset coastal water ecology: belugas (whales in the Gulf of St Lawrence) have been rendered sterile by organochloride pesticides over a number of years. The examples are innumerable but the rapidity of pollution is extraordinary: radioactive fallout from the nuclear accident at Chernobyl was found less than eight days later in particle traps at a depth of 200 m in the Mediterranean. We shall return to the impact of freshwaters on coastal waters in biological terms in the following chapter.

5.6 ECONOMIC ASPECTS OF COASTAL AREAS

Despite all these problems, the accounts for economic activities linked to the shore are impressive. Below are the figures for France.

	Euros (billion)	Employment
Seaside tourism	20	150,000 jobs
Fishing	1–2	52,000 direct and indirect jobs
Shellfish culture	0–3	10,000 direct jobs
Ports	1–5	200,000 direct and indirect jobs
Shipping industry	1–2	35,000 direct jobs
Coastal agriculture	1–8	105,000 jobs

5.7 THE PROBLEMS OF COMMON OWNERSHIP

Three-quarters of the human population live close to the coast and it is in this area that a population double that of today will wish to live. As a result, coastal areas are subjected to changes which are just as significant as those of tropical rainforests; as natural habitats are altered, biodiversity is decreased. All the water used by the human population ends up in coastal waters, with or without being purified. All this occurs in silence, under the surface, with nothing showing above, nor affecting the movement of the waves. Thus beluga carcasses washed up on the St Lawrence coast in Canada are so polluted (via the food chain) by chemical products that they are considered as toxic waste.

This low visibility of anthropogenic effects on coastal waters constitutes its most dramatic characteristic; this impact passes all the more unnoticed since life in the sea is often small and diluted and because aquatic environmental changes are difficult for a terrestrial vertebrate such as mankind, to evaluate. Consider, for example, all the years it took to identify Minamata disease, created by mercury pollution in Japanese coastal waters.

Methods for monitoring coastal water quality require specialised equipment, specialist personnel and a lot of time, and are therefore very expensive. Monitoring is thus limited to restricted sites and for specific purposes: bacterial analysis of bathing waters along busy beaches and bacterial and plankton analysis in mollusc-rearing areas. It is carefully avoided elsewhere (harbour areas), although there are no barriers in the aquatic environment and currents are present which are capable of transporting enormous, undiluted water masses (see Part A). Diseases such as cholera can be carried in marine waters. As for viruses, they have only rarely been studied, although viral particles are extremely abundant in coastal waters.

It should also be noted that media intervention is difficult in the aquatic environment. Filming or photography on land to show the state of an ecosystem is child's play; in the sea, specialised, heavy, expensive and unusual equipment is required, and still this material does not show up the presence of toxic plankton, for example. Information for the general public can only rarely be put forward in image form, which is the most effective route; water degradation is not a media event.

In certain cases, the effects of the impact of human activities are not only hidden, they are not even measurable. We have seen (Section 5.5 above) that infinitesimal concentrations of detergents and metals are sufficient to massively reduce the most fragile stages of marine organisms, but what is the global impact of these substances on the scale of coastal aquatic populations? How can it be evaluated? Without fear of contradiction, it can be stated that coastal waters constitute simultaneously the ecosystem to which man is most attached and that which he manages least. What is the origin of this situation?

In legal terms, French coastal waters are part of the common good; no one person possesses them; they belong in the public domain. In the United Kingdom, they belong to the Crown Estate. Various users may claim ownership illegally: this is often the case with professional fishermen who claim to be the true owners as opposed to amateur fishermen, divers and tourists in general; this is also the case for polluters (individuals or companies) who release their used water or waste into the sea. More recently, certain coastal dwellers have protested against the installation of aquaculture cages in the sea, under the pretext that they spoilt the view. This category of user considers that they had a right "of view" over the occupation of a public domain.

In France, there is no legal representative, member of parliament, councillor or association to look after coastal waters, only groups of users, each of which exploit them for their own profit. In every case, these diverse users are not accompanied by any obligation, no requirement is usually incumbent on an owner, since no-one has legal ownership. The public domain becomes the site for all excesses, without any inconvenient sanctions, i.e. a site of irresponsibility. As we shall see also for access to fishing zones (Chapter 7) this results in liberty applied to common property causing the ruin of all.

In reality, common ownership is a myth, as human behaviour is not compatible with this concept. One has only to open one's eyes to see that the usage is defined from ownership of everything in life today. In another field, the breakdown of the

USSR constitutes an illustration on a grand scale of the problems posed by common ownership.

Elsewhere, statutes affecting coastal waters can be different. In the Pacific, the customary rights of the Melanesians extend to their fishing zones; in Japan the aquaculturists are proprietors of their culture sites, while the fishing rights in the coastal zone belong to local fishing communities. The waters become a property like any other, with rights but also responsibilities. We shall see (Chapter 11) that the best examples of development and management of coastal waters come from Japan.

5.8 SOLUTIONS FOR SAVING WATERS AND COASTS

Because of the multitude of uses of coastal waters and their belonging to the public domain (in France, for example), it is only in Japan that a national or supra-national strategy of development is seen and the management of coastal waters being carried out. In the absence of a grand plan, there exist nevertheless many ideas to put a stop to the real disaster of degradation caused by pollution and overexploitation in coastal waters.

- The principle of "polluter pays" can be applied to all waste ending up in rivers or in the sea, both industrial and domestic waste water; it is known that water treatment is not 100% efficient. Under these conditions, purification can be carried out in the aquatic environment using various types of equipment (see, for example, Chapters 14 and 15) financed by taxes levied under the polluter-pays principle.
- It has been seen that tourism and leisure are the main economic activities in coastal waters. These require both unpolluted waters and the protection of natural sites, which are the basis of biodiversity (by providing habitats for flora and fauna). The success of large aquaria or marine reserves demonstrates the curiosity aroused by aquatic life, as well as the fact that there remains something to be seen and that coastal waters have not been completely emptied of life by an over-equipped fishery or widespread pollution.

The protection of coastal waters for use by leisure activities, and conservation of their flora and fauna simply for us to admire would have appeared unrealistic only a few years ago and would still be difficult today in the absence of any economic justification. But without clean waters and fish, there are no tourists, bathers, divers or pleasure fishermen; all these activities go elsewhere. For tourism, coastal waters thus constitute both ecological and economic capital.

- Like tourism or leisure, aquaculture requires unpolluted water, since these waters are, for example, filtered by reared molluscs. Tourism and aquaculture constitute economic activities which are capable of enhancing the value of coastal waters while protecting them at the same time.

Extending these two types of usage thus constitutes the most efficient means of protecting coastal waters since it concerns the development of economic activities: no state-controlled subsidies or regulations are required.

- The direct acquisition of coastal sea beds by land-based communities also occurs in some countries, to fight against the ravages caused by the status of common ownership. In Brittany (France), the Plan d'Occupation des Sols (POS), which regulates land use by communities on land, has been extended to coastal sea beds up to three miles distant. In the French Mediterranean (Côte d'Azur), a legal statute is being put in place to cover the utilisation of coastal sea beds and waters; the goal is to control the use of these zones within a collective management structure, associating the users with collection of information and making them responsible. These structures would also have a policing function, but with the aim of collecting information, rather than repression. In the same country, privatisation of shellfish culture plots agreed by the State is an established fact, since they are sold between shellfish farmers.
- For fisheries, the main system now dominant in Europe is that of quotas, i.e. the regulation of numbers of fish caught, but many others exist on national or regional scales. Some are legal in nature (fishing prohibitions and various regulations linked to this activity, protection or conservation of zones by regulatory measures); others involve direct action on the environment (restocking with juveniles, management of the sea bed and fertilisation of waters). Privatisation leads to responsibility and requirements which are well-known in aquaculture but not so in fisheries in waters with no physical barriers. The "Law of the Sea" constitutes the main part of the legal basis but, in biological terms, things can be very complicated (e.g. migratory fishes). While in Japan access to coastal sea beds is reserved for well-defined fishing communities, New Zealand has just privatised its fisheries, and the management of salmon fisheries in coastal zones of the Pacific constitutes an evolution towards privatisation (see Section 13.4.3). The proposals for private fisheries based on aquaculture (see also Chapter 13) show that privatisation is another step allowing increased production.

What has changed in relation to the previous situation is that all these propositions include human activity in the functioning of the overall ecosystems; they are integrated with nature and are included in the body of sustainable development.

Thus the protection, development and management of coastal waters become the best ways to increase value in economic terms: ecology becomes the ally of the economy. The entire third section of this book is devoted to these aspects.

5.9 BIBLIOGRAPHY

Aubert M. *La Méditerranée: la mer et les hommes*. Rev. Int. Oceanogr. Med. 1994; 109–112.

Goldberg E. *Coastal zone space. Prelude to conflict.* IOC Ocean Forum 1. UNESCO Publishing, Environment and Development 1994.

Guerrieri M., Lenzi M. and Ugolini R. Situazione e prospettive del l'acquacoltura nella laguna di Orbetello. *Laguna*; 18: 6–11.

Guillaud J.F. and Romana L.A. *La mer et les rejets urbains.* Actes de colloques, no.11. Ifremer Éd., Brest, 1991.

Hale L.Z. Strategies to manage Thailand's coral reefs. *Maritimes* 1992; 36(4): 4–5.

Lacaze J.C. *La dégradation de l'environnement côtier: conséquences écologiques.* Masson Éd., Paris, 1993.

Laubier L. Les océans et l'antarctique. In: *2100 récit du prochain siècle.* Payot Éd., Paris, 1990; 474–495.

Mauvais J.L. *Les ports de plaisance: Impact sur le littoral.* Ifremer Éd., Brest, 1991.

Meinez A., Lefèvre R. and Astier J. Impact of coastal development along the southern Mediterranean shore of Continental France. *Mar. Poll. Bull.* 1991; 23: 243–347.

Morgan E. *Des origines aquatiques de l'homme.* Sand Éd., Paris, 1988.

Naisbitt J. *Les dix commandements de l'avenir.* Sand Éd., Paris, 1982.

OEDC. *Gestion des zones côtières. Politiques intégrées.* OEDC, Paris, 1993.

Paskoff R. *Côtes en danger.* Masson Éd., Paris, 1993.

Pirazolli P.A. *Les littoraux, leur évolution.* Nathan Université Éd., Paris 1993.

Roschild B., Ault J.S., Gouletquer P. and Héral M. Decline of the Chesapeake Bay oyster population: A century of habitat destruction and overfishing. *Mar. Ecol. Prog. Ser.* 1994; 111(1–2): 29–39.

Tamburrino K. Aménagement intégré du littoral méditerranéen. *Rev. Int. Oceanogr. Med* 1991 (101–104: 289–294.

UNESCO. *Coastal studies and sustainable development.* Proceedings of the COMAR. UNESCO Technical Papers in Marine Science, No. 64. UNESCO, Paris, 1992.

6

Freshwater input to the coastal zone

6.1 FRESHWATERS AS VECTORS OF MARINE POLLUTION

Freshwaters from rivers play their part in the overall enrichment of the seas (Chapters 2 and 4), carrying alluvium, suspended matter and dissolved substances, but no distinction has been made about their nature or what proportion is related to natural processes and how much to man. In the open sea, the mean level of organic matter is of the order of 3 mg/l, while it is about 45 mg/l in the coastal zone.

In the preceding chapter, human activity close to the coasts was discussed. To estimate the true impact of human populations on this environment, it is necessary to consider not only littoral populations, but all those living within the water catchment area of the rivers ending at any given marine environment. In the Mediterranean, for example, the "Blue Plan" estimated the population of its water catchment areas at 211 million inhabitants in 1990.

The review on the Mediterranean by Aubert (1994) is particularly interesting, since this almost closed sea, 2.5 million km^2 in area, receives quantified imports of freshwater from its catchment areas and exchanges known amounts of salt water with the Atlantic (the exchanges are about 1.5 million m^3/s from each source, the net import from the Atlantic being only about 41,000 m^3/s). Evaporation accounts for 2900 km^3/year, equivalent to a layer 1 m thick. Rain water represents 98 billion m^3 year (98 km^3/year).

It is also known that one inhabitant excretes about 12–15 g of N/day and 4 g of P which, diluted in the water which this person uses daily, results in an effluent containing 35–90 mg/l of organic and ammoniacal nitrogen (NTK), with a mean of 50–60 mg/l and 10–20 mg/l of P, i.e. 15 mg on average. Both humans and animals also excrete micro-organisms (around 10^{12} per gram of faeces, according to Cormier and Martin, 1991). All this, with or without treatment, finds its way to the residual waters which meet the coastal waters.

Numerous studies have been devoted to these problems; we will refer to some of them, but it is not possible to give an exhaustive presentation of all aspects of coastal water pollution.

6.2 DEFINITION OF MARINE POLLUTION

Man is accused of "polluting", but what exactly is meant by this term? According to the United Nations Organisation:

> *"marine pollution is the introduction by man into the marine environment, including estuaries, directly or indirectly, of substances or energy which can lead to deleterious effects such as: damage to biological resources, danger to human health, hindrance to maritime activities, including fishing, decrease in quality of the sea water from the point of view of its use and reduction in the potential for use in the leisure industry".*

It is easy to characterise those chemicals considered as pollutants of the marine environment, according to the FAO: aluminium, arsenic, beryllium, cadmium, fluorides, acids, hydrogen sulphides, iron, lead, mercury, phosphorus, selenium, titanium, vanadium and zinc. In addition to these pollutants, there are around a million chemical substances, most of which may be released into the sea, either as accidental or routine pollution (Guillemaut-Drai, 1991).

Marine food chains can concentrate chemical substances, even modify them, for example during passage through the digestive tract; this can result in concentrations of toxic substances for the human consumer. Some are biodegradable (hydrocarbons), but the dispersants used to combat oil slicks are not themselves harmless, as they contain detergents (although there is a trend towards the use of colonies of selected bacteria). Pesticides and detergents are other well-known agents of pollution.

Some substances, such as nitrogen and phosphorus, are present in nutrient salts, and are vital to primary production and they cannot be considered as pollutants unless their concentrations reach significantly high levels.

It is impossible to summarise the effects of pollutants here. They vary from fish necroses, found in areas where petrochemical activity is significant, to Minamata disease in man, resulting from toxic mercury concentrated along marine food chains, but often this effect is undetectable; tiny concentrations of copper salts and detergents are toxic to eggs, sperm or larvae of coastal species. The pollution that is known about is, no doubt, only a fraction of the amount which is ignored.

6.3 SOME NUMERICAL DATA

Returning to the Mediterranean, Aubert (1994) estimated that the input of freshwater, polluted to a greater or lesser degree, into the Mediterranean was about 420 billion (420×10^9) m^3/year from water courses and 8–10 billion m^3 from waste waters (50–60 times less volume, but much more polluted). This urban, industrial and agricultural input is mostly discharged without any purification treatment. The mean concentrations of organic and metallic pollutants in these waters are shown in Fig. 6.1.

River waters carry around 350 million tonnes of suspended matter into this sea, organic matter corresponding to 3.3 million tonnes of BOD (biochemical oxygen

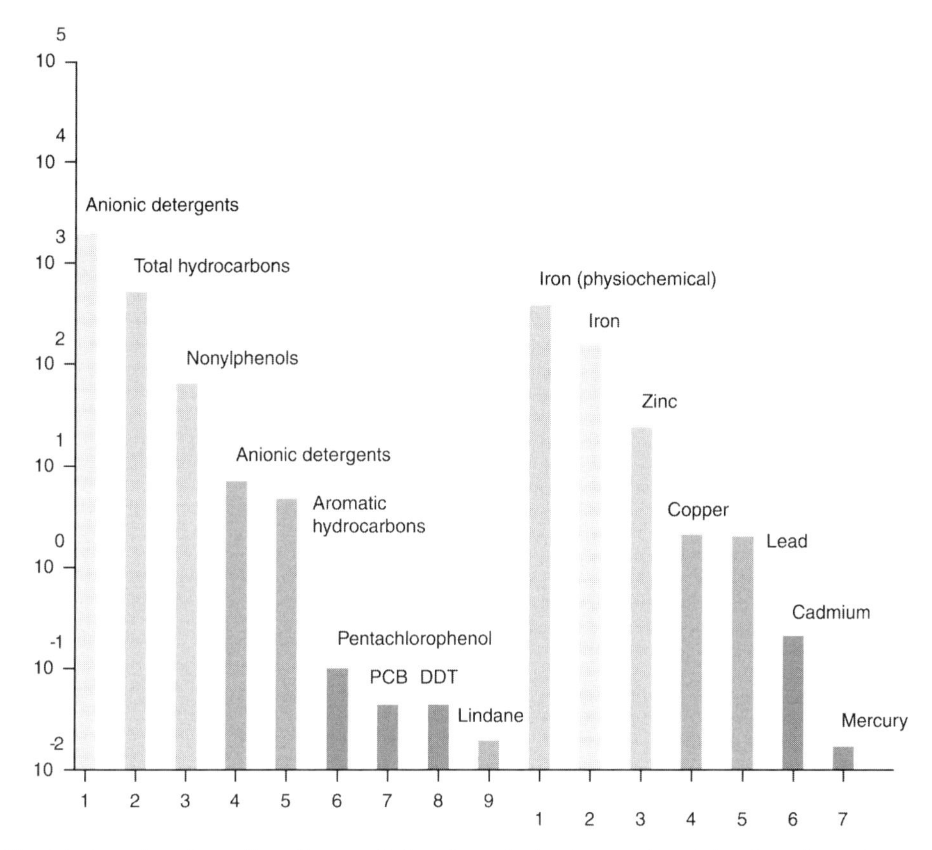

Fig. 6.1. Organic and metallic pollutants. Order of magnitude of contaminated flow from domestic waste water, expressed in mg/day/inhabitant (adapted from Guillot and Romana, 1991).

demand), 8.6 million tonnes of COD (chemical oxygen demand), 1 million tonnes of nitrogen and 360,000 tonnes of phosphorus, as well as 72,000 tonnes of detergents and phenols, 25,000 tonnes of zinc, 4800 tonnes of lead, 2800 tonnes of chromium and 130 tonnes of mercury. Iron and cobalt are also discharged in significant amounts. Diverse petroleum products of various origins (shipping, extraction, industry, urban populations) alone represent around 900,000 tonnes/year. Winds sweeping this sea carry pollutants which often come from a great distance: atmospheric input of N is about 130,000 tonnes/year. Atmospheric inputs of lead are 1–2 times greater than those of the rivers and an order of magnitude greater for cadmium.

Such amounts of N and P, coming from urban effluents and water courses, constitute the main nutrient inputs to the marine coast. These inputs influence plankton production which, as has been seen (Chapters 1 and 4), is limited by nutrients salts or trace elements. The appearance or increase of the frequency of recognised phenomena, such as eutrophication, can be monitored in relation to the nutrient enrichment of coastal waters.

In order to study the impact of freshwaters on marine waters, The European Union, which contains 350 million inhabitants, established in 1988 the EROS 2000 research programme (European River Ocean System), the goal of which is to study the impact on the littoral zone of domestic, agricultural and industrial waste released from the land. An understanding of the biological and geochemical processes and their alteration in the coastal zone is important for the preservation of ecosystems widely implicated in the survival of the biosphere; in the long term, the consequences of global change as a result of human activity such as the rise in sea level must be evaluated, and the potential risks to populations and the effects on global cycles of elements such as carbon must be assessed. This programme has used studies of the polluted western Mediterranean and North Sea as models.

In the western Mediterranean, surface waters are rich in trace metals and low in nutrients, like Atlantic waters. Mathematical modelling shows that the evolution of the plume of Rhône waters and the dispersal of nutrients of continental origin are dependent on meteorological phenomena. Sediments are deposited in the Golfe du Lion (Fig. 6.2). Correlations have been established between the strong winds dominating the Golfe du Lion and the mixing of surface waters, which move in the same direction. One of the most astounding phenomena is the evidence of atmospheric inputs received by the Mediterranean: the Sahara alone deposits 90%, which in Corsica corresponds to a deposition of 16 tonnes/km^2 per annum. Input of nitrates by atmospheric routes to the Mediterranean is more significant than those brought by the Rhône and an order of magnitude greater than total input from rivers. In

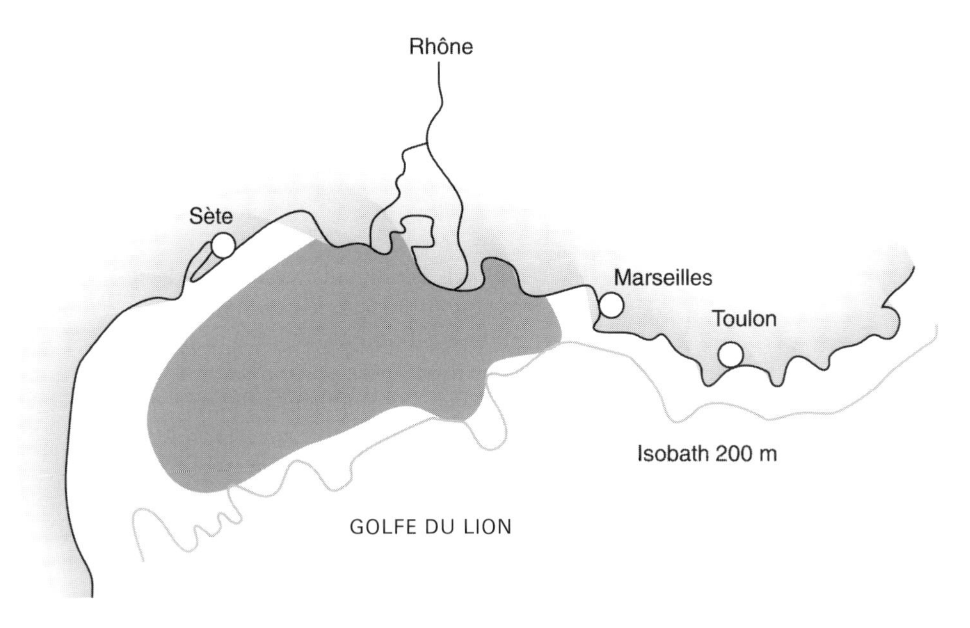

Fig. 6.2. Area where 90% of sediment from the Rhône and the associated pollutants are deposited (adapted from EROS 2000, 1989–1990).

oligotrophic zones, atmospheric N represents 60% of new production. The majority of inputs of lead, zinc, antimony and cadmium to the Mediterranean are of atmospheric origin and linked to human activity. It has also been shown that the seabed on the continental shelf of the Golfe du Lion receives more than 96% of the terrestrial organic waste matter carried by the Rhône. For further details, see the reviews by Martin and Barth (1989, 1990, 1991).

Nitrate concentration in the Rhône has increased greatly during the past 20 years. As a result, coastal waters at the mouth of the Rhône have a planktonic production 4–10 times greater than the open sea; in parallel, bacterial degradation of organic matter is greater at the seabed. An increase in the production of fisheries in certain oligotrophic* coastal waters is attributed to the increase in pollution of human origin.

Globally, inputs of nitrogen from the Rhône into the Mediterranean are estimated at an average of 200 tonnes/day and those of phosphorus at 20 tonnes/day, while waste water of human origin, for a population estimated at 5 million inhabitants, carries 54 tonnes of N and 20 tonnes of P. French urban inputs represent 20% of the riverine inputs.

Phosphorus comes from washing powders and its single treatment at a purification plant would be insufficient. Elimination of nitrogen is also ineffective, while the presence of silica favours the development of diatoms; it is thus only P which is acted upon.

The most abundant organic pollutants thus consist of anionic detergents (flux is estimated at 47 tonnes/year for the town of Toulon). This flux equals 1 g/day/head. After purification at a water treatment plant, the effluent must be diluted 2000 times to regain a P concentration of the order of that found in the environment. Toxicity tests on sea urchin sperm show that the toxicity of the discharge can still be sensed after dilution to 1/10,000. Detergents constitute the greatest ecotoxological risk (Alzieu and Arnoux, 1991).

6.4 EUTROPHICATION

6.4.1 Definition and forms

Eutrophication consists of an excessive enrichment of coastal waters. It has spread since the 1950s and is linked to the widespread use of fertilisers in agriculture and the increase in littoral urban populations. A state of eutrophication is considered to exist when production exceeds $300\,g\,C/m^2/year$.

Agriculture inputs nitrates from the leaching of chemical fertilisers or manure. Urban waste inputs phosphates (washing powders, excreta and industrial discharges) and ammoniacal nitrogen. There are seasonal cycles (Menesguen, 1991), but the trend is upwards as can be noted from the N and P levels in the waters of the Rhine (Fig. 6.3).

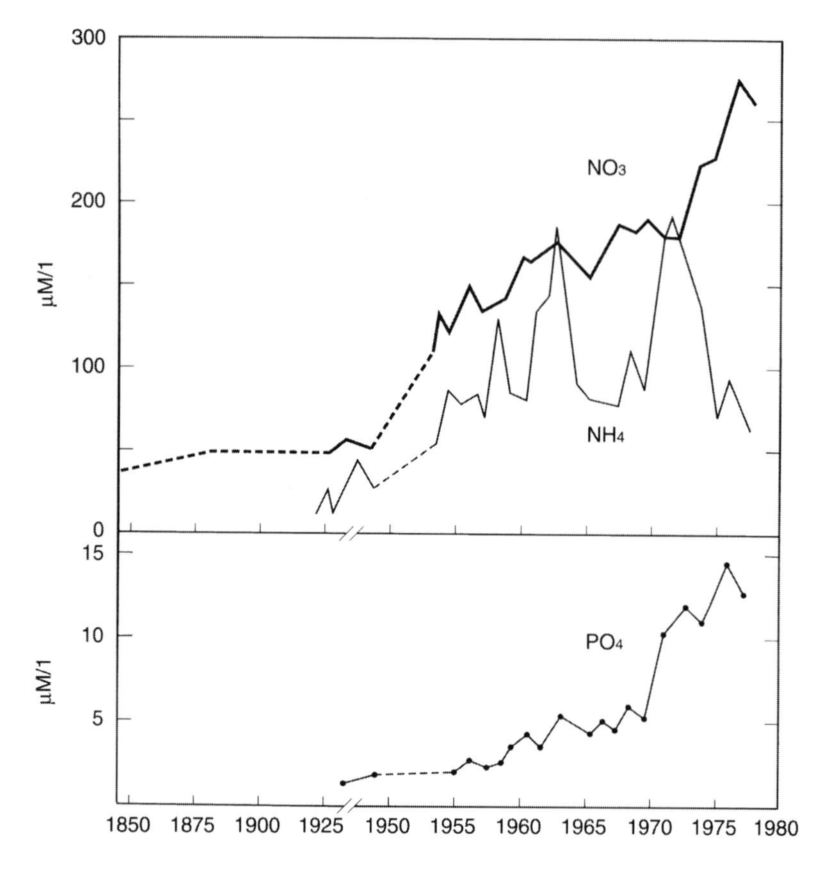

Fig. 6.3. Change in nutrient salt concentrations in the Rhine. Top: nitrogen. Bottom: phosphorus. (From Van Bennekom and Salomons, 1981, cited in Ménesguen 1991.)

Depending on the site, eutrophication can induce development of microalgae or macroalgae.

- Eutrophication leading to macroalgae development mainly occurs in shallow water, both in coastal waters and lagoons: *Ulva* and *Enteromorpha* are two characteristic species. Local Breton communities clear more than 50,000 tonnes/year, but the biomass in the Venice lagoon has been estimated at 550,000 tonnes. The ulva are limited by nitrogen concentration in the water. The remedy for eutrophication is thus the limiting of nitrate inputs.
- Eutrophication associated with phytoplankton results in a bloom at the start of spring (illumination + available nutrient salts). It is rare in the coastal zone in the summer, except in estuaries which can support significant primary production, from spring until autumn at an order of magnitude greater than that of purely marine zones.

6.4.2 Causes

6.4.2.1 Restricted movement of waters

In calm, shallow zones, macroalgae predominate, but disperse and give way to phytoplankton in the deep coastal zone because the latter can proliferate in water masses which have lower concentrations of salts. The imbalance between N and P and silicon would explain the decrease in diatoms in the planktonic bloom and the predominance of dinoflagellates.

However, Menesguen (1991) stated that the sites where eutrophication with macroalgae occur are the beds of bays where the input of nutrients salts is average to low. He also reviewed the hydrodynamic context: it has been seen (Chapter 4) how the natural movements of waters were linked to oceanic production, but the phenomenon is also found in general terms in lakes and in culture. Production is conditioned by the residence time in the zone rich in nutrient salts. Menesguen (1991) describes water movement simulations in the Saint-Brieuc Bay: they show that, despite ample displacements, particles return to their departure point at the end of a tide. Thus there is an almost zero residual circulation of the waters in the beds of these bays, which explains the confinement of ulva to these areas.

6.4.2.2 Enrichment with nutrient salts

The restricted movement of waters does not explain everything. In fact, purification treatment of waste freshwater is poorly adapted to reducing the flux of nutrient materials. The forms of nutrients found in treated water are more rapidly utilisable by algae than those in raw sewage.

In coastal communities, two classic routes of treatment are physicochemical purification by precipitation with iron hydroxide and biological purification using activated sludge; the use of lagoons, which is more ecologically sound, is less widespread because it requires large surface areas.

Biological treatment is more effective than chemical treatment. When there is mineralisation of P and N as a result of the purification process, availability to algae is increased.

The pollutant content of urban effluents is assessed by suspended solids (SS), chemical oxygen demand (COD), biochemical oxygen demand (BOD), total Kjehdal nitrogen (TKN) and total phosphorus (TP).

Aminot and Guillard (1991a, b) studied the density of nutrient elements in the littoral zone, comparing the situation at Toulon (Mediterranean) and Morlaix (Brittany). At Toulon, treatment is physicochemical and the flow of waste water is 13,200 m^3/day (200 l/person/day), while at Morlaix, biological treatment is carried out and the flow is 4400 m^3/day.

In the case of the Toulon outflow, where waste is discharged at a depth of 42 m, 1800 m from the coast, monitoring shows that there is rapid dilution by a factor of at least 100 around the plume which ascends to the surface waters; between 800 m and 1 km away, levels of mineral N are similar to those found in coastal waters, i.e. 0.1–5 μmol/l. The situation is similar for P (between 0.01 μmol/l and 0.04 μmol/l, 1 km from the outflow).

In contrast, at Morlaix, treated water is discharged into the estuary and the same authors have shown that the levels of anionic ammoniacal nitrogen in the waters of the bay are between 0.005 and 0.025 mg/l. For salmonids, the limit is 0.002 mg/l for anionic ammoniacal nitrogen in rearing waters, while the EEC proposes 0.025 mg/l. In an estuary, urban discharges can present a potential risk to salmonids.

Direct inputs of nutrient salts, whatever the type of treatment used, will result in the development of phytoplankton which is considered as a different form of input. Thus, 1 g of N creates 11 g of organic matter and 1 g of P can generate 82 g of organic matter. These calculations (Aminot and Guillaud, 1991) show that nitrogen can be the limiting factor to primary production in an urban effluent. More details can be found in the reports of the Colloquium, *La mer et les rejets urbains* (Guillaud and Romana, 1991).

This phenomenon already occurs in rivers. Despite the operation of treatment stations which have improved water quality downstream of large towns, proliferations of algae already asphyxiate rivers such as the Loire and the Seine along their length. This river eutrophication is characterised by the proliferation of diatoms or chlorophyceae, which leads to a decrease in O_2 in deep waters. Simulations show that decreasing the input of P by 50% decreases the blooms and at the same time improves oxygenation.

6.4.2.3 *Significance of eutrophication*

Waters which are low in salts are poor and non-productive, as we have seen when examining the productivity problems of coastal waters (Part A). Theoretically, rich, eutrophic waters are more productive, but this richness often brings with it production imbalances; in closed or confined environments, the proliferation of macro- or micro-algae can lead to total fauna mortalities, through the exhaustion of oxygen by nocturnal respiration of algae and also the exhaustion of carbon dioxide and large increases in pH.

There is thus an equilibrium to be respected and a compromise to be found between oligotrophy and eutrophy. We shall see that this is one of the routes for management of coastal waters.

6.5 MICROBIOLOGICAL POLLUTION OF COASTAL WATERS FROM FRESHWATERS

Bacteria carried by freshwaters into the sea are mostly (98.5%) attached to particles less than 20 μm in size. The lower density of freshwaters and the small size of these particles result in these bacteria ending up in the surface waters and being diluted there. This dilution and the action of bacteriological processes which occur in sea water have been the subjects of mathematical modelling which allow the prediction of bacterial concentrations at given sites, as a function of current speed and direction.

Bacterial survival in the sea is limited: it is estimated that 90% are eliminated after 2–3 hours, leading to a survival after 12 hours of between 1/100,000 and 1/375,000 (Aubert, 1994).

In the marine environment, bacteriophagic viruses prey on bacteria, but there are also predatory bacteria, such as *Bdellvibrio bacteriovorus* and all the ciliates and flagellates belonging to the marine food chains which sometimes filter half the entire water mass in one day while consuming 10–30 times their own weight in bacteria (Chapters 3 and 4). Phytoplankton (especially the diatoms) are known to release antibacterial substances into the environment, eliminating bacteria. Aubert (1994) described the chemical nature of antibiotics produced by various groups of phytoplankton (p 369–370). Planktonic crustacea, such as the copepods, also graze on bacteria present on particles and floccules suspended in the water (Ogawa, 1977).

A biological equilibrium is thus established based on a collection of interactions, including the predation phenomena which form the basis of food chains, supported by numerous biochemical interactions. The stability of the marine environment rests on this complex assembly.

The surroundings of outflows discharging waste water into the sea are characterised by silting up and the disappearance of marine life, which can extend over several hundred metres. Water treatment plants decrease this phenomenon and can lower the levels of bacteria by around 50%. In this situation, the hydrodynamics of coastal waters again play a vital role in the dispersion and dilution of polluted freshwaters.

Bathing waters (EU standards) must not exceed 500 total coliforms/100 ml and 100 faecal coliforms and streptococcal bacteria/100 ml. The acceptable limits go up to 10,000 total coliforms and 2000 faecal coliforms, with no salmonella being tolerated.

In the Mediterranean, for example, pollution is usually slight except close to large cities (off Barcelona, concentrations of faecal coliforms in coastal waters close to the port were measured at between 30,000 and 50,000/100 ml; Aubert, 1990). This level was only 3000/100 ml at the mouth of the Hérault but was higher around Marseilles before the installation of a large treatment plant in 1987. In Italy, pollution breaks the records in Naples (peaks between 100,000 and 400,000/100 ml), but is also high in Sicily (Messina, Palermo, Syracuse).

Populated areas in Greece are also very polluted while the less populated coasts of Turkey are less so. The little studied Lebanese and Israeli coasts have average pollution. In contrast, Egypt is very polluted. The Algerian, Moroccan and Tunisian coasts are polluted close to large towns.

Bacterial contamination can extend for up to 6–7 km out to sea. Pollution from the north Mediterranean drifts towards to the west, as a result of the currents moving in this direction (see also Fig. 6.2).

In terms of bacterial elimination, the yield at purification plants is about 90% at maximum, which allows bacteria of faecal origin, some of which are pathogenic, to reach the marine environment. This has been reported by Dupray *et al.* (1991a, b) who noted the presence in low numbers (10–100/100 ml) of salmonella bacteria in 74–100% of the treated effluents from five French purification plants using three different processes (physicochemical, activated sludge and lagoon systems, or lagunage). The same authors, this time studying the bacterial evidence of faecal contamination in the littoral zone, noted that sunlight is the main regulator of bacterial survival in coastal waters; they can survive the saline shock, especially when salinities do not exceed 20‰. In the Mediterranean, the transparency of the water leads to the elimination of 90% (T 90) in two hours, but this can take several days in the turbid waters of the Atlantic. Bacteria such as *Salmonella* or *Escherichia coli* can survive in waters rich in organic matter and in sediments where T 90s can be as long as 40 days.

Pathogenic viruses coming from faecal contamination of the water are also present in littoral waters, but they must be distinguished from bacteriophages whose density can reach 2.5×10^8 in unpolluted waters (Chapter 3). Filter-feeding shellfish have often been the source of contamination with hepatitis A and gastroenteritis. Water purification treatments which eliminate bacteria do not reliably eliminate viruses (Schwartzbrod *et al.*, 1991).

According to Poggi (1991), consumers of shellfish are exposed three times more often than others to enteric illnesses. *Enterococcus* also affects bathers when beaches are polluted.

The low effectiveness of the water purification plants in reducing concentrations of faecal bacteria must be emphasised (reduction of 90%), while a proper purification requires a removal of 99.9%. In addition, all enquiries agree in their findings that the majority of plants never function correctly.

After heavy rain, some purification plants discharge the majority of bacterial pollution so that a large part of the raw sewage ends up in the natural environment.

6.6 COLOURED WATERS AND TOXIC PLANKTON

6.6.1 Planktonology

Coloured waters are the result of the proliferation (which can reach several million cells/ml) of phytoplankton, the majority of which are dinoflagellates.

These coloured waters often originate in the open sea but develop close to the coasts, in bays and in zones where there is little water movement, and in lagoons, when long-term anticyclone conditions immobilise water masses which stratify in one place under the influence of sunlight. From a few long-lasting cysts, buried in the sediment, colonisation of the water can take place. By their migratory movements they can dominate other plankton, taking light from the upper layers and mineral salts from the lower.

An increase in the frequency of coloured waters (red tides, which are toxic to shellfish and fish) in the open sea has been noted over a period of 20 years, in parallel to the increase in the concentration of N and P in coastal waters, although it has not been possible to demonstrate a link between the two.

There are many toxic species of phytoplankton with varied mechanisms of development. However, there are common points which can be identified.

- These proliferations occur in spring when the warming of waters or the overlaying of freshwater and salt water lead to stratification, with surface waters lying on top of dense, cold waters rich in nutrients. It has been seen (part A) that, in these conditions, the exhaustion of nutrients (P and N) limits the proliferation of phytoplankton such as diatoms which cannot move in water in the absence of turbulence (current, waves, etc.). Dinoflagellates, which can migrate vertically at a rate of 10 m/day, move in relation to light, but are also likely to live close to the surface during the day in nutrient-poor water; in order to obtain these nutrients, at night they migrate to or below the pycnocline.
- In horizontal terms, they are carried in passive migrations by marine currents. Movements of *Gymnodinium breve* (which causes the illness called NSP) from Florida to North Carolina, where this species has appeared for the first time, have been associated with the transport of warm water masses by the Gulf Stream.
- Some researchers attribute the phenomenon to more detailed surveillance with higher performance equipment and to the development of aquaculture, but other problems are occurring simultaneously. Between 1976 and 1986, while the population of the port of Tolo in Hong Kong increased sixfold, red tides increased by a factor of eight (Anderson, 1994). In Japan, this number changed from 44 to more than 300 between 1965 and 1975; pollution was suspected as the carrier of nutrients to the algae and so rigorous control of the effluents allowed a reduction by half in the frequency of red tides. Despite this, the frequency of red tides is increasing throughout the world.
- This link between nutrients and toxic waters is very controversial, but not without foundation. Monitoring of German waters over a period of 23 years has shown a multiplication by 4 of N and P in relation to silica, a nutrient uniquely vital to diatoms for the construction of their cell walls. These diatoms, which are mostly harmless, constitute the majority in terms of numbers in phytoplankton populations in coastal waters. Populations of diatoms decline in the absence of silica, while those of flagellates can increase.
- Detailed monitoring of the development of the proliferation of *Dinophysis fortii* along the coasts of the Ibaraki prefecture in Japan has been carried out by Iwasaki (1986), in relation to development of temperatures, salinities and movements of water masses. This author showed that the concentrations of this species are the result both of a passive accumulation owing to physical phenomena and the result of the biological activity of cell division. Such phenomena are observed at oceanic fronts between warm and cold waters; internal tidal waves and coastal upwellings seem to be responsible for the transport of swarms from the open sea to zones where bivalves are cultured. *Dinophysis* thrives best in less saline ($S < 34.3‰$) and warmer

(T° >11°C), optimal conditions being around 15–16°C and 33‰ salinity, at a depth of about 10 m.

- From examination of sediments, proliferations of *Gymnodium* in Norway (which have been associated with salmon mortalities) have been attributed, not to pollution, but to increases in water temperature. This parameter is rarely taken into consideration, when nuclear power plants discharge water warmed by 7° to 13°C, at a rate of 100–200 m^3/s depending on their capacity. Red tides caused by *Chrysochromulina polylepsis* have caused significant salmon mortalities in Norway.
- Mediators excreted by bacteria of telluric origin (bioptine, thiamine, vitamin B12) are necessary in the medium used for microalgae culture and are thus likely to favour their proliferation. Inhibitors of dinoflagellate growth have been extracted from sediments by Japanese authors.
- Study of the variation in dissolved amino acid concentrations during a diatom bloom has shown that phytoplankton blooms leave their trail in terms of levels of various amino acid concentrations in the waters as well as in terms of the C/N ratio.
- Phytoplankton itself excretes mediators which slow the development of other species and these substances, often proteins, are broken down by certain pollutants such as pesticides, heavy metals or surfactants (toxic to marine fish larvae at a concentration of 0.0034 ml/l). This can play a role in the development of dinoflagellates.
- Giannacourou (1995) cultivated dinoflagellates in a 20 m^3 mesocosm. They develop in waters which move very little, but their presence is especially associated with high pH. It should also be noted that the percentage of peridinians in Mediterranean ports is 50% higher than that of exterior waters (Aubert, 1994).
- Abundant plant production resulting from nutrient richness is not stabilised by zooplankton predation. In addition, the zooplankton is sensitive to the pesticides brought by agricultural runoff.
- Dinoflagellate spores overwinter in mud on the seabed; recolonisation of the water column can therefore be rapid.

These species are linked to relatively high temperatures and, in Europe, proliferations extend from spring to the start of autumn. Development occurs at the thermocline more than at the surface or on the bottom; developments in the open sea have also been noted. Other effects of blooms are the deoxygenation of the water when the bloom starts to decline. Burrowing species such as *Mya arenaria, Ensis, Cerastoderma* and *Tapes decussata* all ascend to the surface of the sediment when there is a *Gyrodinium* bloom.

The increase in frequency of proliferations of toxic phytoplankton is also due to the fact that these toxic species are transported by ships in their ballast water and by infected shellfish. Anderson (1994) cited the example of the examination of the ballast mud of the large cargo ship "Tasmania" which revealed the presence of 300 million dinoflagellate cysts per tank. The appearance of a toxic dinoflagellate in the plankton has been associated with the development of the timber industry, transported by ships which empty their ballast tanks before loading with timber.

6.6.2 Toxicological aspects

Potentially, there can be over 18 different toxins present in a shellfish (revealed by an Elisa test). Obviously, the toxins affect those who ingest them, but these plankton also cause itching and burning sensations in bathers or divers.

The toxicity of a given species can vary in relation to ambient conditions. Laboratory rearing of *Aureococcus anophagefferens* shows that cultures at 24°C are more toxic than those reared at 20°C. Cultures reared in less intense light conditions are less toxic; finally, toxicity disappears when density falls below $40–70 \times 10^4$ cell/ml (Tracey *et al.*, 1988).

Diarrhoeic toxins (Diarrhoeic Shellfish Poisoning or DSP) include okadaic acid. This acid is considered to be the major component of toxins in France and Spain, Holland and Sweden; however, research on the toxin has been hindered as it is not known how to rear *Dinophysis acuminata*. The European method based on testing on mice lacks specificity, sensitivity and precision (Shumway, 1990). The Japanese have developed a very sensitive chemical test (UBE industries) which allows identification of traces of toxin from a plankton sample. Both in Japan and Europe, many toxin profiles have been identified.

Accumulation in mussels can last for five months after the *Dinophysis* has disappeared. Concentrations of less than 100 *Dinophysis*/ml often produce high concentrations of DSP in mussels. It is true that poisoning can lead to death and the clam *Venerupis semidecussata*, contaminated with *Procentrum minimum*, caused 114 deaths in Japan in 1942.

In contrast, certain species of bivalve such as the American hard-shell clam, *Mercenaria mercenaria,* do not concentrate toxins, as they bury themselves deep in the seabed when toxic plankton are present; similarly for oysters, which close up.

Crustaceans do not accumulate toxins and can thus still be sold during the most intense blooms. In this case, mussels can be used to feed lobsters (conversion rate of 10, but commercially viable because of the price of lobsters).

The "Butter Clam" industry in Alaska which was producing five million pounds of shellfish in 1917 has practically disappeared and today all Alaska's beaches are considered unsafe for the consumption of shellfish. The promising development of mussel rearing in Sweden was stopped by dinoflagellate blooms and serious threats exist to French mytiliculture.

6.7 CONCLUSIONS

In this brief chapter, other significant pollution, such as thermal pollution, has been passed over, as it is considered elsewhere (reports of numerous Electricité de France (EDF) Colloquia on the impact of nuclear power stations).

Some types of pollution are reversible (bacterial pollution), while others are irreversible, such as chemical or radioactive pollution. The latter are obviously the most dangerous for coastal waters.

In a context where ecology must compromise with the economy, the development of coastal waters constitutes a rampart for the defence of the environment. In order to conserve the natural character of tourist resorts or to favour the production of marine species for commercial ends, the absence of pollution is a *sine qua non*; while it seems to be a difficult state to obtain in the absence of economic justification, it becomes vital when such stakes exist.

6.8 BIBLIOGRAPHY

Alzieu C. and Arnoux A. Rejets urbains et micro polluants: synthèse. In: Guillot J.F. and Romana L.A. (Eds) *La mer et les rejets urbains*. IFREMER Éd., Brest, 1991; 187–189.

Aminot A. and Guillaud J.F. Apports en matière organique et en sels nutritifs par les stations d'épuration. In: Guillot J.F. and Romana L.A. (Eds) *La mer et les rejets urbains*. IFREMER Éd., Brest, 1991a; 11–26.

Aminot A. and Guillaud J.F. Devenir des éléments nutritifs en zone littorale. In: Guillot J.F. and Romana L.A. (Eds) *La mer et les rejets urbains*. IFREMER Éd., Brest, 1991b; 27–34.

Anderson D. Eaux colorées et phytoplancton toxique. *Pour la Science*, Oct. 94; 204: 68–74.

Aubert M. Les risques en matière de pollution microbiologique de la Méditerranée. *Rev. Int. Oceanogr. Med.* LXXXXVII–LXXXXVIII: 6–20.

Aubert M. *La Méditerranée: La mer et les hommes*. Rev. Int. Oceanogr. Med., Vols 109–112, 1994.

Cormier M. and Martin Y. Rejets urbains et salubrité du littoral; synthèse. In: Guillot J.F. and Romana L.A. (Eds) *La mer et les rejets urbains*. IFREMER Éd., Brest, 1991; 149–151.

Dupray E., Baleux B., Bonnefont J.L., Guichaoua C., Pommepuy M. and Derrien A. Apports en bactéries par les stations d'épuration. In: Guillot J.F. and Romana L.A. (Eds) *La mer et les rejets urbains*. IFREMER Éd., Brest, 1991a; 81–87.

Dupray E., Baleux B., Bonnefont J.L., Guichaoua C., Pommepuy M. and Darrien A. Le devenir de bactéries en zone littorale. In: Guillot J.F. and Romana L.A. (Eds) *La mer et les rejets urbains*. IFREMER Éd., Brest, 1991b; 89–99.

Giannacourou A. *Réseaux trophiques planctoniques utilisables pour l'élevage larvaire de la daurade* (*Sparus aurata L., 1758*). Doctoral Thesis, Montpellier University; 1995.

Guillemaut-Drai M. *Pollutions chimiques et conséquences sanitaires*. Rev. Int. Oceanogr. Med., Vols. 101–104, 1991; 205–213.

Guillot J.F. and Romana L.A. (Eds). *La mer et les rejets urbains*. IFREMER Éd., Brest, 1991.

Iwasaki J. The mechanisms of mass occurrence of *Dinophysis fortii* along the coast of Obaraki Prefecture. *Bull. Tohoku Reg. Fish. Res. Lab.* 1986; 48: 125–136.

Martin J.M. and Barth H. (Eds). Water Pollution Research Report 13. Commission of the European Communities, Brussels, 1989.

Martin J.M. and Barth H. (Eds). Water Pollution Research Report 20. Commission of the European Communities, Brussels, 1990.

Martin J.M. and Barth H. (Eds). Water Pollution Research Report 28. Commission of the European Communities, Brussels, 1991.

Menesguen A. Présentation du phénomène d'eutrophisation littorale. In: Guillot J.F and Romana L.A. (Eds) *La mer et les rejets urbains*. IFREMER Éd., Brest, 1991; 35–52.

Ogawa K. The role of bacterial floc as food for zooplankton in the sea. *Bull. Jap. Soc. Scient. Fish.* 1977; 43(3): 395–407.

Poggi R. Impact sanitaire des contaminations microbiologiques. In: Guillot J.F and Romana L.A. (Eds) *La mer et les rejets urbains*. IFREMER Éd., Brest, 1991; 115–131.

Schwartzbrod K., Jehl-Pietri Boher S., Hugues B., Albert M. and Béril C. Les contaminations par les virus. In: Guillot J.F. and Romana L.A. (Eds) *La mer et les rejets urbains*. IFREMER Éd., Brest, 1991; 101–114.

Shumway S.A. A review of the effects of algal blooms on shellfish and aquaculture. *J. World Aquac. Soc.* 1990; 21(2): 65–104.

Tracey G., Steels R. and Wright L. Variable toxicity of the brown tide organism *Aureococcus anophagefferens* in relation to environmental conditions of growth. In: Granel E. *et al.* (Eds) *Toxic Marine Phytoplankton*. Elsevier Science, 1988; 233–237.

7

The exploitation of aquatic living resources: fisheries and aquaculture

7.1 GENERAL SITUATION

Aquatic ecosystems are utilised by numerous human activities whose strategies are very different; fisheries and aquaculture traditionally exploit living resources, but leisure activities such as pleasure fishing, underwater fishing, gathering and collection of shellfish and the importation of foreign marine plants and animals (such as corals for tourism) can be added to traditional methods of capture. The construction of ports, marinas and artificial islands and the use of corals or sediments for construction also constitute ways of utilising these ecosystems which must be taken into account. All these activities occur within a complex context, simultaneously ecobiological, geographical, economic and legal (regulatory). However, professional fishing and aquaculture still constitute the two main forms of exploitation of aquatic living resources.

Fisheries and aquaculture are both based on the production of living matter in the oceans. These are immense (362 million km^2), and include an enormous biomass, since the total weight of living matter represents between four and seven times that of terrestrial organisms, the most abundant animals on the planet being planktonic copepods. Primary production has most often been estimated at between 150 and 200 billion tonnes/year, but these estimates vary between 55 and 400 billion tonnes, depending on author.

Another figure very often cited by fisheries science experts is that of the upper limit of renewable fisheries resources being 100 million tonnes per annum. Note that this figure does not even represent a thousandth of oceanic primary production and this is due to the complexity of aquatic food chains—fishing activities rely on a myriad of exchanges in aquatic food chains. This level of fishery catches has been described elsewhere, with four trophic levels and a yield estimated at 15% at each transfer (Troadec, 1989a).

International organisations include aquaculture production in their fishery statistics. Actual production has been estimated at 18 million tonnes/year (22 million tonnes, including algae) but, while fisheries production reached a historic peak (close to the calculated limit) in 1989 and has been stagnant or in decline since, aquaculture production has increased by 5–10% per year over the past 20 years (production being about 6 million tonnes in 1976). It should be noted that aquaculture also supports sport fishing, which is an important sector of the leisure economy.

Fisheries and aquaculture provide more animal protein than any other resource, including cattle, sheep or poultry farming (80 million tonnes of meat and 20 million tonnes of eggs, according to Barbault, 1991).

Fish consumption is very poorly reported; according to the FAO (1994), northern countries consume 27 kg of fish per inhabitant per year in contrast to 9 kg for southern countries. On average, world consumption was 17.8 kg of fish/inhabitant in 1990; fish is of great importance in developing countries since 60% of the inhabitants in these countries obtain at least 40% of their protein from fish.

Fish contributes to 16% of animal protein consumption, but this figure varies from country to country. Weber (1994) cited the figures below from data from the FAO for 1988:

North America	6.6%
Western Europe	9.7%
Africa	21.1%
Latin America and the Caribbean	8.2%
Middle East	7.8%
Far East	27.8%
China	21.7%

Within Europe itself, this consumption shows great differences between countries, varying between 5 kg/head/year in Switzerland and Yugoslavia to more than 30 kg/head/year in Spain and Portugal (Barnabé, 1990b).

Subject to the constraints of both ecology and economics, the exploitation of marine living resources develops through a series of crises marked by events, such as the fierce demonstrations by fishermen in Rennes in 1994. This case is not unique, as we shall see. The trend is to mainly blame economical dictates, but this does not explain everything and biological and ecological aspects must not be underestimated; methods for posing questions and resolving problems also merit attention.

7.2 THE CRISIS AND DECLINE IN FISHERIES

7.2.1 "Fisheries science": a theory in crisis

Fisheries techniques are as old as humanity. Observation, experimentation and empirism have allowed coastal populations to live a life dependent on aquatic organisms, whether these be fish, crustacea, shellfish or algae. Exploitation of coastal seas, followed by that of the open sea when mechanical engineering allows it, is thus carried out by fishermen, then aquaculturists and, more generally, all those with a practical knowledge of the environment, even through leisure activities.

It was only after the Second World War that science, and especially technology, became really concerned with fisheries. Current fishery techniques resulted from this progress; they have benefited from improvements in several areas (underwater acoustics with sonar, radio-location and progress in nets and fishing equipment with synthetic fibres and other materials, etc.). Fisheries science was set up to scientifically manage exploited fish populations: its goal was to allow the regulation of catches by adjustment between recruitment (juveniles produced by spawning, incorporated into the catchable portion of the species population) and sampling by fishing. Thus fisheries theory, intended to be predictive, was born at the end of the 1950s. What stage is it at now, 40 years later?

Without going into the detail of the analytical models on which they are based (see, for example, Laeavastu, 1967; Gulland, 1969; Kesteven, 1974; Holden and Raitt, 1974; Postel, 1976), in order to maximise fish production, fishing effort and the age at which fish can first be caught (age at recruitment) must be simultaneously regulated. The optimal yield of a stock is thus reached when the biomass of the stock has more or less the same value as that of an unexploited stock. Environment and recruitment are considered to be stable with time. In practical terms, this is never the case.

As stated by Troadec (1989a), the objective of fisheries management was very honourable, as it was concerned with optimal exploitation of fished populations by aiming fishing effort at well-defined targets in the population in order to catch a maximal biomass without affecting reproductive potential.

In fact, as stated by this author as well as the FAO (Marine Resources Service, 1994), fisheries theory has never been able to predict the intensity of recruitment of species exploited by the main fisheries. Thus, to propose a balanced management plan for exploited stocks: "The relationships between recruitment levels and the spawning stock are not reflected in the data" (von Walhert and von Walhert, 1989). In other words, the facts do not correspond to the theory, which should be the main purpose of fisheries science.

This "inability to predict" is a serious handicap for an economic activity. Predictions have been made by specialists, the imprecise nature of which can now be measured: Ricker (1972, cited in Hanson, 1974) stated that fisheries in the year 2000 would be of the order of 150–160 million tonnes while Chapman (1972, cited in Hanson, 1974) estimated this production at 400 million tonnes. From this we would predict that in the year 2000, fisheries would contribute 75% of human protein requirements, while in fact they constitute only 16%.

As is stated by numerous authors in the publication edited by Troadec (1989b), the absence of theoretical support passed rather unnoticed while expansion possibilities existed for fisheries, but energy crises and the exploitation (or overexploitation) of all catchable stocks could not continue to hide this failing. Fisheries science is thus a discipline in crisis (which does not indicate that fisheries theory is without value). It is limited to the study of adult populations and their capture by fishermen. Little work has been done towards improving understanding of the mechanisms of biological production within aquatic ecosystems, and even less towards the study of its renewal, previously considered to be constant.

From another point of view, fisheries is, *sensu stricto*, a gathering activity; fishermen utilise a common natural resource and are in competition for this resource, but they leave the renewal of the resource to natural processes. In the price one pays to the fisherman, there is his labour, fuel, depreciation of the boat, etc., but the fish is free, as its capture is not paid for.

7.2.2 Fisheries decline and overinvestment

Stocks constituting traditional fisheries are depressed, while their management has been considered as a model for development: catches of cod changed from 700,000 tonnes in the 1970s to 50,000 tonnes in 1992 (Fig. 7.1). Fisheries therefore moved to catching lower value fish such as small pelagic species (sardines, for example). These fish, situated at a lower level on the food chain, can no doubt sustain a higher fishing effort, but the monetary value of catches did not increase. No high-value fish species has seen their catches increase. Emphasis has recently been placed on deep water fishes such as the *Hoplostethus atlanticus* (orange roughy), but these are slow- growing species, whose capture requires gear which is expensive to operate.

In global terms, the FAO (Marine Resource Division, 1994), noted a fall in marine fisheries production in 1990, the first since 1976 (Fig. 7.2). A member of this organisation showed that lasting development such as that proposed at the Rio Conference in 1992 cannot be reached under a régime where there is open access

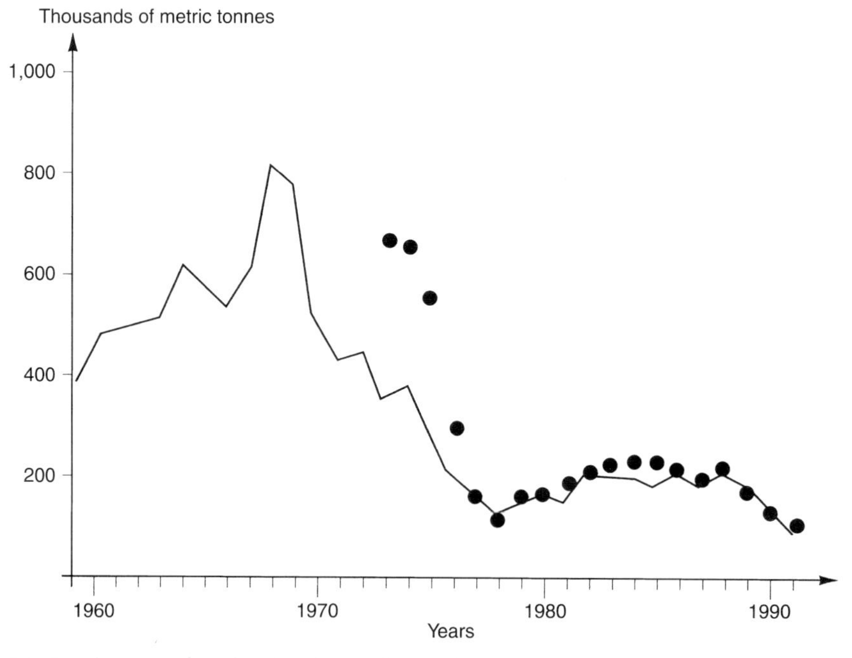

Fig. 7.1. Development of catches (continuous line) and total allowable catches (TAC) (dots) of the cod fishery in the North Atlantic (FAO, 1994).

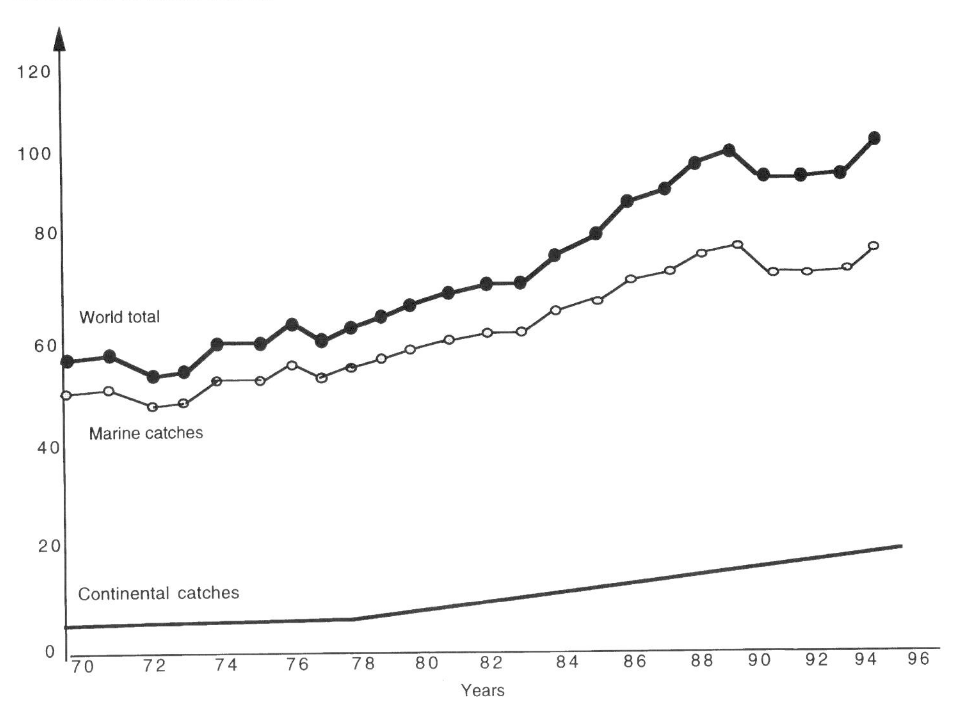

Fig. 7.2. Development of worldwide fisheries catches (FAO, 1997).

to resources, both in exclusive economic zones (EEZ) and elsewhere. The FAO declared that not only have fishing limits been reached, but that this was in fact the case for many fisheries for a decade before the peak in production in 1989.

All the experts are unanimous: all exploitable stocks are exploited and already 70% of stocks are overfished. Production peaked in 1989 at 100.23 million tonnes and has decreased since (97.46 million tonnes in 1990 and 96.95 million tonnes in 1991). These figures include aquaculture production (which is in excess of 545,000 tonnes, while those of fisheries have declined by 5 million tonnes). Production which is purely marine (including aquaculture) was 81.75 million tonnes in 1991 in contrast to 82.8 million tonnes in 1990 and 86.4 million tonnes in 1989. Fisheries, however, still depend on the first levels of marine food chains.

Greenpeace (1994) also blames overfishing for the starvation of whales, dolphins, seals and seabirds and observers have noticed more and more mass migrations of marine animals dying of hunger. Species which allowed increased catches in the 1980s and which represented 29% of total catches by weight in 1989, only represented 6% in value.

French fisheries provide a good example of the downturn in catches and employment: they tripled between 1946 and 1975, going from 245,000 to 695,000 tonnes; they reached 733,000 tonnes in 1988; but returned to 640,000 tonnes in 1995. In 1956 there were 51,500 fishermen, falling to 25,000 in 1976, and now there are less than 16,000, perhaps even around 10,000 (Billard, personal observation), equivalent to a loss of more than 15,000 jobs in 20 years. The number of boats fell from 32,400 in 1956 to 6600 in 1995. Despite this, the national French consumption of fish has increased (30 kg per person per year) and imports of seafood products represent a negative trade balance (FF13 billion in 1993). According to Chaussade (1994), national production provides 35% of the amount consumed, but only 17% of its value.

Despite this overexploitation, the size of industrial fishing fleets continues to increase, which represents a problem from an economic point of view as well as for the resource; the FAO (1994) reported convergent results from two documentary sources which estimated the operating costs of the world's fishing fleet at more than $22,000 million (≈FF130 billion) of revenue, without even considering the depreciation of the capital. This phenomenon results from the subsidies given by the countries to their fishermen. This same body predicts that, despite this over-investment, there will be 25 million tonnes less food derived from fish in the year 2000. Fishing is characterised by an inability both to regulate fishing effort and contain overinvestment.

None of these figures take into account the enormous numbers of reject fish (not utilisable or too small for commercial purposes). New and Csavas (1995) estimated that 27 million tonnes of reject catches, along with, for example, 100,000 tonnes of undersized cod caught in non-selective nets, could be transformed into meal which could be used in aquaculture.

International organisations which are sensitive to these degradations of resources are starting to react. In 1992, the UNO adopted a resolution prohibiting the use of drift nets in international waters and their prohibition in EU waters has been total since 1995. In the 1980s, Greenpeace (1994) estimated the length of drift nets set each night in the north Pacific to be 50,000 km.

Taking more severe action, Canada closed the cod fishery in 1992 (to try to allow the recovery of overexploited stocks), causing the loss of 20,000 jobs. In 1995, negotiations of a new agreement between Morocco and the EU were at an impasse, as Morocco wanted to decrease catches; this would have especially limited Spanish fishermen who caught 80% of their catches outside their own territorial waters. The closure of other fisheries, or their abandonment because of non-profitability, is certain to occur. This is a serious problem as there were an estimated 12 million fishermen in the world in 1992 and 200 million people who depended on fishing as their means of subsistence.

Overexploitation cannot be attributed to fisheries theory, but fishermen profit from the inaccuracy of scientific predictions in estimating recruitment. At the moment, regulations are based on these predictions and, already little inclined to respect constraints, here the fishermen find a new argument for getting round them.

7.2.3 Recruitment and its determination

Most marine species are characterised by very high reproductive potential resulting in the spawning of hundreds of thousands of eggs. This high fecundity must be viewed in the context of low survival under conditions of population equilibrium; there is thus "ecological regulation" and this capacity appears to be enormous. Since the turn of the century, many hypotheses have been advanced on this subject.

These eggs hatch into planktonic larvae which metamorphose into juveniles capable of leading a life identical to that of the adult. For all these species, the pelagic larval stage constitutes a regulatory phase of population size and has been termed a "critical stage".

Hydrodynamic factors which can move eggs and larvae play a determining role: turbulences, currents and stratification of the water can either disperse or concentrate the larvae and their prey, as well as their predators. Adapting to the physicochemical environment is a primary requirement for these pelagic stages which are particularly sensitive to pollution (a few nano g/l of metals or detergents, for example).

Thus, concentrations or swarms of fish larvae and also of their prey (copepods) may occur at interfaces between water masses or by simple passive gathering between turbulences. Fish eggs and then the young larva live first on yolk reserves inherited from the mother (*vitellus*), but when this is finished they must transfer to exotrophy (food caught in the environment), which corresponds to a critical period of mortality. As has been shown in aquaculture research, the acquisition of living plankton food of the right size and composition at first feeding determines survival.

Predation carried out by small pelagic fish and also jellyfish, chaetognaths and simply older larvae (cannibalism) further complicates the phenomenon. It is very likely that every hypothesis put forward affects survival by a different percentage but, as stated by Bakun (1989), "a strong relationship must exist between recruitment and climate".

Birkeland (1992) actually reported very large numbers of correlations between recruitment of marine species and inputs from rivers or rains, making clear the problems of nutrient richness of waters. By introducing data on the intensity of upwelling in preceding years to an African fisheries model, Fréon (1991) was able to obtain a good correlation between predicted and actual catches.

Experimental studies on larval rearing in mesocosms (Divanach *et al.*, 1992; Barnabé *et al.*, 1995) have shown that, under very favourable trophic conditions, larval survival of the sea bream *Sparus aurata* up to one month old can reach 100%. The idea of critical stage must be revised: it is not the life cycle stages which are critical, but the conditions required for their survival. This fact brings back into play the hypothesis implicating inherited genetic characteristics and their capacity for survival. This is not always the determining factor. These experiments and many others have confirmed that biotic factors of the environment determine the success of larval survival; this has been demonstrated in the wild by Dufour and Galzin (1993) for reef fish.

For Hunter (1981), although larval mortality had been well established, there was no definitive evidence that this variation on rhythm of mortality determined the size of age classes recruited. In the case of anchovy larvae (*Engraulis mordax*), which he studied, he stated that it was not possible to detect changes in recruitment potential from larval abundance, whatever their stage, up to a length of 15 mm.

The identification of mechanisms regulating recruitment would therefore require study of the entire larval stage and all the juvenile stages.

Since the main problem stems from the variability in renewal of fished populations, marine and fisheries ecologists have more recently become interested in the determination of recruitment, i.e. research into the causes of variations in abundance of juveniles produced by reproduction, which are integrated every year (where reproduction has an annual cycle) into the existing population of the species. This has been the object of a national research programme in France, the "Programme National de Recrutement" (PNDR) and an international programme, GLOBEC International. The continuing PNDR project has brought a great deal of new knowledge about the species to which it was devoted, but it has also shown the extraordinary range of responses that one species can give to environmental variations. Special reports by the PNDR devoted to this subject have been published since 1991.

Like GLOBEC International, PNDR is first interested in hydrodynamics, before taking into account physiological or behavioural problems. Ten years from its start, this programme was still not shedding any light on the future intensity of recruitment (Anon, 1994).

Note also that programmes concerned with larval stages require a precise base of knowledge about the ecosystem itself and also require an understanding of a large part of the biological cycle of the species (fecundity, stock, etc.). A range of data must therefore be available and an assembly of interactions understood before the slightest prediction can be made.

An example of an unforeseen phenomenon likely to have an effect on recruitment is provided in the studies by Nissling (1994). In the Baltic Sea, successful spawning of cod is limited to deep areas where salinity is between 10 and 17. These waters can remain in the same place for a long time and are often low in oxygen. The survival of eggs and larvae improves with increased salinity and dissolved oxygen (minimum threshold of 2.5–3 mg O_2/l). The Baltic is widely polluted by human activities which can alter the oxygen concentrations of the water, thus determining recruitment; but which fisheries recruitment model takes this factor into account?

The variability of oceanic marine populations has been the subject of many models, but the difficulties involved in the investigation of such a vast, three-dimensional environment have not helped the subject to advance. As stated by Bakun (1989), the long duration of phases of larval drift can render in vain, or at least relatively unimportant and unprofitable, a type of management which only considers local problems. This same author and Lasker (1989) made a well-documented point about explicative hypotheses put forward by fisheries scientists, one insisting on the importance of hydrodynamic phenomena and the other reviewing exhaustively all the hypotheses put forward since the turn of the century. We shall not examine these in detail, but it is necessary to see if fisheries resources are limited at the recruitment level; food availability for the growth of juveniles, which are then incorporated into the population, is no longer considered to be the limiting factor.

The state of knowledge about recruitment was effectively summarised by Miller (1994), who reviewed the colloquium, devoted to this subject for flatfish, held at Texel in 1993. The question was asked: "Why, after more than a century of concerted effort, do we still not understand (we cannot predict) the variability in abundance of fish age classes?" He suggested two possible answers: 1, we lack the required data, and 2, we still do not have the structure to interpret existing data.

We do not believe that answering these questions would suffice. The influence of hydrodynamic phenomena has been widely demonstrated in recruitment, and the problems posed by these phenomena and their sensitivity to initial conditions render them unpredictable since they are regulated by the deterministic law of chaos (Revaud d'Allonnes, 1994). The hydrodynamics of oceanic waters are piloted by the climate, but terrestrial atmosphere is also regulated by the same laws of chaos. Elsewhere, the subject of climate has evoked the term "butterfly effect" to illustrate sensitivity to initial conditions and the "impossibility of predicting the general pattern, even in the short term" (Lorenz, 1972 in Lorenz, 1995). In other words, the climate currently remains unpredictable.

Climate and hydrology constitute two interwoven processes of unpredictable phenomena, directly implicated in recruitment. It is not only the level of complexity or the absence of sufficient data that limit the understanding of recruitment determination, it is the nature of the phenomena themselves which excludes prediction. One can no more talk about the determination of recruitment than determination of climate (e.g. predict the trajectory of a cyclone) or the determination of hydrodynamic phenomena (e.g. the divergence of two buoys released simultaneously from the same point in a vast, turbulent ocean). As stated by one expert, "All numerical predictive models turn out to be doomed to more or less imminent divergence" (Revaud d'Allonnes, 1994). What, therefore, can be said about recruitment determination?

The question must be asked, as it has been by certain international organisations, as to whether it would not be simpler simply to change strategy.

Supposing one comes across well-defined situations from elsewhere from which to predict recruitment—to what practical use in fisheries could this knowledge be put? Understanding or modelling a state of shortage does not provide the solution to overexploitation or the fisheries crisis, no matter how far one looks. The propositions described in the last part of this book constitute a response in this sense, and a new alternative applicable to fisheries.

7.2.4 Fisheries management

As recruitment is unpredictable and unmanageable, it is "fisheries management", based on fisheries theory, which has mobilised the fisheries scientists. However, although it is now termed "management", it is not management in its true sense; it is more a matter of *attempted* management. A system of management is traditionally based on regulations permitting the controlled exploitation of resources (Van Tilbeurgh,1992).

For fisheries, access regulations are required for a resource which, by definition, is accessible to the whole world. Limits to fishing effort are regulated by limits to power or size of the boats, size of net meshes and fishing quotas for fleets of a certain size. This type of management often remains inefficient on account of "the infinite ingenuity with which fishermen circumvent the regulations" (Larkin, 1989). Management of fisheries thus reverts to regulations, or, as defined by Troadec (1989a, p. 29) "the planning (of fisheries) is reduced to their conservation".

The FAO Marine Resources Service (1994) emphasised that this catastrophic situation raises fundamental questions concerning the efficiency of strategies utilised to manage marine fisheries over the past two decades. We appear to be at a critical stage in the research for practical alternatives for the management of coastal waters and for replacing the standard system of open access to the resource which has dominated until now.

This is not a new question. The OECD (1993) reported the proposal of Harding in 1968: the management of a common resource or "tragedy of the commons", ending with "liberty applied to common property causes the ruin of all". Details and references on economic analyses carried out on the regulation of fisheries since the 1950s can be found in the above publication.

For the fishery resources in their EEZ, many countries have adopted a system of individual transferable quotas (ITQ) within the body of a total allowable catch (TAC). This mechanism constitutes the management system and current conservation of limited fishery resources.

With regard to international waters, the United Nations Convention on the law of the sea (UNCLOS) was put in place on 16 November 1994 and should bring order to political discussions on fishing. Not everything is regulated, but it is a judicial instrument which has required 27 years of dialogue and which has come to life 12 years after its signature. International waters have thus come into common ownership and the fisheries in these waters fall under this jurisdiction.

Paradoxically, in some zones (Black Sea, Golfe du Lion), fisheries production is increasing and the Marine Resources Division of the FAO (1994) attribute this fact to the input of nutrients produced by river basins into a sea otherwise poor in production. This is an example of the involuntary action of man on a large marine ecosystem. Nothing proves that it will be beneficial to the ecosystem in its entirety: since the 1970s, the frequency of coloured waters (proliferations of phytoplankton, many species of which are toxic) has clearly increased and seems to evolve in parallel to inputs of fertilising substances to coastal waters as a result of human activities. The concentration of mercury in several fish in the Mediterranean exceeds the acceptable level of 0.7 ppm; in the long term, the accumulation of discharged chemical products into closed seas threatens the consumption of certain very popular fish (Caddy and Griffiths, 1991).

In developing countries, projects for developing artisanal fisheries carried out over the last two decades have almost all failed (Chaussade, 1994); the idea of development itself has been challenged.

A final example of false "prediction" meriting a mention is that of krill (defined here as various species of small pelagic shrimps in boreal waters). Estimates of potential catches based on fisheries theory have reached 400–500 million tonnes. It is thought that the quasi-extermination of whales by overfishing has allowed a krill population explosion. Trials carried out by the Russians and Japanese have shown the difficulty of catching these minuscule shrimps. Large trawl nets with fine mesh require engines of several thousand horsepower; the long distances to the fishing grounds in boreal waters and the fact that the catches often contain toxic species which are difficult to separate from the krill (jellyfish), have never allowed the expansion of this low-value fishery. Populations of birds and fish, previously in competition with the whales, now consume the krill.

Currently, catches of krill achieved by the Russians and Japanese come from the Antarctic and were about 380,000 tonnes in 1990. As krill production is low in relation to its biomass, the resource may be sensitive to overfishing and should be managed. A Russian–Japanese partnership in the form of a joint venture is in the process of fishing for krill for use in feed for reared salmon. As for fish meal, the fishery finds an outlet in aquaculture!

7.2.5 Fishery perspectives

On the basis of analysis of landings, the FAO published an interesting analysis of marine fisheries potential in 1997. First, it was noted that, in the current fisheries system which is generally characterised by small size at first capture and significant discard, the total maximum catch corresponds to about 82 million tonnes, which is a value close to the average landings of 1990–1994 of around 83 million tonnes.

Another approach was carried out by the same organisation, with the same goal but using catch analyses from each ocean (ecosystem method). This study provided a different result, since the total estimates here reached 100 million tonnes.

Using these two data series, the FAO calculated the relative growth rates of landings for each ocean, and the global trend. Figure 7.3 (FAO, 1997) shows the

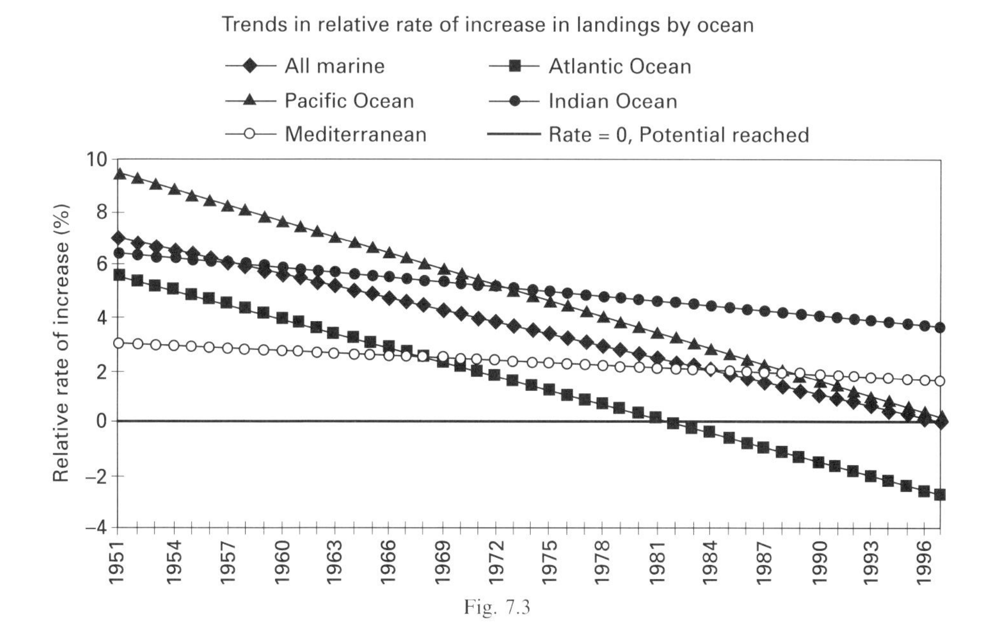

Fig. 7.3

decrease in this rate for all the large ecosystems; the Mediterranean, which has the slowest rate of decline, owes this to the input of nutrients (nitrogen and phosphorus produced from agriculture and waste water) by the rivers which flow into it—the pattern of fisheries production decreases less quickly than elsewhere as a result of pollution.

The separate optimisation of each of the large fisheries in each ocean would allow the maximum potential of 125 million tonnes to be reached. The FAO however, makes its data relative: from its most accurate estimates this organisation estimates that an increase in fishery landings of 10 million tonnes is possible, which would result in 93 million tonnes of potential marine production. The prospect of fisheries production reaching 125 million tonnes remains highly uncertain.

The overall recommendations of this international organisation for fisheries management concern: the increase in age at first capture, the prohibition of exploitation of juveniles, the increase in net mesh sizes and the temporary or permanent closure of fisheries in areas where young fish are concentrated (nurseries). Examples of the 100% restoration in 18 months of the sustainable exploitation of fisheries resources in the Philippines and Cyprus are given, as well as the long-term restoration of the Arctic cod.

The other main problem concerns unwanted bycatch. This catch is thrown back into the sea (and thus does not appear in figures for landings) and represents between 18 and 37 million tonnes (on average 27 million tonnes) per annum. This tonnage is made up of fish with no commercial value and also a large proportion of juveniles. Thus, the total fisheries catch is in reality around 110 million tonnes, for 83 million tonnes landed.

The reduction in these discards, along with the rehabilitation of degraded fisheries and the sustainable exploitation of resources which have been badly exploited, constitute the conditions for substantial increase in fisheries potential.

However, many unknowns remain. The widespread fisheries for small pelagic planktonophages alter the predation of these species on the eggs and larvae of pelagic or benthic species which can be either ichthyophages or planktonophages. These pelagic planktonophages also serve as prey for larger predators. The impact of the various interactions of the fishery on these food webs is very difficult to predict.

7.3 EXPLOITATION OF AQUATIC RESOURCES BY AQUACULTURE

7.3.1 The hatchery-nursery and management of recruitment

Aquaculture can be defined as the art of breeding and rearing aquatic plants and animals. Man intervenes in the natural dynamics of a species (expressed through the processes of reproduction, growth and population dynamics, etc.) by controlling them either partially or totally. In specific cases (e.g. intensive trout culture), water only provides a physical support for the organisms which live in it, their food being provided from elsewhere. The system is artificial in the extreme, often requiring input of oxygen into the water.

In terms of rules and regulations, for international organisations such as the FAO, it is the individual or corporate ownership of a cultivated stock and the culture area which serve to distinguish aquaculture production statistics (aquatic organisms harvested by a legally entitled individual or a person who has owned them during their entire rearing period) from fisheries statistics (production exploitable as a common ownership resource, with or without licence).

In the definition above, breeding precedes rearing, since the aquaculture of any single species firstly requires the acquisition of the subjects to be reared; this phase occurs in the hatchery-nursery which is the structure at the base of aquaculture.

Whatever the site or species under consideration, the objectives of the hatchery remain similar: completion of sexual maturation of captive broodstock is required, followed by reproduction and then fertilisation of the eggs and their incubation. However, the maintenance and feeding of the fragile and minuscule larvae which hatch from the eggs are also the function of the hatchery. Producing live planktonic prey for these larvae requires the reconstruction of a true plankton food chain.

In the hatchery-nursery, the absence of predators, maintenance of adequate physicochemical parameters and the provision of the right size, amount and quality of plankton dictate the survival rate which, coupled to the prolific nature of reproduction in marine species, allows the mass production of juveniles. Billions of salmonid and mollusc juveniles, and millions of marine fish and crustacean juveniles, are now produced in the hatchery-nursery setup.

Hatchery-nursery systems which form the source of juveniles for aquaculture, provide the route for managing "recruitment" which is so sorely lacking in fisheries; they also have the capacity to provide cohorts of juveniles to reinforce natural populations or to set up new populations in marine ecosystems.

7.3.2 Fisheries, aquaculture and products from the sea

Despite its advantages, aquaculture cannot, except by a miracle, respond in the medium term to the demand for seafood products. The reasons for this are simple: first, some species of high commercial value produced by aquaculture are often reared using low market-value fisheries products (in Japan, 27% of fisheries catches are used to feed reared species) and, second, it is not possible to rear a species in high demand just anywhere. The main products imported into Europe cannot be reared or caught there since these are salmon, tropical prawns, cod and tuna—species which are not adapted to the temperatures of many European waters.

This paucity of aquatic products explains why there are only few problems of competition to fear between aquaculture and fisheries, in as much as the species proposed are often different and fisheries cannot respond to the demand.

This lack of high-value marine species does not guarantee that problems of overproduction do not exist in aquaculture. However, they are due to the success of rearing systems as was the case for the salmon in northern Europe, the market price of which fell from £10 to £5 per kg between 1991 and 1993. In other sectors, some prices fell over the same period (mussels) and others rose (oysters), as market characteristics are complex (currency fluctuations, availability of commercial product, cash flow requirements of producers, etc.).

We shall not dwell upon other advantages such as the improved conversion of food ingested by fish, their greater percentage of flesh in relation to total weight (in comparison to farmed terrestrial species) and even the currently-perceived benefits of foods based on seafood products for circulatory diseases, rheumatism, diabetes and even the brain (Castell, 1988).

7.3.3 Productive capacity and economic value of aquaculture

Some current levels of animal production in aquaculture represent, in terms of yield, around 10 times the highest-yielding agriculture production in terms of primary production, on identical temporal and spatial scales. In the rias of north-west Spain, suspended rearing of mussels results in production of 200–250 tonnes (wet weight)/ha/year, which compares favourably with the 2 tonnes of chickens or 0.3 tonnes of cattle which can be produced from one hectare of maize or prairie respectively. The utilisation of the volume of a mass of water, not only its surface, allows the exploitation of natural production which occurs at significant depths and over an area which greatly exceeds the managed surface, due to the movement of marine currents. The comparison has no ecological significance in the strictest sense, since the biomass of filter feeders uses primary production which may have been generated elsewhere; this works to the advantage of the producer.

In the most developed systems, such as integrated agriculture and aquaculture as practised in China for example, production reaches 9 tonnes of fish/ha/year and more than 30 tonnes of duck from the same area of water. What is more, the final use of this nutrient-rich pond water for irrigation of agricultural productions increases the yield from these by more than 10 tonnes/ha (Schoonbee and Prinsloo, 1988).

The economic impact of aquaculture products is also distinct from that of fisheries (on a downturn) by the fact that production is increasing (between 1983 and 1992, aquaculture production increased by 60%). On the basis of current production in Norway (Fig. 7.4), Torissen (1995) predicted an aquaculture production of salmon greater than 1 million tonnes for 2005 and Larkin (1989) predicted that aquaculture production would exceed that of fisheries before 2020.

In monetary terms, the estimates of the FAO in 1994 showed that the total value of aquaculture products had tripled from 1984 to 1994, going from $12,000 million to $32,200 million dollars; crustacean production increased by 205%, fish production by 105% and that of molluscs, 55%. Fisheries production has only increased by 15% since 1984, but the value of catches has increased by a factor of 2 as a result of price increases ($82,000 million). Aquaculture production contributes directly to human food while 29% of fish caught is destined for processing (oils and animal feeds). Only 22 million tonnes of caught fish are sold fresh. Aquaculture therefore supplies more than half of the consumption of fresh products (Nash, 1995). In Japan, the world leader in fisheries and aquaculture, aquaculture production is 1.4 million tonnes, representing 13%, by volume, but 27% by value, of fisheries produc-

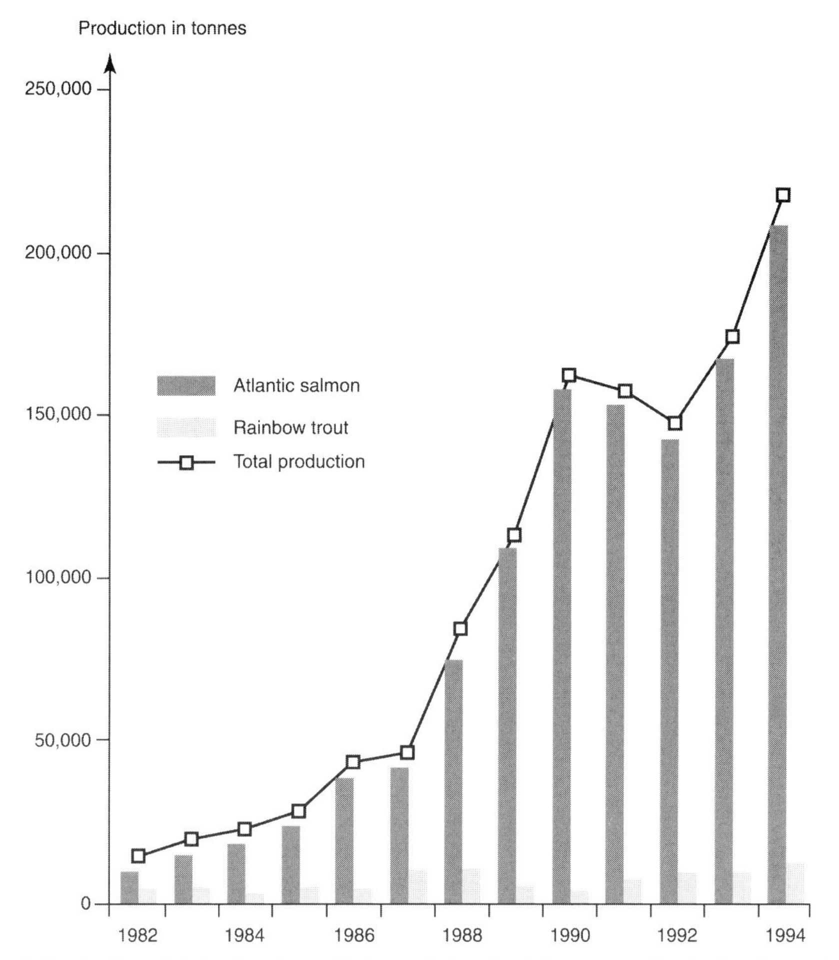

Fig. 7.4. Production of Atlantic salmon (*Salmo salar*) and rainbow trout (*Onchorhynchus mykiss*) in Norway. (After Torrissen, 1995.)

tion. This trend is found world-wide and in 1989, for example, aquaculture production increased by 11% in tonnage, but by 18% in value.

7.3.4 Diversification of production

Around 150 aquatic species are reared, including 89 fish, 23 crustaceans, 35 molluscs, 4 algae and various other species such as frogs, terrapins and sponges. For each of 34 species, annual production exceeds 20,000 tonnes.

Thousands of hatcheries have been constructed, mainly since the 1980s. These range from the small family-run unit to the industrial structure capable of producing millions of reared individuals (Cupimar, a Spanish farm, produces 12 million sea bream, *Sparus auratus* juveniles). The areas used for these rearing systems amount to millions of hectares and Nash (1995) estimated the volume of rearing pens in use in the world at 10 million m^3.

Despite the relative homogeneity of its objectives (biological production), aquaculture involves very diverse activities. Production for food is often associated with it, but there are many other objectives: the production of fish for leisure fishing (trout, salmon, pond fish) or juveniles for the reconstitution of natural stocks (scallops); the introduction of new species for commercial fishing (clams from the Philippines); the production of ornamental fish for aquaria. All the above demonstrate the power and diversity of these, often new, techniques throughout the aquatic environment. In Singapore, for example, the market for aquarium fishes represents $50 million. In the United States, the turnover of this business is estimated at $40 million (wholesale).

Other, no less important, roles seem to have fallen to aquaculture. Its social role in the maintenance of populations has often been cited, as it creates and maintains jobs along the shore and in rural areas. This is true, for example, for the Tahitian pearl culture which provides jobs in the far away Tuamotu islands, and the cultures of seaweeds, prawns, mussels and oysters which are carried out in many wetlands or coastal zones throughout the world. Taking France as an example, marine fishery produces a little over 700,000 tonnes per annum for around 10,000 direct jobs (R. Billard, pers. comm.), while aquaculture, with a production of around 200,000 tonnes (33% of aquatic production), provides 23,000 jobs.

The economic value to aquaculture of low-value fisheries products is well known, since aquaculture uses low-grade fish, either frozen or ensiled or after transformation into fishmeal. In 1993, Japanese aquaculture thus absorbed 1,473,000 tonnes of frozen "trash fish" and 270,000 tonnes of fishmeal, while European aquaculture of sea bass and sea bream consumed around 50,000 tonnes of fishmeal (Doumenge, 1995). In total, this amounts to around 4 million tonnes of captured fish which are made into fishmeal and 2 million tonnes of "trash fish" which are utilised by aquaculture and which contribute to their maintenance and profitability.

There are many examples, but the new areas in which aquaculture appears to be able to play a major role are those of coastal water development, water purification and, in wider terms, environmental sciences.

7.3.5 Aquaculture, recycling and environmental protection

Certain types of aquaculture can take advantage of pollution and recycle it into products which are useful to man. Pig manure is used in daphnia production, domestic waste water is treated in lagoons (there are around 2400 of these in France) and fish culture in ponds is essentially a recycling system. The Asiatic countries thus have an excellent method of using a pond ecosystem for fish production and the purification of waste from terrestrial stock rearing (Doumenge and Marcel, 1990; Jhingram, 1990). The situation is similar in eastern Europe, in a temperate climate (Marcel, 1990).

These are real examples of ecosystems developed by man. While the techniques used in this management are empirical, they result in the production of several million tonnes of animal protein (around 12 million tonnes of carp in China).

Man is using these production processes to control large ecosystems, including those in big American lakes (Barraclough and Robinson, 1972).

In other situations, the ability to concentrate marine species can be used in anti-pollutant applications: large quantities of mussels are used to filter the Mersey estuary (Liverpool) and have given it "a quality which it has not had for a long time"; tropical mussels are used to clean up water from shrimp rearing ponds in south-east Asia (these filter feeders are not destined for human consumption). One food chain can thus serve as a double-edged sword in the protection of the environment: it can be used to get rid of pollution but, conversely, its sensitivity to pollution can be utilised to protect and conserve aquaculture activities.

Aquaculture techniques thus offer an extraordinary and unique opportunity; they can generate new products while assuring large-scale recycling of waste destined for the aquatic environment, previously considered to be a disposal problem.

7.4 THE SPLIT BETWEEN SCIENTISTS AND PROFESSIONAL USERS OF THE SEA

Traditional shellfish and carp culture existed long before scientists took an interest in them; production is ahead of the research which has followed close behind. This does not appear to have been a handicap: the oyster is at the top of cultivated or fished species in France with 150,000 tonnes representing FF1.5 billion.

The so-called "new" aquaculture has developed on the basis of the progress of ecology and biology and here the reverse of the above is true: hatchery production has followed after research and did not exist before laboratory experiments. This is a significant event as reared molluscs do not seem to have been as important as fish to scientists in the past. However, the situation has changed and traditional activities are integrating scientific progress (remote capture of molluscs, triploid oysters, etc.).

Van Tilbeurgh (1994) devoted an in-depth sociological study by means of a working group with the single aim of understanding oyster culture, partly made up of producers and partly of biologists from IFREMER*.

Brought into being by the occurrence of an outbreak of disease that destroyed a significant section of the Brittany oyster population, this study showed the distance which exists between the theories constructed by the biologists and those of the professionals. To summarise, citing Van Tilbeurgh:

> *... faced with oyster mortalities and needing to identify a means of treatment to overcome these, the scientists described an explanatory theory based on overexploitation; since then, the objective of the study for the scientists has been to demonstrate that the trophic capacities of the water act as a limiting factor in the oyster culture ecosystem, with epizootics occurring in situations of overuse of the basin. Thus, until now, in an attempt to check different epizootics, this laboratory (Ifremer), has instituted levels for the natural system. These rest on the adaptation of oyster culture techniques to the constraints imposed by the scientists ...*
>
> *... The majority of oyster farmers do not support this theory of possible overexploitation. For the user, production must go along with nature, using its contin-*

uous, renewed variability. The oyster farmer thus uses environmental variability for production; this method of working depends on the observation of nature ...

... The absence of total proof of the theory of overexploitation by the oyster farmers is evidence of the absence of total agreement between the different bodies of knowledge. Through the development of management policies in oyster culture ponds, the ecologists can only respond through their data system (space and time) to problems put to them (existence of an overexploitation threshold) in relation to the methods at their disposal (statistical approach to the environment to maintain the status quo). These two types of knowledge seem to operate in two distinct spheres.

There is no point in arguing who is right or wrong. The variability in oyster culture for over a century has been reported by Héral (1990), and Doumenge (1992) reported the variability affecting the fishery for pearl oysters in the Indo-Singhalese strait of Mannar. There the abundance of natural banks of pearl oysters fluctuated so much that their presence was considered as a gift from the gods; there were 95 fishing campaigns over three centuries, interrupted for periods of between 16 and 37 years of failed production. These were natural banks which were only accessible to divers for a few weeks of the year. The variability in natural or cultivated populations is thus demonstrated.

From another point of view, and applicable to oysters, prawns or fish, the natural tendency of the producer is towards intensification, with the associated consequences of lightning epidemics which destroy rearing systems based on new technology even before they have established themselves. Numerous examples of this classic phenomenon can be found in Barnabé (1990c).

Thus a gulf appears between the methods of producers and the scientists who have the job of following this production, as a result of having different objectives. "For the scientists, an experiment serves to confirm or disprove a theory. It is of no significance except in relation to theoretical knowledge. For the oyster farmer, an experiment serves as proof, not of theoretical knowledge, but of previously-observed facts" (Van Tilbeurgh, 1994).

This division between representations of the same phenomena made by different socioprofessional communities is felt to be an "abuse of power" and sometimes has destructive repercussions. In 1990, in the Étang de Thau, the prohibition of shellfish sales decreed by Ifremer, which appeared unfounded to the shellfish farmers, led them to destroy this organisation's laboratory at Sète. Details of this sad affair, testimony to a dramatic lack of understanding, are given in Piétrasanta (1993). The rebellious nature of the fishermen's riots at Rennes in 1994 revealed a similar lack of understanding between the professional view and institutional perception.

What must be remembered is that the scientists' opinion prevails over that of the professional since the former guides the decisions of protection authorities and results in regulations, although the validity of this opinion is not always demonstrated. We shall not pronounce on the problem of responsibility for shellfish culture basins, but the failure of fisheries theory provides the argument; over several tens of years, it has constituted a universally-accepted and applied dogma, even though it does not relate to the reality of the data.

While the professional fishermen and fish farmers are saying to the scientists "pay attention to our problems", the "research prospectors" at Ifremer, the organisation in charge of these questions (Ifremer, 1994) may appear to be light years away from these preoccupations. As a result, the working group which provides the science on the exploitation of aquatic resources is no longer considered credible by many professionals.

This is not true everywhere. In Japan, the synergy between "Fisheries Universities" and professional cooperatives results in an extraordinary dynamism in the production sector, which progresses constantly despite the creation of marine EEZs, which has reduced the access to distant fishing grounds. Japan's artisanal fishery and aquaculture are expanding rapidly and this is also the case in the whole of south-east Asia and in China, and also in Italy, Spain and Greece.

7.5 CONCLUSIONS

While fisheries' limits are reached because fished stocks are finite in character, the productive potential of coastal or oceanic ecosystems is very far from being so.

The implicitly-accepted law of the sea is reflected in criteria based on the geography of the resources for access to them: there is open competition for resources which are currently fully exploited (they can sustain this competition as long as exploitation is not complete); a sort of national or supranational (EU) law has appeared, while the fishery is characterised by an incapacity to regulate fishing effort and contain its overinvestment.

Within this framework, artisanal fisheries which are, by definition, coastal ones, hold the trump cards: fewer intermediate consumers, higher gross and net added values, lower cost of capital investment and higher numbers of direct jobs. The riches created go into the national economy and its annual production of 24 million tonnes is destined for human consumption, mostly local.

Although it is concerned with both fisheries and aquaculture, the scientific work and assistance in production activities is not in phase with the problems of producers. This lack of understanding is serious and contrasts with the situation in countries such as Japan: the previously-mentioned synergy between "Fisheries Universities" and professional cooperatives results in an extraordinary dynamism in the productive sector which is constantly making progress. Japan's artisanal fishery and aquaculture are expanding rapidly throughout the whole of south-east Asia, as they are in the business of management, not only of the strategic vision which has failed in France, but also of current projects and fruitful collaboration.

It is thus necessary to change strategy, which is what we propose in the following chapters. Our proposals are based on the work on developing the rearing of sea bream and sea bass in our laboratory (Barnabé and René, 1972, 1973; Barnabé, 1974, 1976, 1990a), fisheries studies (Barnabé, 1973, 1976), eco-ethological studies of fish in the littoral zone (Barnabé and Chauvet, 1992; Chauvet *et al.*, 1991), and also on the coordination of studies examining aquaculture and aquatic resources (Barnabé, 1990c, 1991). In order to expose what is happening elsewhere in the area of exploitation of aquatic resources, we have chosen the term "development", as it also implies "management" and comes under the perspective of "sustainable development" as it has been called since the Rio Conference in 1992.

7.6 BIBLIOGRAPHY

Anon. *Compte rendu séminaire PBD-Globec, Nantes, July 1993*. Unpublished paper, February 1994.

Bakun A. L'océan et la variabilité des populations marines. In: *L'homme et les ressources halieutiques*. IFREMER Éd., Plouzané, 1989; 155–188.

Barbault R. *Écologie générale*. Masson Éd., Paris, 1991.

Barnabé, G. Contribution à la connaissance de la croissance et de la sexualité du loup (*Dicentrarchus labrax L.*) de la région de Sète. *Ann. Inst. Oceanogr*., Paris 1973; 49(1): 49–75.

Barnabé G. Mass rearing of the bass *Dicentrarchus labrax L.* In: Blaxter J.H.S. (ed) *The early life history of fish*. Springer-Verlag, Berlin, 1974: 749–753.

Barnabé G. *Contribution à la connaissance de la biologie du loup* Dicentrarchus labrax (*L.*) *Poisson Serranidae*. Doctoral thesis. Technical Science University, Languedoc, Montpellier, 1976.

Barnabé G. Rearing bass and gilthead bream. In: Barnabé G. (ed) *Aquaculture* Vol.2. Ellis Horwood, Chichester, 1990a; 647–683.

Barnabé G. Some details of aquaculture production. In: Barnabé G. (ed) *Aquaculture* Vol.2. Ellis Horwood, Chichester, 1990b; 1095–1100.

Barnabé G. (Ed.) *Aquaculture*. Vols 1 and 2. Ellis Horwood, Chichester, 1990c.

Barnabé G. (Ed.) *Biology and ecology of cultured species*. Ellis Horwood, Chichester, 1994.

Barnabé G. and Chauvet C. Évaluation de la faune ichtyologique dans la réserve sous-marine de Monaco. *Assoc. Moneg. Prot. Nat.* CR 1990–1991,1992; 51–59.

Barnabé G. and René F. Reproduction contrôlée du loup *Dicentrarchus labrax* (Linné) et production en masse d'alevins. *CR Acad. Sc*., Paris 1972; 276D: 2741–2744.

Barnabé G. and René F. Reproduction contrôlée et production d'alevins chez la daurade *Sparus auratus* Linné 1758. *CR Acad. Sc*., Paris 1973; 275D: 1621–1624.

Barnabé G., Giannakourou A. and Ben Khemis I. *A comparative study of cost effectiveness with growth performance of sea bream* Sparus aurata *larvae reared under two different climatic conditions in France and Greece, using three types of mesocosms*. Confidential final report. Contract no. Aq. 429, Commission of the European Communities, Research Program in the Fisheries Sector (FAR), 1995.

Barraclough W.E. and Robinson D. The fertilization of Great Central Lake. II: Effect on juvenile sockeye salmon. *Fish Bull.* 1972; 70: 37–48.

Birkeland C. Difference among coastal systems. The controlling influences of nutrient input and the practical implications for management. In: *Coastal studies and sustainable development*. Proceedings of the COMAR. UNESCO Technical Papers in Marine Science No.64, UNESCO, Paris, 1992; 8–16.

Caddy J.F. and Griffiths R.C. *Tendances récentes des pêches et de l'environnement dans la zone couverte par le Conseil Général des Pêches pour la Méditerranée.* Ét. et Rev., CGPM, FAO (Rome) No.63, 1988.

Castell J.D. Fish as brain food. *World Aquaculture* 1988; 21–22.

Chaussade J. *La mer nourricière, enjeu du XXI siècle.* Chaussade Pub., Nantes, 1994.

Chauvet C., Barnabé G., Harmelin J.G., Bianchoni C.H. and Miniconi R. Quelques aspects de la structure sociale et démographique des populations de mérou *Epinephelus guaza* des réserves corses de Scandola et des îles Lavezzi. In: Boudouresque J.C., Avon M. and Graves V (Eds) *Les espèces marines à protéger en Méditerranée.* C.R. Colloque International. GIS Posidonie Pub., Marseilles, 1991; 277–290.

Divanach P., Barnabé G. and Connes R. *A comparative study of cost effectiveness with growth performance of sea bream* Sparus aurata *larvae reared under two different climatic conditions in France and Greece, using three types of mesocosms.* Periodic report no.2. Contract no. Aq. 429, Commission of the European Communities, Research Program in the Fisheries Sector (FAR), 1992.

Doumenge F. Nacres et perles. Traditions et changements. *Bull. Inst. Oceanogr.* 1992; 8:1–52.

Doumenge F. L'interface pêche/aquaculture; coopération, coexistence ou conflit. *Norois* 1995; 42(165): 205–223.

Doumenge F. and Marcel J. Aquaculture in China. In: Barnabé G. (ed) *Aquaculture Vol.2.* Ellis Horwood, Chichester, 1990; 946–963.

Dufour V. and Galzin R. Colonization pattern of reef fish larvae to the lagoon of Moorea island, French Polynesia. *Mar. Ecol. Prog. Ser.* 1993; 102: 143–152.

FAO, Marine Resources Service. *Review of the state of world marine fishery resources.* FAO Fisheries Technical Paper 335, Rome, 1994.

FAO Marine Resources Service. *Review of the state of world marine fishery resources: marine resources.* FAO Fisheries Circular 920, Rome, 1997.

Fréon P. L'introduction d'une variable climatique dans les modèles globaux de production. In: Cury P. and Roy C. (eds) *Pêcheries Ouest-Africaines.* Orstom Pub., Paris, 1991; 395–424.

GEP. *L'exploitation des ressources vivantes aquatiques: Rôle et perspectives de la recherche.* Collection Études 1989. Min. Rech. Technol. DGRT, Paris, 1989.

Greenpeace. *Dossier: Les dents de la mer.* Greenpeace 4/1994; 3–25.

Gulland J.A. *Manuel des méthodes d'évaluation des stocks d'animaux aquatiques.* Manuel FAO de science halieutique, FAO, Rome, 1974.

Hanson J.A. Open sea mariculture in utilitarian perspective. In: Hanson J.A. (ed) *Open sea mariculture.* Dowden Hutchinson & Ross Inc., Strousberg, USA, 1974; 17–58.

Héral M. Traditional oyster culture in France. In: Barnabé G. (ed) *Aquaculture Vol.1.* Ellis Horwood, Chichester, 1990; 342–380.

Holden M.J. and Raitt D.F.S. *Manuel de science halieutique. Part 2: Méthodes de recherche sur les ressources et leur application.* FAO, Rome, 1974.

Hunter J. Feeding ecology and predation of marine fish larvae. In: Lasker R. (ed) *Marine fish larvae.* University of Washington Press, Seattle, 1981; 33–77.

IFREMER. *Prospective de recherche (4 fascicules)*. IFREMER Éd., Issy-les-Moulineaux, 1994.

Jhingram V.G. Aquaculture in India. In: Barnabé G. (ed) *Aquaculture Vol.2*. Ellis Horwood, Chichester, 1990; 988–1016.

Kesteven G.L. *Manuel de science halieutique. Part 1: Introduction á la science halieutique*. FAO, Rome, 1974.

Laeavastu T. *Manuel des méthodes de sciences halieutiques (10 fascicules)*. FAO, Rome, 1967.

Larkin P. Mariculture and fisheries: Prospects and partnerships. Symposium on the ecology and management aspects of extensive mariculture. ICES, Nantes, EMEM, No 58, 1989.

Lasker R. Les déterminants du recrutement. In: *L'homme et les ressources halieutiques*. Ifremer Éd, Plouzané, 1989; 115–157.

Lorenz E.N. Un battement d'aile de papillon au Brésil peut-il déclencher une tornade au Texas. *Alliage* 1995; 22: 42–45.

Marcel J. Fish culture in ponds. In: Barnabé G. (ed) *Aquaculture Vol. 2*. Ellis Horwood, Chichester, 1990; 593–626.

Miller J.M. An overview of the Second Flatfish Symposium: Recruitment in flatfish. *Neth. J. Sea Ecol.* 1994; 32(2): 103–106.

Nash C. Introduction to the production of fishes. In: Nash C. and Novotny A.J. (eds) *Production of aquatic animals*. World Animal Science, C8. Elsevier, Amsterdam, 1995; 1–20.

New M. and Csavas I. Will there be enough fish meal for fish meals? *Aquaculture Europe* 1995; 19(3): 6–13.

Nissling A. Survival of eggs and yolk-sac larvae of Baltic cod (*Gadus morhua*) at low oxygen levels in different salinities. In: *Cod and climate change*. ICES Mar. Sci. Symp. 198, 1994; 626–631.

OECD. *Gestion des zones côtières: Politiques intégrées*. OECD, Paris, 1993.

Piétrasanta Y. *L'écharpe verte. Combats pour une nouvelle écologie*. Albin Michel Éd., Paris, 1993.

Postel E. La théorie des pêches: Dynamique des populations exploitées. In: Bougis P. et al. (eds) *Océanographie biologique appliquée*. Masson Éd., Paris, 1976; 116–157.

Revaud d'Allonnes, M. Turbulence océanique, l'effet coquilles Saint-Jacques. *Science et Avenir* 1994; Hors Série 98: 24–29.

Schoonbee H.J. and Prinsloo J.F. The use of polyculture in integrated agriculture/aquaculture production system aimed at rural community development. *J. Aquat. Prod.* 1988; 2(1): 99–123.

Torrissen O.J. Norwegian salmon culture: 1 million tons in 2005? *Aquaculture Europe* 1995; 19(4): 6–11.

Troadec J.P. Les pêches et les cultures marines face à la rareté de la ressource. In: *L'homme et les ressources halieutiques*. Ifremer Éd., Plouzané, 1989a; 1–46.

Troadec J.P. (ed) *L'homme et les ressources halieutiques*. IFREMER Éd., Plouzané, 1989b.

Van Tilbeurgh V. What is meant by management of ecosystems? *Océanis* 1992; 18(4): 479–486.
Van Tilbeurgh V. *L'huitre, le biologiste et l'ostréiculture: Lecture entrecroisée d'un milieu naturel*. Coil. Logiques Sociales, L'Harmattan Ed., Paris 1994.
von Walhert G. and von Walhert H. Extensive rural mariculture in the tropics: Experience and issues. In: *Aquaculture, a biotechnology in progress*. European Aquaculture Society, Bredene, Belgium, 1989; 97–102.
Weber P. Sauver les océans. In: *L'état de la planète*, 1994. Worldwatch Institute Éd., la Découverte, Paris, 1994; 65–92.

Part C

Management of coastal waters

8

Ecological planning

8.1 THE SCIENTIFIC BASIS OF AQUATIC DEVELOPMENT

The term "development" has been described in the introduction to this book as the application of scientific and technical ideas and principles to animal or plant populations, as well as their habitats, with a view to improving the reproduction or establishment of particular species or to ensure their durability and good health, for example. We have also seen that this should include the human dimension and that this would involve spatial organisation in order to improve the population's living conditions, develop economic activities and place value on natural resources while avoiding the perturbation of natural ecosystems.

Sustainable development can be considered as ecological engineering and is based on scientific theory: it rests on the principles of ecological and environmental sciences, developed in the past. These principles have now matured and are ready for application (Mitsch and Jorgensen, 1989).

Ecological engineering (or ecotechnology) is a development tool and is based on the scientific report that natural ecosystems are self-organising entities. The use of technological methods for the development of ecosystems must therefore be based on a deep understanding of ecology. The objective is to bring all of this together.

According to Odum (1989), ecological engineering consists of the manipulation of the environment by man, by using low inputs of complementary, or auxiliary, energy in order to control the systems in which the main energy leading the system comes from natural resources. This principle is the management of self-organisation; this implies that conserving the functioning processes of ecosystems is effective, because ecotechnology uses natural structures and processes. The starting point is a natural ecosystem, but the new ecosystem, which develops as a result of the conditions introduced by engineering, differs little—the components introduced by man and the ecological characteristics combine within the self-organised structure of the ecosystem, resulting in a process which maximises performances. A simplistic example can be given: increasing the temperature of a pond by recycling or thermal protection (glass) can allow accelerated growth of the populations within. Odum's term "partner with nature" is thus justified, but it is still necessary to choose ecosystems which are adaptable to human requirements.

Functioning of ecosystems depends on basic energy, or solar radiation (Frontier and Pichod-Viale, 1991). These authors also emphasise the significance of auxiliary energy, the importance of which they illustrate with the fact that a plant uses 36 times more energy to transport water from the ground to its leaves than for photosynthesis. In aquatic environments, the emphasis is placed on turbulence, a particularly efficient form of auxiliary energy, to which we have devoted a large part of Chapters 2 and 4. Complementary energy provided by man is termed "artificial auxiliary" energy. In the aquatic environment, the creation of water movements, for example, constitutes an artificial auxiliary energy.

Another method of development has been indirectly formulated by von Walhert and von Walhert (1989) with regard to extensive mariculture*. They noted that this is one of the ways of making progress in ecology and fisheries; the setting up of mariculture installations results in a structural change in habitat which is not easy to interpret in a quantitative way using ecosystem theory, but which can be understood in the context of ecological thinking based on the concept of the ecological niche, i.e. the relationships between species within one environment. Within this concept, waste products from a species (CO_2 or nitrogen, for example) are not considered as such, but as the starting point for new primary production. Recycling thus appears as one form of management and affects many sectors (e.g. water purification and extensive aquaculture). Edwards (1980) devoted a well-documented review to aspects of organic waste recycling by fish, emphasising the significance of this technique in terms of safeguarding the environment and economising on fuel.

It should also be noted that sustainable development only exists with the addition of complementary energy to the functioning of a natural ecosystem. Thus it is distinct from intensive aquaculture, which reconstructs an artificial system with humans completely in control, providing all energy and feeding requirements. Economic efficiency is therefore very different in the two cases.

In aquatic environments, ecosystem processes seem to be sufficiently understood in general terms for it to be possible to envisage the manipulation of natural phenomena for mankind's useful ends, without perturbing natural ecosystems. Humans thus becomes a partner with nature, rather than being set apart from it.

The development or management of aquatic environments by ecotechnological means is undergoing rapid development throughout most of the world and numerous institutions have centres or organisations which are very active in this area. Many aquatic environments have been ecologically managed for various purposes, from lakes and water courses to large marine ecosystems ("large marine ecosystems" of greater than 200,000 km^2, Sherman, 1990), some examples of which will be given later.

The fact that fisheries' limits have been reached (see Chapter 7) does not indicate that the potential production of coastal ecosystems has been reached. It is interesting to compare the potential for a fishery, with estimated catches of 2 kg/ha/year on average in exploited waters, and the potential of even the most extensive form of aquaculture, which produces about 100 kg/ha/year and up to 200 tonnes/ha/year in shellfish culture areas.

Inclusive management, such as the integrated modification of the functioning of natural ecosystems, constitutes a common tool for these types of productive activities, but the economic development or management of coastal waters is not limited to edible species; it concerns many other aspects of the aquatic environment.

Ecological management is also entirely different to biotechnologies based on the manipulation of genetic structures: it does not introduce new species and its actions are reversible by nature.

8.2 SCIENCE AND SUSTAINABLE DEVELOPMENT

It can be noted firstly, following Lefeuvre (1995), that ecology and the environment are the "poor cousins" of the life sciences, which in France are focused on molecular biology, the representatives of which are in the majority at all levels of decision making. It is estimated that for the past few years, ecology has only represented 3.65% of the life sciences budget. In 1993, Lefeuvre reported that the total number of research grants allocated to all ecological laboratories was less than that allocated to a single cellular and molecular biology laboratory (i.e. 29). This French example is a good indicator of the situation occurring throughout Europe.

More recently, the environment has become a fashionable subject as social pressures are applied for its management and protection, but this does not mean that the situation has changed in research organisations or universities. This social demand whets the appetite of numerous disciplines more concerned with getting funds than on investing in environmental problems.

In Europe, development of the aquatic environment is practised rarely: in some organisations in France, the research is either too pure or too applied and in others, outside their remit. In contrast, the activities of the large scientific organisations in France are concentrated on increasingly sophisticated "monitoring" of marine ecosystems in order to best define their precise functioning.

For fundamental research, "monitoring" constitutes the endpoint, the only objective being to understand and not to act. In its report on biodiversity and the environment, the Académie des Sciences de l'Institut de France (1995) defined the classical stages of research using three words: assessing, functioning, modelling. The area of aquatic ecology is favourable for endless monitoring, since a new instrument or modelling technique, or simply the passage of time, requires the redoing of what has been done before in order to confirm or deepen understanding of certain matters. Noting the presence of pollution and understanding its causes is a prerequisite, but would not constitute an end in itself. In these areas, research is often reduced to records. When these records are published in reputable scientific reviews, the desired result is considered to have been achieved.

This does not, however, satisfy requirements worldwide: thus the Intergovernmental Oceanographic Commission (UNESCO), under the leadership of its Secretary General, G. Kullenberg (1993), noted that science should address itself to the integrated management of coastal waters. He wrote: "Even when they are based on a considerable amount of data, the most in-depth studies often end with the phrase:

research should be continued ..." This is not surprising, from a scientific point of view, but it is exasperating for managers presented with a very technical, ambiguous document and the requirement to make fast decisions. Having paid a large sum for a detailed report, they expect a clearly expressed opinion on the methods for improving fishery stocks or means of reducing pollution. Those responsible undoubtedly forget that all laboratory directors, in parallel to their research and from necessity, pursue another goal which is to obtain new grants for further research.

In terms of the theory of knowledge, as stated by Peters (1980), the test for ecology is not logical consistence, aesthetic appeal or mathematical precision, but the accuracy and usefulness of its predictions. Both experimental and theoretical ecology have forgotten this. The same author (Peters, 1986) demonstrated that predictive ecology constitutes a theory providing useful tools for describing, manipulating and controlling our environment: it sacrifices precise description and details for generalities and the application of predictions, since "we must manage our planet in the short term to reduce the effects of pollution and respond to the requirements of increasing populations".

The "passivity" of current ecological science is similarly denounced by Gliwicz (1992) who stated that the role of the ecologist "is confined to saying what must not be done, rather than proposing solutions, which is left to the chemists or engineers".

With regard to modelling, De Angelis (1988) clearly emphasised the problems and difficulties unique to aquatic populations. The data for use in modelling have to be very detailed, concerning both the ecosystem (spatial heterogeneity, population structure, dependence on environmental conditions, etc.) and the physiology or the individual behaviour of the organisms. A complete data set is necessary to establish a valid model, but is rarely available, and the author rightly cites marine fish larvae recruitment hypotheses as an example.

With regard to the protection of the marine environment, Rougerie (1995) in a critical review of research on the marine environment in the south Pacific over the past 30 years, noted that while our understanding of large island and marine ecosystems has greatly increased over this period, their overall state of health has regressed. Thus, he noted a paradox between the improvement of knowledge, the accelerated use of resources and the degradation inflicted on the environment, constituting the central stumbling block on which rests all hopes for balanced development. For the same author, the highly-publicised increase in sea level is nothing in comparison to the human population explosion which grows at a rate of 95 million inhabitants per year (equivalent to the population of Mexico), the main part of which is spreading through ecosystems in the intertropical zone.

In light of such reports, an underlying question is posed more and more: what purpose does research serve? What does it bring apart from knowledge?

In France, the 1994 report of the Conseil Supérieur de la Recherche et de la Technologie (CSRT), reporting to the Ministry for Higher Education and Research, emphasised that new directions should be given to research. Whilst more and more pressure certainly appears to be in favour of fundamental research, it should be oriented towards economic or social objectives; "take an interest in us, fellow citizens appeared to be saying to the researchers". The CNRS followed suit through the proposals of its director (Aubert, 1995). The journal *Alliage* devoted several articles to this subject, including a well-argued one by Lévy-Leblond (1995). This can be summarised by saying that pure research does not lead automatically to beneficial consequences for the economy; the gap between research and development becomes obvious and he cites the example of South Korea which succeeded in economic lift-off without any significant research, and emphasises the lack of recent defining innovations regarding the main obstacles encountered by biology and physics.

A detailed analysis of the problems of the usefulness of fundamental research had also already been carried out by Durand (1992) who remarked on the lack of a relationship between a country's scientific research effort and its industrial and economic success. He gave the examples of Germany and Japan and also the "Tiger economies" (Korea, Taiwan, Singapore and Hong Kong) to demonstrate that it is not necessary to be leading international scientific competition (dominated by the USA and the UK) to be at the forefront of technological and industrial developments. For France, this author attributes the over-investment in high-level research to the weight given to fundamentalist researchers in the corridors of power and to the science lobby.

He advocates the redistribution of some of the grants to basic research towards technological advancements. This direction was taken into account by the Minister for National Education, Higher Education and Research, who announced technological projects in higher education.

This is what is already happening elsewhere: in Europe, Germany directs research money towards technology and definite objectives; in the USA, military research grants are directed towards civil objectives since "war becomes economic"; the power of the CIA is now directed towards economic objectives!

The management and protection of aquatic environments are there for the taking for science: a scientific concept such as biodiversity can be put into practice, as there is a continuum between basic and applied research. Thus it can be hoped that attitudes will change or at least develop. At a colloquium devoted to the state of the environment in the south Pacific, an ecologist specialising in the study of reef systems (Salvat, 1995) proposed, on the basis of scientific reports, the introduction of new species to repopulate species-poor lagoon areas, while emphasising that such introductions carried out carefully would not bring harm to the ecosystem and would, on the contrary, allow the development of economic activities which would be well integrated with ecosystem function. This is a rarely found constructive attitude.

8.3 OTHER ASPECTS OF THE DEVELOPMENT OF ECOLOGICAL SYSTEMS

In comparison with aquaculture and fisheries which are centred on the commercialisation of a biomass of a species, either collected or reared, ecological development does not have the production of commercial items as its sole purpose. Recycling of aquatic pollution generated by human activity, conservation of natural areas and management of productive ecosystems all constitute other aspects which are also important, but whose economic value is not obvious enough to give them a market value. The seed is sown, however, and the attributing of an economic value to man's environment is an idea which comes up again and again, despite the difficulties in such an evaluation. In some cases, the costs have been evaluated (see Chapter 9).

Ecological development joins with, and thus remains inseparable from, biodiversity, as it is known that this diversity is itself a guarantee of the equilibrium and healthy functioning of ecosystems. It is all the more appreciated by people in their leisure time: the attraction of the countryside results from the equilibrium and diversity of living organisms within it (see Chapter 5). The same applies underwater, to coral reefs and other submarine seascapes visited by divers, tourist submarines or glass-bottomed boats. We shall see in the following chapter that conservation and the protection of nature constitute the first form of management.

8.4 GLOBAL DIMENSION OF DEVELOPMENT

8.4.1 Ecology and human activities

We have approached sustainable development from its ecological perspective but the changes involved could be very different, since human populations have, throughout history, added many other facets to the initial activity of obtaining food: clothing themselves, moving about and living in communities have created complex systems whose functioning is divided between numerous players. Natural ecosystems are no longer the only ones in play, although they still form the basis for all the rest. The creation of a road on land, a harbour in the sea or the discharge of human waste all cause disruption to the natural environment.

While exploitation of the sea by fishing requires ports and roads or even the fertilisation of lagoon systems with manure, the course of the road, location of the port and quantity and type of fertiliser are matters of choice. The interests involved and their complexity render objectivity illusory and there are many choices to be made—for example, how to define "quality of life". Within the human community, this choice is not only a matter for the scientists involved in the relevant social disciplines, but for individuals; the political wishes of the community are therefore challenged and a consensus must be reached. This consensus can then focus on management goals, improving and optimising that which already exists and not creating something from scratch. Ecotechnological progress needs to be integrated into the process and is based on the dynamics of natural ecological processes.

The scientific manager thus only intervenes in order to elucidate the collective choice and to propose the methods for attaining the goal. This progress would thus fulfil society's desire for knowledge and allow the actions of the decision makers, with research becoming integrated. While it is true that within the environmental sciences, ecology is only one approach, it is central to the social sciences.

8.4.2 Impacts of non-ecologically based planning

Scientific ecology is a recent phenomenon, while the management of coastal zones has a historical dimension; from the ports of Ancient Greece and Carthage to Fos, Antifer or Rotterdam, examples of harbour management are innumerable and must have had an ecological impact which has only recently been taken into account (see Chapter 5).

Man has achieved the greatest known work with water-based projects: from the Canal du Midi to the Suez and Panama Canals, gigantic constructions put in place for navigation. More recently, in Holland, the big construction company, Marios, installed a series of enormous dykes to contain sea waters; in 1995, the largest mobile man-made construction, the petroleum and gas platform "Troll", weighing a million tonnes (16,000 m^2 in area, 370 m high and 225,000 m^3 of concrete) was put into the water. It was installed 80 km off the coast of Norway and its base sunk 12 m below sea level, at a depth of 303 m. These structures tower 45 m above the surface of the water.

These achievements, both recent and historic, are now integrated into our environment "for better or for worse", it could be said. Their often disastrous impact on the ecology of coastal waters constitutes negative and widespread evidence of the impact of developments carried out by man.

Aquaculture is not excluded from this category. In many tropical countries, the mangroves which stabilise the coastline have been transformed into rearing ponds for shrimp or fish without taking specific environmental concerns into account. Nevertheless, with time, there is an integration of extensive aquaculture installations within coastal zones. Thus, the majority of traditional rearing sites for mussels or oysters, or fish-rearing ponds now form part of the landscape; this is the case, for example, for the oyster-beds (claires)* of the Marennes Oléron or the "valli"* in the Pô plain.

It has already been said (Chapter 5) that the edge of the sea has an irrational attraction for humans, who choose coastal waters for both leisure and tourism. For several decades, new forms of littoral management have come into existence, bringing together harbours and accommodation ("marinas") as well as numerous new coastal towns. Their impact is analogous to that of other harbours or towns, but they involve the installation of new human activities—walking, bathing, sport fishing, pleasure-boating, etc. (with the pollution created by these activities)—which add to the historical activities and to the growth in human population to increase the impact on littoral ecosystems. Only ecological management can compensate for the perturbations created by the installation of these new populations and their requirements (quality of water for drinking and swimming, conservation of natural ecosystems and fish for sport fishing), since natural ecosystems are changed completely.

8.4.3 Legal aspects: appropriation of resources

For international organisations such as the FAO, it is the individual or corporate ownership of a cultivated stock and the culture area which distinguish the statistics concerning aquaculture production (aquatic organisms collected by an individual or a person legally having ownership during the entire rearing period) from those concerning fisheries (production exploitable as a resource with common ownership, with or without a licence).

Since 1982, the law of the sea has resulted in the setting up of economic exclusive zones (EEZ), which extend the sovereignty of the states 200 miles out into the open sea. In practice, it limits the freedom of access to neighbouring fishing zones of coastal states. This is all the more important since the majority (90%) of fisheries resources come from the littoral zone or continental shelves situated in the EEZ of the coastal states.

As a result, marine resources range from having no owner, i.e. resources which are accessible to all, to that of common resources, i.e. resources for which the users can be defined at a national level (Troadec, 1989). Within this framework, the economic development demanded by national politics and the legal system already offers to state (or country) managers the marine zones contiguous with their land, which are the richest in terms of biological production. This is not a new development; in Polynesia and New Caledonia, the customary right attributes the ownership to tribes or families, not only of land, but also of aquatic domains (lagoons and well-known fishing areas such as tuna grounds).

8.4.4 Economic aspects

Until now, marine products have been thought to have no other value than their market value. In the context of the market economy, two questions dominate the debate: Can what is caught or produced be sold? Can what is desired by the market be caught or produced?

The establishment of the sale price has only the law of supply and demand as its basis (the system of selling fishery products by auction is a typical example). Aquaculture products, and especially those from managed environments, are in a different position although they do not entirely escape this situation. The product from rearing or culture is harvested on demand, in theory allowing a better price to be obtained (by supplying when demand is highest). Supply can be changed with time, and there should be no low prices associated with large fish catches, nor lack of income when there are no fish. This aspect should be emphasised, as once the supply of a given product is assured and regular, the retailers can gain customer loyalty. As a result, the market for less well-known fish (because they are rarely caught) or ones of exotic origin can be developed. This type of production also allows the closing of contracts between producers and distributors for definite periods, which is a factor affecting price stability, in a sector where variability in supply from fishing is unpredictable.

Economic constraints also determine the preferred choice of actions:

- In terms of biological production, nothing is to be gained by producing or catching a species which cannot be sold. For this reason, the farm owner or manager must determine where there are gaps in the market which can be targetted, before working out ways in which they can be developed. They have an advantage over the fishermen, who cannot choose what they catch; in shrimp fishing, the bycatch (fish which cannot be sold), often represents more than 90% of the catch. Greenpeace (1994) reported that the rejects from shrimp fishing was of the order of 5 million tonnes of pelagic fish and 17 million tonnes for all marine species (equivalent to a quarter of the world's fishery catch).

The economic framework is variable. The collapse in the price for preferred species of fish caught and reared in 1993 was not predictable; on the contrary, on the basis of the scarcity of the resource, an official at Ifremer (Troadec, 1989), predicted a price explosion which did not take place for three reasons:

- as a result of the free circulation of products in Europe following the coming into force of the single market (and the illegal import of foreign products, made easier in Europe by this device);
- the devaluation of the Spanish peseta and the Italian lire, lowering the price in these countries, which are big importers of fish;
- as result of the production explosion for farmed marine fish such as the bass and sea bream, which has doubled almost every year, saturating the market. Now, the problem of "mad cow disease" has pushed the trend towards fish consumption.

As time goes by, improvement of techniques helps towards decreasing the production costs for aquatic species, both for juveniles and the growing on of animals. This technological progress allows the production of a large number of individuals in one place by improving survival or growth rate, allowing a larger number of production cycles. In economic terms, this also holds true for the multiplication of production units which creates a certain amount of competition and also to over-expansion which requires new investors to achieve economies of scale thereafter. There is a relentless pursuit of size in aquaculture infrastructure: European hatcheries of marine fish thus present an over-capacity for production, increased all the more by technical progress.

8.4.5 New socio-economic considerations

The attraction of the coast results in new economic activities centred around tourism and leisure; they are in the process of changing the economic value of this area. In France, Spain and elsewhere, sandy beaches frequented in the past by a few groups of seasonal fisherman, today provide the essential requirements for new tourist complexes. It is not only land ownership or harbour or urban developments which constitute capital, but the environment. Littoral ecosystems become capital in a financial sense since, without beaches, coastal waters and marine sea beds, these towns and harbours would have no meaning—the ecological disaster of the drying up of the Aral Sea and the ruins created by it demonstrate this point. There are also the tourist spots installed along the coast of the Red Sea, in quasi-desert sites, but with marvellous sea beds attracting divers from around the world.

This capital constitutes a positive value, often linked only to the presence of coastal waters and to a geographical criterion, without taking into account the "ecological value" (for example waters used for sailing) to which we shall return in the next chapter.

Tourism and leisure can thus be listed, without much hesitation, in the context of "durable development" as defined at the beginning of the next chapter; their introduction, although not exclusively beneficial (Chapter 5) can, with precautions taken against pollution, turn out to be "neutral" for the environment. This is very different from the situation we have described for fisheries (Chapter 7).

These are types of use for space which, without degrading ecosystems, bring new activities to the littoral zone while increasing its intrinsic value. They can coexist with historical activities, such as fishing, which often provide the basis for traditions (folklore, type of boat or buildings used).

We have emphasised the economic importance of the tourist sector (Chapter 5) which now constitutes the main global industry in terms of turnover, ahead of the petroleum industry. Making the tourism and leisure industry one of the targets for development in coastal waters is thus not incompatible with the demands of "sustainable development".

8.5 CONCLUSIONS

We have noted the development of man's ideas concerning the aquatic environment; the exploitation and management which prevailed in the 1970s has given way to "sustainable development".

Within this framework, the development of coastal waters requires a multidisciplinary approach including fisheries, aquaculture, marine biology and ecology and economic and social sciences, since the overpopulation of coastal zones without doubt poses the most serious threat.

With regard to coastal waters, until now the use, if not ownership, of the marine littoral area has been allocated to fishermen, but this area has now become the framework for new activities; tourism, leisure—whether pleasure-boating, diving or sea water therapy, etc.—cannot exist in a sea emptied of fish by industrial fishing or used as a dustbin by the activities of man on land. Within the framework of exploitation of oceanic resources, traditional activities such as fisheries and aquaculture must be made part of an integrated approach to the coastal ecosystem, including both commercial species and others and the potential of production systems managed using new technologies and waste recycling methods, etc. Above all, space must be made for new categories of users, which in itself will cause problems (Goldberg, 1994).

Ecological management can integrate the often divergent aspirations of multiple users, since they all agree in placing respect for the sea and natural processes at the centre of their concerns. This is a new phenomenon which contrasts with the trends seen in the recent past (overexploitation by fishing, pollution by the oil industry and use of the sea as the ultimate refuse bin).

In such a context, the role of the scientist is not to stop with reports, but to relate their results to reality and society's demands, becoming a powerful player with many rights in relation to the political decision maker; this is a new definition for research, a new strategic vision which must be put forward for evaluation.

As stated by Tamburrino (1991), the objective being aimed for is vast and very ambitious: "... we must conserve as much as possible, but above all think, imagine and achieve without limit, with the main goal of creating new values, new realities and new harmonies"; this cannot be done without thought for the ethical implications.

Our modest approach concerning the management of coastal waters is both descriptive, reporting what is happening elsewhere, and prospective with the aim of exploring the possible future in order to support strategies.

As stated by Salvat (1995): "As well as searching and finding, one must valorise". It is only under these conditions that the coastline, a frontier in the past, will become one of the articulations of the future.

8.6 BIBLIOGRAPHY

Académie des Sciences. *Biodiversité et environnement*. Report No. 33. Tec & Doc Lavoisier, Paris, 1995.

Amanieu M. *Ecothau. Programme de recherches intégrées sur l'étang de Thau*. Univ. Sc. Tech. Languedoc, Montpellier, 1985.

Aubert J. Editorial. Les Clubs CRIN. *La Lettre* 1995; 19.

Bannister R.C.A., Harding D. and Lockwood S.J. Larval mortality and subsequent year-class strength in the plaice (*Pleuronectes platessa*). In: Baxter J.H.S. (Ed) *The early life of fish*. Springer Verlag Berlin, 1974, 21–37.

Brown A.C. and McLachlan A. *Ecology of sandy shores*. Elsevier, Amsterdam, 1990.

Carter R.W.G. *Coastal environments*. Academic Press, London, 1991.

Chaussade J. *La mer nourricière, enjeu du XXI siècle*. Chaussade Pub., Nantes, 1994.

De Angelis D.L. Strategies and difficulties of applying models to aquatic populations and food webs. *Ecological Modelling* 1988; 43: 57–73.

Durand T. Prix Nobel et développement économique. *La Recherche* 1992; 249: 1410–1414.

Edwards P. A review of recycling organic wastes into fish, with emphasis on the tropics. *Aquaculture* 1980; 21: 261–279.

Frontier S. and Pichod-Viale D. *Ecosystèmes: Structure, fonctionnement, évolution*. Masson Éd., Paris, 1991.

Gentelle P. *La lettre des Clubs* CRIN 1994; 14.

Gliwicz Z.M. Can ecological theory be used to improve water quality? *Hydrobiologica* 1992; 243–244: 283–291.

Goldberg E. *Coastal zone space. Prelude to conflict*. IOC Ocean Forum 1. UNESCO Publishing, Environment and Development, 1994.

Greenpeace. *Les dents de la mer*. Greenpeace 4/1994; 3–25.

Ifremer. *Prospectives de recherche*. (2 parts) Ifremer Éd, Issy les Moulineaux, 1994.

Incze L. and Schumacher J.D. Variability of the environment and selected fisheries resources of the eastern Bering Sea ecosystem. In: Sherman K. and Alexander L.M. (Eds) *Variability and management of large marine ecosystems*. AAAS Selected Symposium 99, Washington 1985; 109–143.

Iversen E.S. *Farming the edge of the sea*. Fishing News (Books) Ltd, Farnham, 1972.

Kullenberg G. Gestion intégrée de la zone côtière. Que faire? UNESCO, *Bull. Sc. Mer*. 1993; 68.

Lacaze J.C. *La dégradation de l'environnement côtier*. Masson Éd., Paris, 1993.

Lefeuvre J.C. Écologie et environnement, les mal aimés de la science française. *Aménagement et Nature* 1995; 116: 33–47.

Lévy-Leblond J.M. A quoi sert la recherche? *Alliage* 1995; 22: 2–6.

Michael R.G. *Managed aquatic ecosystems*. Ecosystems of the world, no. 29. Elsevier, Amsterdam, 1987.

Mitsch W.J. and Jorgensen S.E. (Eds). *Ecological engineering: An introduction to ecotechnology*. John Wiley & Sons, New York, 1989.

Needham T. Canadian aquaculture: Let's farm the oceans. World Aquaculture 1990; 21(2): 76–80.

Nihoul M. *Conclusions du conseil scientifique du programme Ecothau*. Ecothau. 2nd CR d'activités, 1987–1988. Univ. Sc. Tech. Languedoc, Montpellier, 1987.

Odum H.T. Experimental study of self-organization in estuarine ponds. In: Mitsch W.J. and Jorgensen S.E. (Eds) *Ecological engineering: An introduction to ecotechnology*. John Wiley & Sons, New York, 1989: 291–340.

Olsen S. Coastal management: The nature of the challenge. *Maritimes* 1992; 36(4): 1–2.

Panneau J.M. *La Nouvelle République*, 06 04 1990.

Peters R.H. Useful concepts for predictive ecology. *Synthese* 1980; 43: 257–269.

Peters R.H. The role of prediction in limnology. *Limnol. Oceanogr.* 1986; 31(5): 1143–1159.

Riley J.D. The distribution and mortality of sole eggs *Solea solea* (L) in inshore seas. In: Baxter J.H.S. (Ed) *The early life of fish*. Springer Verlag Berlin, 1974, 39–52.

Rougerie F. Réflexions sur l'oceanologie hauturière et récifale dans le Pacifique Sud (1965–1994). In: *CR Colloque: Quelle recherche en environnement dans le Pacifique Sud: Bilans thématiques (compléments)*. MESR, Paris, 28–31 March 1995.

Salvat B. La biodiversité et son maintien en Polynésie française. In: *CR Colloque: Quelle recherche en environnement dans le Pacifique Sud: Bilans thématiques (compléments)*. MESR, Paris, 28–31 March 1995; 37–46.

Sherman K. *Large marine ecosystems global units for management: An ecological perspective*. ICES, Biol. Oceanogr. Com., 1/24,1990.

Tamburrino L. Aménagement intégré du littoral méditerranéen. *Rev. Int. Oceanogr. Med.* 1991; 101–104: 289–294.

Troadec J.P. Les pêches et les cultures marines face à la rareté de la ressource. In: *L'homme et les ressources halieutiques*. Ifremer Éd., Plouzané, 1989; 1–46.

Van Tilbeurgh V. What is meant by management of ecosystems? *Océans* 1992; 18(4): 479–486.

von Walhert G. and von Walhert H. Extensive rural mariculture in the tropics: Experiences and issues. In: *Aquaculture, a biotechnology in progress*. European Aquaculture Society, Bredene, Belgium, 1989; 97–102.

9

Conservation—the first form of development

9.1 WHY CONSERVATION?

The conservation of living resources as part of "sustainable development" was defined in 1980 by the International Union for the Conservation of Nature (IUCN) as the management of man's use of the biosphere in such a way that current generations gain maximum advantage from living resources while ensuring their sustainability in order to satisfy the requirements and aspirations of future generations. This organisation added that the objective can be attained while maintaining vital ecological processes, preserving biodiversity* and looking after the sustainable use of species and ecosystems.

This concept is fundamental, since it is accepted with a unanimity rarely shown by scientists, administrative or religious authorities and it is not synonymous with the integrity or maintenance of equilibrium. Conservation should be accepted as an aspect of sustainable development.

Natural ecosystems such as forests or coral reefs came into existence and underwent evolution long before the appearance of *Homo sapiens* and, as a species, humans did not play a part within their natural functioning. In contrast, the proliferation of the human species is such that their negative actions on the environment are often so harmful (see part B) that one can now imagine entire species or ecosystems being destroyed by their actions.

This standpoint is one where the ecological scientists have followed the aesthetes, hunters and pioneers who, together with the public powers and groups which administer national territories, have played a decisive role. Long before the political ecologists and other groups in society, they made the decision makers aware of the stakes involved for human society, i.e. the protection of nature, conservation of species and the environment in which they live—at least in developed countries (e.g. creation of Yellowstone Park in the USA in 1872). Elsewhere, events such as war, famine and the need for fuel have often meant that the environment takes second place. Profit is not the least of these reasons: the deforestation of Brazil or Indonesia is there to remind us; in the Philippines, 90% of the forest disappeared since the start of the twentieth century. Deforestation has also destroyed 70% of the forests in the United States and Europe over the last two centuries. In Madagascar, Ghana and the Ivory Coast, 70% of the forests have disappeared in a period of 20–30 years, leading to the disappearance of about 15 species of lemur in Madagascar (Levesque, 1994). The latter example clearly shows the relationship between habitat and the survival of a species.

This crisis of conscience is recent. Our hunting and farming ancestors only considered species in terms of their immediate usefulness; the disappearance of numerous hunted mammals and birds proves this. They had the excuse, which can no longer be used, of not having the knowledge.

The criteria for "sustainable development" today include reference to management in general and thus conservation in particular, but there is something else which must act as an anchor point for sustainable development. This involves Principle 15 or the "precautionary principle" of the Declaration of Rio (June, 1992): "Where there are threats of serious or irreversible damage, lack of full scientific certainty shall not be used as a reason for postponing cost-effective measures to prevent environmental degradation".

Sustainable development and the precautionary principle thus constitute the new paradigm within which the management of coastal waters can be written.

9.2 CHARACTERISTICS OF CONSERVATION IN COASTAL WATERS

The damage affecting marine ecosystems and the coastal zone is less visible but just as significant. This subject has given rise to many works: either specifically about the terrestrial shoreline (Lacaze, 1993; Pirazzoli, 1993) or both the terrestrial and marine parts (Mauvais, 1991; Aubert, 1994; Boudouresque, 1995).

Addessi (1993) showed that human recreational activities have had a long-term effect on intertidal communities of the rocky seabeds off San Diego, California, between 1971 and 1991; densities of all species have decreased in the most frequented areas. The hidden fauna has similarly changed. The only observed increase in density has been that of small gastropods. According to the FAO, in the Mediterranean, 15 species of mollusc and 3 species of crustacea have disappeared.

With regard to the management of coastal zones, the OECD (1993a) proposed integrated management through its recommendations for sustainable development. This involves management rather than development, but there are many common problems; a list of the main international conventions on the regulation and management of coastal zones can be found in this work and that of Boudouresque (1995).

This mainly economic international organisation reported that the management of the coastal zone cannot be envisaged without taking ecosystems as the starting point, i.e. by favouring the ecological approach and recommending reconciliation between ecology and the economy. It tackles very well the management problems arising from state politics, but raises the issue that the management of ecosystems is incompatible with governmental or administrative structures, which often block the system. Some cases of management which have got past these blockages and resulted in an integrated policy are reviewed in a special report (OECD, 1993b).

In coastal waters, conservation is approached in different ways, depending on whether it involves resources exploited by a fishery (fish, molluscs and crustaceans) or the protection of the environment for the benefit of marine life.

In biological terms, of the 71 phyla encompassing all forms of life, 43 are in the oceans, while only 28 are terrestrial. Five of these include 95% of marine species. The range of marine species is much more diverse than terrestrial ones; in one environment (Roscoff), there can be more than 2500 species of macrofauna and 600 species of algae (Lasserre, 1992).

Large groups, such as the plankton and filter feeders, have no equivalent on land. Coral reefs shelter a great diversity of species. Fish account for about 28,000 species, about three-quarters of which are marine; in numbers, they represent one-half of all vertebrate species.

The inventory of natural substances which come exclusively from the sea is far from being complete, but already products from sponges have been used in the treatment of leukaemia and coral fragments in bone transplants and dental repairs; as have compounds from the shark which are anti-infectious or active on the circulatory system. Antibacterial substances have been identified in a sponge originating from Tunisia. Macroalgae provide hypertension drugs, anti-inflammatories can be found in a mussel from New Zealand and obesity is treated with alginic acid (from marine algae)—as stated by Weber (1994) "... the oceans are an important new frontier for research". Castell (1988) reported the many benefits to the human organism of certain polyunsaturated fatty acids which are exclusively marine in origin.

Several medicines of marine origin have become essential: cephalosporines from a marine fungus, vidabirine (antiviral) and cybarine (antitumoral) extracted from sponges, as well as various anaesthetics. Marine animal toxins have been the subject of numerous studies in defining the power of anaesthetics but, overall, less than 1% of marine species have been studied in terms of their pharmacological potential.

9.3 COMPATIBILITY BETWEEN CONSERVATION AND MANAGEMENT

The strictly scientific reasons for conservation which prevailed at the start have now been replaced by less idealistic preoccupations:

- Two-thirds of our medicines contain active ingredients of natural origin; thus the conservation of the biological diversity of natural ecosystems is also conserving this formidable reservoir of future medicines which have been little exploited. The current step towards natural products and political ecology contributes to the reinforcement of this conservation requirement, especially in developed countries.
- Humanity must conserve its natural resources of food (fished species), primary materials (wood) and manage the biological inheritance without

which it cannot survive: domesticated plants and animals are products of wild species. The long-term survival of species depends on their genetic diversity (Barbault, 1993). Moreover, natural populations constitute a gene pool both for current domestic species and those which may exist in the future.

One example is Razon wheat, obtained by INRA* by hybridisation of a cultivated and a wild species, which is more resistant to parasites without loss of productivity. There is also the example of a marine littoral fish which has been domesticated for the past 15 years, the sea bass, *Dicentrarchus labrax*. A stock of broodstock from fish caught in the sea in the late 1970s has been built up at the Marine Biology Station at Sète (France); this stock has been regularly reinforced with animals reared on the same site, from individuals selected on the basis of their growth rate (mass selection), and also from fish caught in the wild. The quality of the descendants of this broodstock, in comparison with progeny of reared fish, has been recognised by the majority of fish farmers in the Mediterranean, and hundreds of millions of eggs have been provided to production units in almost every bass-producing country.

Conservation has become obligatory for mankind's survival, but it must be emphasised in ecological considerations arising from development, as we have seen above: intervention with natural ecosystems is thus inseparable from conservation. For more details about these areas, see Barbault (1993, pp. 215–217), Wilson (1988) and Purves *et al.* (1994, pp. 1126–1145).

An original example of increasing value by conservation can be given. In Martinique, the explosion of Mount Pelée in 1902 sank all the ships anchored in Saint-Pierre Bay; an experienced diver, Mr Météry, patiently searched for and located the many wrecks lying at depths of 15–85 m. The interest of divers in these historic remains has resulted in the creation of numerous diving clubs and the installation of a base for "L'union des centres de plein air" (UCPA) which can welcome and accommodate more than 80 divers on site. A tourist submarine, the "Mobilis", is moored at Saint-Pierre to allow non-divers to discover the most beautiful wrecks. These installations have obviously led to investment, created jobs and also increased prices for rented accommodation, but here is a form of increased value which is compatible with sustainable development. Now the protection of remarkably beautiful coral reef areas is envisaged: their conservation is not incompatible with their ultimate increase in value by subaquatic tourism.

The compatibility between conservation and management in the domain of coastal fisheries and aquaculture has been the subject for recent studies and recommendations by major international institutions. In 1997, for example, the FAO* published, in the series "Technical directives for a responsible fishery", a pamphlet entitled: *The precautionary approach applied to capture fisheries and to the introduction of species.* This was followed in 1998, in the series "FAO Guidelines", by a report entitled *Integrated coastal area management and agriculture forestry and fisheries*, which devoted an entire section, to the "Integration of fisheries into coastal area management". As stated by the Editor: "These guidelines examine issues specific to the agriculture, forestry and fisheries sectors, and suggest the processes, information

1

Visualisation of water stratification. Warm fresh spring water flows into a cold, turbid, salt water pond. The freshwater lies above the salt water and their different turbidities render stratification visible.

(*Fondame spring near Salses in the Perpignan region*)

2

Plastic refuse is abundant in coastal waters. It is very light and is moved about by swell and currents in the shallow waters, finally being deposited on the shore.

(*Plage de la Corniche, Sète (Hérault), depth 4 m*)

In shallow waters (−3 m), several tens of metres from the beach, soft seabeds are deserted. Any hard substrate, such as the rock shown here, constitutes an attachment substrate for small mussels. Some small, light-coloured fish fry stand out against this dark background. This scenario is reminiscent of an oasis.

(*Plage de la Corniche, Sète*)

Even in sheltered waters, soft substrates are not very favourable to marine life: on this seabed in Nouméa (New Caledonia), a semi-buried discarded beer can provides support for these queen scallops. Their shells are in turn colonised by invertebrates—the lack of suitable habitats is demonstrated.

When larger objects are placed on sandy seabeds, such as this cement pipe (1.9 m in diameter, 3 m long). mussel biomass and numbers of fish fry increase.

(*Plage de Marseillan* (*Hérault*), *depth 10 m*)

5

6

Within the same pipe, conger eels, mussels and small fish cohabit at very high population densities. However, the surrounding sandy seabeds remain like deserts.

Another inhabitant of the same habitat, the forkbeard *Phycis blennoides*. (*Marseillan Plage, depth 16 m*)

8

The lobster *Homarus vulgaris* also uses this hiding place.
(*Marseillan Plage, depth 13 m*)

9

Another crustacean often abundant on mussels is *Scyllarus arctus.*

(*Marseillan Plage, depth 16 m*)

10

At other sites, the flat oyster *Ostrea edulis* constitutes the dominant population of filter feeders.

(*Jetty at Sète harbour, depth 8 m*)

11

This 50 cm diameter buoy submerged at a depth of 5 m in May, was colonised by flat oyster (*Ostrea edulis*) spat at the beginning of July.

(*Shellfish culture zone at Agde* (*Hérault*), *photograph taken in August*)

12

The same buoy photographed 10 months later. Mussels have now also colonised the substrate. The oysters are smothered by this competition for attachment and access to open water which is their source of food and oxygen. The populations succeed one another in time and are in competition for the same substrate.

Even in tropical waters, all objects which are submerged in mid-water, such as this buoy and especially its mooring, are covered in flora and fauna which in turn attract fish.

(*Caribbean coast of Martinique, Saint Pierre, depth 3 m*)

13

Coral reefs provide habitats for juvenile fish in tropical waters.

(*Nouméa lagoon, New Caledonia, depth 6 m*)

14

15

This car wreck (front wheel axle unit) attracts juvenile fish in the absence of a coral reef on this sandy seabed.

(*Caribbean coast of Martinique, Fond Corré, depth 7 m*)

Coral reefs are frequented by the multitude of adult fish (here, soldier fish).

(*Caribbean coast of Martinique, Prêcheur, depth 12 m*)

17

Plankton harvesting equipment, filter bag containing the harvest (copepods), achieved in a few hours from a productive salt marsh.

18

Artificial reef frequented by the sea bass, *Dicentrarchus labrax*.
(*Photograph by Gy F. Bachet.*)

Artificial reefs are often made up of very large structures, such as this 158 m^3 module (the diver shows the scale).

(*Photograph by Gy F. Bachet.*)

When waters are rich in nourishing seston, the biomass and production of mussels in suspended culture become enormous—each 10 m long rope produces more than 100 kg in nine months.

(*Shellfish culture zone at Agde, depth 22 m*)

requirement, policy directions, planning tools and possible interventions that are necessary for integrated coastal area management".

9.4 BIODIVERSITY AND BIOLOGICAL "INVASIONS"

Biodiversity, or biological diversity, represents the variety and variability of living organisms and the ecological complexes to which they belong. This term also refers to the assembly of biological and genetic resources, i.e. our biological capital.

This word has become fashionable and the biodiversity of the marine coast has not escaped this trend. As for biological capital, the importance of marine biodiversity does not need to be emphasised: the disappearance of one or several species corresponds to a loss in biological capital. On the other hand, according to the theory of ecosystems, the most stable ecosystems show the greatest biological diversity; this is the case for tropical rain forests, coral reefs or mangrove swamps. Young ecosystems are more productive, but their biodiversity is lower. This is also true for artificial ecosystems, such as sewage treatment ponds, to which we shall return in Chapter 11.

The other problem, that of the function of biodiversity, is far from being resolved: the majority of ecologists think along the same lines as Heip (1995), that any reduction in biological diversity will seriously damage the ability of ecosystems to maintain a viable biosphere when physical changes are occurring. On the large scale, biodiversity corresponds to a reservoir of potential adaptations to physical changes, for example climatic ones. In ecosystems with great biodiversity, there is a certain amount of redundancy, as one function can be accomplished by different groups of species. If conditions change (temperature, oxygen levels, etc.) some species would no longer be effective, but others would be more so.

The fact that the role of biodiversity is not fully understood must not impede vigilance over its protection in relation to the "precautionary principle" decided upon by the Rio Conference in 1992.

The preservation *in situ* of the biodiversity at many thousands of sites worldwide, representing 5 million km^2, costs on average €44 per km^2, but this varies from €11 in Tanzania to €2740 in France.

Marine "invasions" result from transport, and then the expansion of species from one region to another. These transfers are not always voluntary and are the consequence of the development of international transport by air and sea (ballast water in ships). In a few hours, aquatic organisms are transported from one side of the world to the other, these movements being totally impossible without man's intervention. Carlton (1995) cited several key examples, such as the invasion of the American ctenophore, *Mnemiopsis leidy*, which overran the Azov Sea and the Black Sea. It destroyed anchovy stocks which, in the past, had been the basis of a very productive fishery in Russia, Ukraine, Romania and Turkey. Similarly, the cholera virus was passed from South America to the Gulf of Mexico by ballast water. Thus several thousand species move from place to place on the planet every day.

The source of these invasions is the shipping industry, with the transport of ballast water, to which can also be attributed the explosion of the phenomena of toxic algal blooms. Other factors responsible for these invasions are the transport of many aquarium species and, lastly, aquaculture "which only causes a slight flux of prawns, bivalve molluscs and fin fish" (Carlton, 1995).

Food chains are altered and very diverse communities are often replaced by one single species. However, certain marine cultures would not exist without these transfers. In Europe, the Japanese oyster *Crassostrea gigas*, and the Philippine clam, *Ruditapes semidecussatus*, have become the main cultured species (in excess of 200,000 tonnes), since indigenous species have been decimated by disease. Thus, from the point of view of the ecologist or farmer, the balance is different. On land, the majority of cultivated species in Europe are of exotic origin.

Health authorities limit the transfer of species, but with regard to ballast waters, an *ad hoc* group of the International Maritime Organisation is carrying out investigations, i.e. the affair is adjourned *sine die*.

9.5 THE ECONOMIC VALUE OF NATURE

As stated by Purves *et al.* (1994, p 1142), natural ecosystems fulfil numerous vital functions without which we could not survive. Ecosystems absorb carbon dioxide and other gases, produce oxygen, purify water, regulate water courses and constitute landscaped and recreational space. The economic value of these functions is difficult to establish, but some estimates have been carried out. Sustainable development, which wishes to reconcile environment and development, must thus attribute an economic value to the protection of the environment, seen as a development tool.

Firstly, a value for direct consumption must be attributed to the products of hunting, fishing or gathering; in many tropical countries, more than three-quarters of food proteins come from these activities. The productive value corresponds to the turnover of exploitation by forestry or agriculture or fisheries. The recreational or tourist value comes from the turnover created by leisure activities (bathing, sailing, etc.). The ecological value is more difficult to estimate in economic terms.

There are, nevertheless, some examples. The cost of a system which would combine the purification of waste water and fish production over a hectare of the tropical area of Louisiana is estimated at €153,000. The storing and purification of water, soil fixing and forest fertilisation carried out by a hectare of vegetation in Georgia (United States) has been evaluated at €1526 per year (Purves *et al.*, 1994).

It is interesting to compare these figures with those for the forest cited by the French Revenue (7-10-94, p 7): one hectare of forest produces on average (over a long period) 9 m^3 of wood/year. Two-thirds of this wood is from trunks, which are sold at a price of €30–38 per m^3. Branches fetch between €3–6 per m^3. Fir is a little more expensive (€38–45) and oak €60–68; depending on the age of the trees, a hectare would be worth between €4500 and 7600. The annual ecological value is directly related to the value of the land.

In the United States, the OECD (1993a) stated that 87% of the value of commercial or sport fisheries depended on species whose life cycle was either totally or partially within aquatic habitats situated close to coasts, and that these activities represent more than \$10 billion. The same organisation reported that an improvement in water quality of between 10 and 20% in Chesapeake Bay improved pleasure boating, fishing and bathing activities, generating supplementary income estimated at \$10–100 million. Several publishers have produced books on this theme.

In Europe, a large number of ecological and management studies have been carried out on the Wadden Sea. This sea is 70 km long and 30 km wide and is connected to the North Sea, from which it is separated by the low islands of northern Holland. The waves there are less than 3 m high and the tide varies from 1–1.5 m, but this is sufficient to move 2 billion m^3 of water per tide, with currents reaching 220 cm/s. It is the largest intertidal zone in Europe. Its waters are turbid (up to 50 mg/l of suspended matter); this coastal sea has been partially managed (dykes, salt marshes, marinas) and around 4000 ha of its area have been used for mussel rearing on the seabed. Its multiple uses make its natural value in terms of a breeding ground for birds and a nursery for North Sea fisheries immense (Essink, 1992), but this author does not give any precise figures.

In coastal waters, a single resource may be exploited in many ways and this is true for fish. For a long time, the sale price of fish caught during an angling championship has been compared with that of the commercial price for the same species caught by fishermen (Barnabé, 1974): the fishing effort during a championship is known precisely, as are the expenses incurred on the fishing trip (accommodation, food, tackle, etc.). A fish caught by the angler costs around four times more than the same fish bought at the fishmonger, but it is very difficult to put a price on leisure activities.

In any case, using the professional fisherman is not the best way to estimate the value of a fish. One method has been used for the grouper in Port-Cros National Park: "Price of a caught grouper, sold by a professional fisherman: 15 kg × €9.9 = €150. Price of a grouper living in its environment, observed by 500 divers per year: €75 × 100 per dive = €7500. Average life expectancy of an adult grouper: 20 years. Total economic value of an adult grouper: 50,000 × 20 years = €153,000. This comparative approach uses simple integration of the cost of diving. Indirect local economic consequences can also be added (accommodation, restaurants etc.)". The grouper's conservation thus has an irrefutable economic significance.

A global estimate of the services provided to man by nature has been carried out by American and European ecologists and economists. They have applied the classic rules for evaluating the costs and benefits of services provided by different ecosystems—provision of drinking water, role in climate regulation, protection against erosion, pollination of cereals, etc.

It is obviously not possible to make estimates for all the reserves involved, but this assessment speaks for itself. A monetary estimate is given:

7,423	billion € for oceanic ecosystems
11,143	billion € for littoral ecosystems
4,330	billion € for wetland ecosystems

3,353	billion € for tropical rainforest ecosystems
762	billion € for other forest ecosystems
1.5	billion € for lake and river ecosystems
1	billion € for prairie and farmed ecosystems

The total comes to €29,560 billion per year, or double the world's annual Gross National Product (GNP). The services of aquatic ecosystems and those of wetland areas represent €24,405 billion, or 83% of the total. This is an indirect evaluation, but it demonstrates the fundamental impact of ecosystem function on human life.

With regard to coastal seas and other resources such as fisheries, an indirect measure exists in the form of reference to, for example, the price of real estate along the shore. There are also intangible goods, such as the "unimpeded" view over a seascape, which has already given rise to legal controversies. In Canada and France, the owners of properties overlooking the sea have taken out proceedings preventing the installation of aquaculture units on the sea shore (rearing cages, only the superstructure of which is visible above the water). Similar developments have also been objected to in Scotland.

The economic value of nature is thus on the rise again.

9.6 TYPES OF CONSERVATION AND PROTECTION

Nature conservation can take many different forms: it can integrate both the organisms and the environment in which they live (biotope) or a larger entity on which they depend for their survival (ecosystem), or in contrast be reduced to the conservation of one species or even its genes in a freezer.

Species may be preserved outside their natural environment, for example in zoos, botanical gardens, aquaria, conservatories of domestic species or gene banks, where reproductive products are conserved by cryopreservation in liquid nitrogen at −196°C. In France there is a bureau of genetic resources, located at the National Museum of Natural History in Paris which is concerned with the conservation of animal and plant species, but does not operate in the marine domain.

The best known form of *in situ* preservation is that carried out in well-defined geographical zones such as parks and reserves. In fact, the loss of biodiversity (i.e. of species numbers) is most often the result of the degradation of the environment in which these species live. Conservation of biodiversity is thus included in the framework of sustainable development (responding to current requirements without affecting the capabilities for future generations).

In Europe, this protection is based on the setting up of regulations or prohibitions and very rarely on true management. This is more evident in the physical rehabilitation of beaches or threatened coastlines, or in the construction of harbours; with the exception of the installation of some artificial reefs or the very limited reintroduction of *Posidonia* species, there are few operations whose goal is the restoration of marine ecosystems. The situation is very different elsewhere and, in the following chapters, we shall provide numerous examples of management techniques which are working towards conservation and protection.

Legal regulations thus remain the main tool for conservation of the marine environment. Details of the arsenal of laws put in place are not given here, since they vary from country to country; examples reported by the OECD (1993a, b), Weber (1994) or numerous publications by UNESCO* (UNESCO, 1994, for example) can be consulted. Boudouresque (1995) described in detail the types of legal protection for species and biotopes in the Mediterranean, since the protection of a species *in situ* is often achieved by protecting its habitat; this is the idea of the "Habitats Directive" of the European Union, but it only concerns a few marine species.

Various forms of nature protection and conservation can be distinguished and protected sites constitute the most common form. These respond to four types of objective (Boudouresque, 1995):

- to establish reserves for threatened species or ecosystems;
- to have undisturbed areas available for education and/or research;
- to build up overexploited populations of species with a view to exploiting them *in situ* or in adjacent areas;
- to organise areas suitable for underwater tourism.

According to the same author, there are five levels of protection in protected marine sites:

Level 1: All forms of human activity are authorised, by legislation; it is thus possible to predict the general protection of a threatened species (e.g. *Posidonia*).
Level 2: Professional fishing is allowed with free access to the area, with only certain types of sport-fishing prohibited (most often underwater spear-fishing by divers).
Level 3: Professional fishing allowed, but with restrictions; access to area is controlled; all forms of sport fishing prohibited.
Level 4: Professional and sport fishing prohibited; controlled access.
Level 5: All human activities (including presence) prohibited; this is the law of "sanctuary".

Prohibitions concerning professional fishing, sport angling and fishing using an underwater spear-gun or all collection of animals or plants are well understood, but diving, bathing, anchoring or any access by boats can also be prohibited. Often the success of protected areas is such that access has to be limited: in the Mèdes islands (Costa Brava), as in Port-Cros National Park (French Mediterranean), the number of divers has had to be limited at the most "spectacular" sites. The number of authorised divers is around 500/day at smaller sites.

These regulations are put in place within the diverse administrative structures which manage protected sites. In France, the main structures are nature reserves, national parks and regional nature reserves. The "Conservatoire du littoral" aims to control linear urbanisation of the land bordering the sea and has, for this purpose, acquired around 10% of it, i.e. 500 km of the total 5500 km of French coast.

In France, there are also regulations aimed at areas requiring special protection:

- Fishing reserves are zones where professional fishing is temporarily prohibited for five years in order to allow the recovery of exploited fish species.
- Fishing establishments are permitted areas (for groups or individuals) to be used for a specific activity (installation of reefs, management, aquaculture, etc.).
- Certain specific regulations have as their goal the control of taking of individual adults outside spawning periods and the protection of juveniles. Others are used to stabilise the size of certain populations. Once again we are faced with the problems of fisheries management, to which we shall return. Other practices, such as rat extermination, constitute selective destruction, which could also be advocated in the marine domain (elimination of the conger eel in certain places, for example).

The operation of geographically distinct zones is not the only management tool: utility taxes and fees, the principle of polluter pays, the attribution of property rights and the prohibition of sale and even transport of species constitute economic instruments which are often much better than zonation.

9.7 RESERVES AND NATURAL PARKS

9.7.1 World-scale reserves

These are the best known forms of nature conservation.

Under the aegis of UNESCO, an international network of reserves has been set up to "improve the wellbeing of human populations by the rational use of natural or modified ecosystems". Within the framework of the large-scale programme "Man and Biosphere", the programme of reserves in the biosphere led to the constitution (by 1994) of 324 reserves in 82 countries, varying in size from 500 hectares to 70 million hectares. The World Heritage Convention, signed by 136 countries, established a list of 411 sites of universal value, of which 90 are natural and 16 both natural and cultural. In the marine environment, the Great Barrier Reef must be mentioned, which extends for over 2000 km along the north-east coast of Australia and covers 348,000 km^2, nearly two-thirds of the area of France. The Great Barrier Reef Marine Park is managed by the Great Barrier Reef Marine Park Authority, located in Townsville; the regulations are not the same in all areas of the park, but this institution, which has been functioning since 1975, has numerous achievements to its credit. The conservation of the Wadden Sea can also be cited (divided between Holland, Germany and Denmark), which has given rise to numerous scientific studies.

The creation of a vast marine park in the western Mediterranean has been proposed by representatives of international organisations (EU, UNESCO, ICES) and by French, Italian and Monagasque experts, following a colloquium on cross-border marine parks organised in Nice in 1992. This is now an international marine park stretching from Corsica to the shores of Provence, to Monaco and Liguria, as well as an international park around the Strait of Bonifacio (between Corsica and Sardinia).

With regard to the open sea, we have seen that the United Nations Convention on the rights of the sea ("UNCLOS") was put into action on 16 November 1994 and should bring order to political discussions about fisheries. Not everything is regulated, but this concerns a legal tool which required 27 years to be put together and only came into play 12 years after it was signed! International waters are becoming the common property of humanity.

9.7.2 Conservation and reserve size

The size of a reserve plays a determining role in its effectiveness for several reasons. Studies of island ecology have shown that the diversity of populations on an island are a function of its surface area. On the other hand, Barbault (1993) reported that in order to maintain sufficient genetic diversity within a population, it must consist of at least 500 individuals. The other aspect is linked to the peripheral zone or edge of a reserve: the larger the reserve, the smaller its peripheral zone, in relative terms. The smaller it becomes, the more it can be assimilated into a peripheral zone.

Parks and reserves associate the idea of wide open spaces with protection and are not concerned with small areas. Very roughly, flora and fauna are associated with their environment; the entirety in play here is the ecosystem*. In Europe, these zones are still modest in size (several tens of hectares to several hundreds of km^2), but in very large, sparsely-populated countries such as Australia, 20% of the land's surface is involved and marine coastal zones are not forgotten—the Great Barrier Reef National Park extends to more than 2000 km in length.

In central America, Costa Rica has transformed 25% of its land into nature reserves. The American chemicals manufacturer, Merck, puts aside funds to explore the pharmaceutical or agrochemical potential of natural substances (in relation to the Convention on biodiversity signed in 1992 in Rio, which declared the principle of national sovereignty over biological resources). With regard to marine waters, UNESCO launched an international programme on marine biodiversity and a coastal programme ("COMAR": The Coastal Marine Programme) explained by Steyaert and Suzyumov (1992).

The situation is very different in the Mediterranean where Boudouresque (1995) calculated that all the protected coastal sites represented only 2420 km^2 or 1.9% of depths less than 50 m. If one excludes what are called "paper parks", i.e. zones which are theoretically protected but without real protection, this figure falls to 0.09% of depths less than 50 m.

International organisations can label these protected spaces: the World Heritage of UNESCO or the Biosphere Reserves of the same organisation (320 reserves in 81 countries) are the best known. The world's oceans (except EEZs) and the Antarctic were the first geographical areas to be officially declared as the common heritage of humanity, implying an equable sharing between nations and the emergence of responsibility for the planet.

It is regrettable that while international treaties exist for the protection of whales or for the classification of numerous marine and terrestrial zones, there is no treaty related to the unique protection of coastal waters. Even at the Conference in Rio in 1992, Agenda 21 left the coastal zone under the national sovereignty of each country.

9.8 CONDITIONS FOR THE SUCCESS OF PROTECTED ZONES

The OECD (1993a) gave a detailed report of the conditions for success of a protected marine zone:

- centralisation of power (one single organisation manages the zone);
- consultation (the organisation consults all parties involved);
- research and monitoring, organised by the leader of the organisation;
- capacity for effective control;
- own budget, to prevent the favouring of lucrative activities to the detriment of habitat protection.

It must be emphasised that very often the regulations adopted are not put into action as there is a lack of power to enforce them on land, and even less so in the sea. Thus, the Galapagos Marine Park (50,000 km^2) is often subjected to invasions by Japanese fishing boats (Constant, 1994).

9.9 SOME ASPECTS OF THE RESERVE EFFECT

The beneficial role of reserves, parks and all other forms of nature conservation cannot be underestimated. Genetic and ecological conservation mean that it is only in parks and reserves that one can find species which do not exist elsewhere; it may be possible for populations which survive in these protected areas to regain their place within the ecosystems from which they came. In addition to their utilitarian role, it is here uniquely that they can still be observed for educational or aesthetic purposes; at the dawn of a leisure-oriented civilisation, this may often be more important than their economic significance.

The different aspects of the reserve effect have been reviewed by Boudouresque (1995):

- The reappearance of rare species, or an increase in their abundance, is the best known. The population of groupers at Port-Cros was estimated at 91 individuals in 1994, 160 in 1996 and 300 in 1999, while they have disappeared from neighbouring areas, which is also true for many other species. In Monaco, we have noted (Barnabé, 1995) that the reserve is an order of magnitude richer in the number of individuals than the adjoining area outside the reserve.

- There is a balanced structure of populations, with an abundance of large adults. In all reserves, large fish are more abundant than outside them. In Monaco, the biomass of fish inside the reserve is two orders of magnitude greater than that of a similar area outside (due to differences in both size and abundance) (Barnabé, 1995). This characteristic allows reproduction in fish which change sex (groupers, sparids) in the middle of their lives, and also increases the fecundity of others; fecundity does not increase linearly with length, but in relation to weight. This also avoids genetic drift, which is noted in overexploited species (the most precocious spawners are selected by fishing since the largest fish are the most caught).
- Buffering effect: seasonal variations are less marked in reserves.
- Cascade effect: the abundance of certain fish, for example large predators, will cause a decrease in the number of their herbivorous invertebrate prey, resulting in less browsing of algal mats. Thus the setting up of a reserve has a very marked effect on the entire ecosystem.
- The most important impact concerns the reproduction of marine species; as we have seen, the majority of them have (Chapter 3) a very large reproductive potential and their biological cycle includes planktonic larval stages which, carried passively by currents, serve the function of dissemination to colonise often distant places. The lobster has many planktonic stages and the planktonic phase of its biological cycle lasts a year; these larvae can find themselves far from their starting point when they are ready to position themselves on the seabed. Therefore the impact of a marine reserve is not only limited to its immediate environment.

Many species have reproductive behaviour which cannot be expressed in fished areas. The displays of the sea bream, sexual behaviour of a group of groupers and the reproduction very close to land of many other species (some laying their eggs in the gravel on the beach), result in gatherings which are favoured targets for fishermen who preferentially search for these collections of adult (hence large-sized) individuals in a restricted coastal area. Other examples include invertebrates (molluscs and crustaceans). Some edible gastropods form masses weighing several tens of kilograms at the time of spawning and fishermen actively seek these out. It is estimated elsewhere that three-quarters of fished species spend at least part of their life cycle in the coastal zone.

Allowing reproduction thus favours recruitment, even if other factors render it uncertain; as stated by Boudouresque (1995), a protected space will "export" eggs and benefit commercial fisheries. These eggs and the later larval stages of exploited species are almost the only ones not to be subjected to capture, because of their small size. Above a size of only a few centimetres, fisheries activity damages the juveniles via various means: the cod-ends of trawls have a mesh which changes shape under the influence of the boat's speed, closing (to form a lozenge shape which is more or less "flattened"); these meshes can get clogged during the trawl and thus limit the escape of smaller fish. In certain areas, the capture in limited quantities of small commercial species such as *Atherines*, or smelt, is used as justification for nets with small mesh sizes*, which also catch juveniles of all other species.

There are many examples of this premature harvesting: thus in the Monaco and Nice regions, there are still some fishermen who specialise in the capture of larvae of fish species, which are sold under the name of "poutine". In Spain, elvers (glass eels) are sold for consumption, but they are very expensive because they have become very rare as a result of overfishing. A substitute made from fish paste has appeared—even the dark eyes of the elver are reproduced—and they taste excellent.

Boudouresque also reported the example of the Tabarca reserve in Spain, created in 1986, where more high-value fish are caught in the peripheral zone of the reserve than were previously caught in the collective reserve + peripheral zone: the catches of sea bream increased threefold between 1984 and 1989 and those of dentex, tenfold.

9.10 AN ARTEFACT OF THE RESERVE EFFECT IN COASTAL WATERS: CONCENTRATION

We should add to the above list an unusual and unexpected effect which we have noticed at the many reserves which we have visited. The starting point is the chain effect: the presence of large predators obviously is owing to the absence of catches by fisheries in protected zones and also from the fact that their prey (small fish or invertebrates) are not fished and are thus more abundant than outside the reserve.

How can a reserve of limited surface area concentrate, and especially feed, the significant gatherings of fish which are seen there? Here again, the answer lies in the plankton, with planktonophagic fishes at the base of this concentration effect. Since they are not fished in the reserve, they are abundant and take their food from the plankton, the only resource not overexploited by man. This food is largely produced, but not consumed, elsewhere as there is a lack of fish outside the reserve, and is carried into the reserve by currents. The biomass of planktonophages is thus not exclusively linked to local trophic resources; there can be over-concentration linked both to the exportation of planktonic production not consumed in zones outside the reserve and the absence of fishing of these planktonophages in the reserve. At a higher trophic level, predators of these small fish are concentrated in the reserve for feeding (active concentration) and can stay there since they are not fished (passive concentration).

Reserves thus do not present a reference to the original natural state, but exaggerate the differences between the protected and unprotected zones. Concentrations of fauna found there cannot serve as reference points to the natural state prior to predation by humans: fauna richness in reserves results mainly from the poverty in fauna of the contiguous zones. While protected marine zones cannot act as reference points to a natural state, this richness clearly emphasises the impact of human activities outside protected zones.

9.11 MANAGEMENT OF FISHERIES: AN AMBIGUOUS CASE OF CONSERVATION

We have dealt with fisheries in Chapter 7 but, as stated by Troadec (1989), "fisheries management corresponds to conservation" since it allows natural processes to maintain populations at their optimum yield. We shall limit ourselves to this aspect here.

The regulation of fisheries based on the setting of quotas or TAC (total allowable catch) is not true management, but it should be noted that there has been an evolution towards limitations allowing "sustainable exploitation"; this is the case for the krill in the Antarctic, catches of which are limited to 1.5 million tonnes. Considering sustainable exploitation for resources whose variability is very large is rather tricky, but constitutes a measure which can prevent the exhaustion of stocks. In Australia and Israel, the government collects taxes representing between 11 and 60% of the catch value for the use of fishing zones. The coastal and marine observatory of the English Channel and the North Sea has proposed the creation of reserves for marine life which could be established around artificial reefs, as a measure against the fall in price of certain species. This would create zones populated with reproducers likely to increase recruitment. For the fishing industry, the problem of stocks must be resolved by reducing the over-capacity of the European fishing fleet. In these terms, overexploitation concerning biological resources has impacts in both economic and social terms which are very different from the simple protection of nature.

Fisheries management must reconcile two objectives: the conservation and the sustainable exploitation of resources. This is not simple in itself because of the variability of recruitment (Chapter 7), but when these resources are common to several countries or when stocks overlap (i.e. populations which, during their lives, frequent economic zones belonging to different countries or where the laws are different), this becomes problematical and a source of conflict (France–Spain for the tuna, Europe–Canada for the turbot, etc.).

In Japan, fisheries co-operatives are, in a way, the owners of the coastal fishing zone which they exploit (see Doumenge, 1990). This implies not only the common property rights over the fish, but also an obligation to ensure its survival and management. We shall see in the following chapters that this conscious responsibility linked to common ownership has resulted in true management. From an initial fishing activity, using management tools, the Japanese have succeeded in integrated management of the resource. The situation is similar in Italy, where the common ownership of lagoons ("valli") has led to their management by fisheries co-operatives.

In contrast, an attempt to transfer this technique to France at a comparable site (Étang de Leucate) ended in failure because of the lack of collective responsibility displayed by the fishermen who retained individual behaviour. Development and management of fisheries must therefore become part of the culture before it can become a reality.

The FAO conference in 1995 (FAO, 1997) adopted an International Code of Conduct prescribing the use of a precautionary approach to all forms of fishery and to all aquatic systems, whatever their jurisdiction, thus recognising that the majority of problems suffered by fisheries arise from a lack of foresight in their management regime, taking into account the degree of uncertainty which characterises this area (see also Section 7.2).

One of the main problems is firstly to establish the social value of a fishery within society and to divide the resources between its different uses and users (commercial fishing and sport fishing, for example). The problem of limiting access to resources remains central (see Chapter 7), but the variability and uncertainty in fisheries dynamics must be taken into account in their management, which becomes a precautionary approach.

Despite these contingencies, the route towards better fisheries management is clear; it requires three, apparently simple, conditions:

- a thorough knowledge of the fisheries and the stocks which they exploit;
- agreement on a well-defined policy;
- enforcement of fisheries regulations.

The perspectives described by the FAO for the future of fisheries have been reported earlier (Sections 7.2–7.5).

9.12 CONSERVATION ASPECTS OF AQUACULTURE

Many tropical fish are bred in captivity to satisfy the demand for freshwater aquarists. In Florida, this type of rearing in freshwater involves around 3000 farms, with an annual turnover of $70 million. Some marine species are also reared, but the majority of seawater aquaria are still stocked with animals taken from the natural environment.

An interesting case of rehabilitation using marine parks has happened in Chesapeake Bay. The Maryland Oyster Action Plan made provision for the construction of several oyster rehabilitation zones by using aquaculture to co-ordinate the requirements of multiple users (McCoy, 1994). The story is similar for the pearl oyster in Japan, French Polynesia and Australia, and pearls are now produced almost exclusively from cultured pearl oysters.

Many aquatic species used in the laboratory come from farms and many more could in the future, with the use of similar techniques.

With regard to over-fished species, or those in the process of disappearing, fisheries based on aquaculture constitute a solution which is already operational in some places.

9.13 THE MANY DIFFERENT ROUTES FOR DEVELOPMENT

The many forms of development, from the most draconian imposition of complete reserve status to unlimited human exploitation, were described during a colloquium devoted to the Wadden Sea (Netherlands Institute for Sea Research, 1992).

- The ecologists believed that it was impossible to return the reserve to its virgin natural state, and recommended "a natural reserve with zero usage", i.e. prohibition of human activity. This is all the more paradoxical since dyke building has been carried out on this sea for at least a thousand years, there are numerous salt marshes and, in short, its development and use by man has long since transformed the natural environment, and yet they are trying to take it away.
- Others have remarked that, despite the effective preservation of almost all of this sea by placing it in a reserve, these measures limit its degradation but do not have a net positive effect on its restoration or return it to its previous state. This was described by Wolf (1992) who emphasised that the state of the Wadden Sea is largely influenced by that of contiguous ecosystems, notably the North Sea.
- Meijer and Holsink (1992) envisaged nature conservation within the framework of leisure usage in this same sea, and noted that 57,282 pleasure boats were counted there in 1988. For Bakker and Coljjn (1992), who constitute "the voice of the state of Holland", the objective to be attained is an ecosystem capable of being used and managed in a sustainable way; to do this, they proposed a management strategy based on an ecological process for evaluating the state of the ecosystem (AMOEBA: Abstract Method for Overall Examination of the Biological Ambience).
- De Vlas (1992) envisaged the maximal utilisation of the Wadden Sea by man within the framework of sustainable development. He proposed a complete management catalogue from bird and marine mammal watching and other tourist and leisure activities to classic fisheries, development of bait collection and mussel culture, introduction of new species, use of wind power or sand extraction, (Fig. 9.1). These proposals can no doubt be criticised, but they reconcile ecology and economics, in contrast with nature conservation projects which depend on grants from the community to achieve their objectives and thus are not considered as true partners by the public powers.

It is understandable that the final recommendations of such a symposium correspond to a compromise, the main components being:

- zonation (definition of zones with different levels of human activity);
- the creation of nature reserves where biotic and abiotic processes carry on without human intervention (zero use nature reserve);
- limitation of fishery activities likely to destroy habitats, such as catching of worms buried in the sediment, and also collection of cockles or mussels;
- limitation of hunting or its prohibition (hunting from a boat);
- limitation of pleasure boating to its current level;
- adaptation of the tourist industry in line with nature conservation objectives (speed limits for boats, etc.);
- reduction by 90% of phosphorus and 50% of nitrogenous inputs, as well as more detailed studies of ecotoxicological aspects.

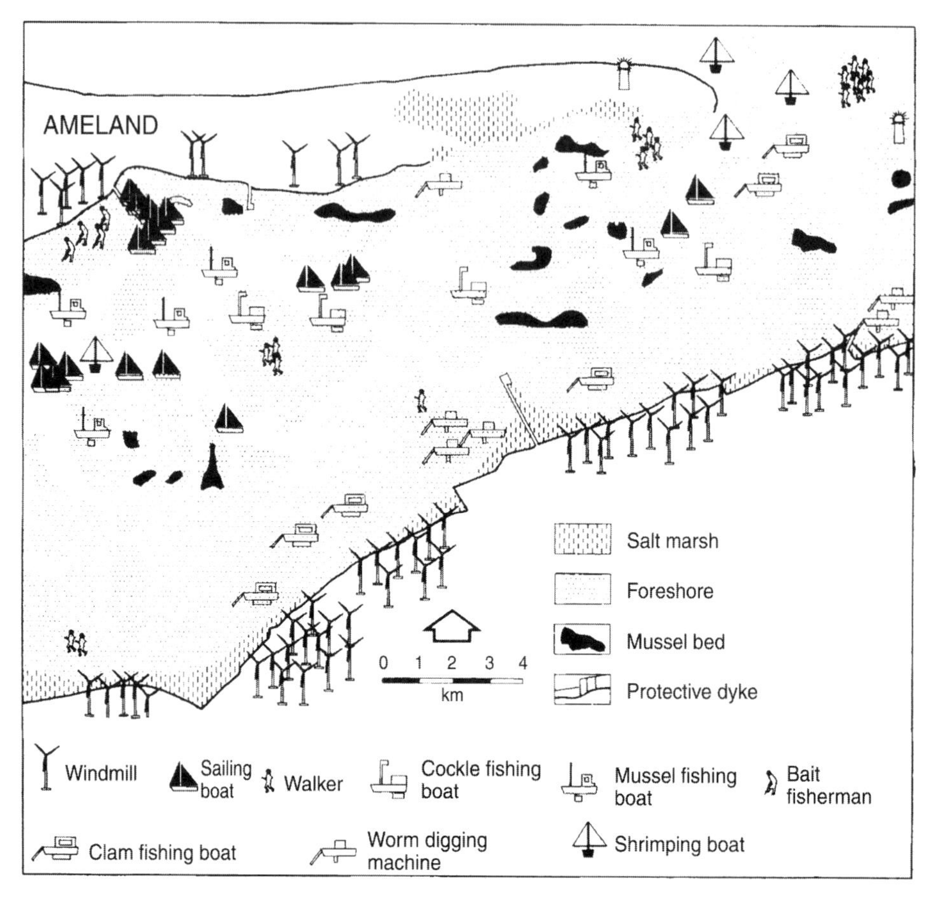

Fig. 9.1. The Wadden Sea, Ameland region, in a situation of maximal sustainable development.

9.14 CONCLUSIONS

Nature conservation, in all its different forms and guises and whatever its aim, is the first form of management; it effectively implies a human action which is an act of voluntary protection, able to be modified in both space and time.

Although it has its imperfections and deficiencies, the creation of protected marine areas, managed in an ecological fashion, has provided a demonstration with consistent results. However, these results are only part of the story. The degradation of the marine or terrestrial environment continues, as does the disappearance of species, i.e. the loss of biodiversity constitutes the evidence and things are getting worse. According to various estimations, between 5 and 20% of species currently in existence will have disappeared from the Earth's surface and the oceans during the last decade of the twentieth century.

Protected zones are all the more effective when they are very large but, despite this, there is no truly isolated area, since all ecosystems are interdependent on each other. Thus the state of the Wadden Sea, itself protected, always depends on that of the North Sea, which is itself influenced by imports of pollutants from the Rhine, or exchanges with the English Channel. In biological terms, we have seen how populations in protected marine areas, far from representing an "original" state, can also be influenced by the forms of exploitation in the surrounding area. Interdependence thus requires the generalised protection of the environment.

Reports based on ecological monitoring have provided the basis for current recommendations by scientists. Some have resulted in the creation of protected areas (main conservation measure). In this way, prohibitions or limitations on usage are put in place. At the time of writing, there are no proposals from the scientists, other than regulation by prohibition (see Section 9.13, above). Such proposals are rarely integrated with the practical preoccupations of the decision makers, confronted with the developmental demands of civil society. Ecology is based on evolution, while the current doctrine of the majority of ecologists is confined to maintenance of the status quo.

In such a context, it is necessary to return to the regulator, both to obtain legal decisions and also to obtain finances from central funding for their application, which is always costly. This state of dependency is certainly prejudicial to the protection of the environment.

Some ill-considered developments can also lead to ecological disaster: conservation, protection and, in more general terms, respect for nature, must act as a guide in all forms of management.

Sustainable management can be reconciled both with nature conservation and man's utilisation of ecosystems (considered as integrated with nature and not divorced from it). We shall see that management methods are very varied: their choice arises as much from politics as from science, but management which integrates ecology and economics also has the advantage of being sustainable in a socio-economic sense.

A real change in the paradigm must be achieved. Ecological science must become a partner, a force in proposition making and innovation and not a limitation. Sustainable management by the integration of human activities into the functioning of ecosystems constitutes this new paradigm.

Beyond conservation and its simple current rules, many forms of sustainable management of coastal waters are practicable and the following chapters describe several methods.

9.15 BIBLIOGRAPHY

Addessi L. Human disturbance and long-term changes on a rocky intertidal community. *Ecol. Appl.* 1993; 4(4): 786–797.

Aubert M. *La Méditerranée: La mer et les hommes.* Rev. Int. Oceanogr. Med., Vols 109–112, 1994.

Bakker M. and Coljin N. A target ecosystem for the Wadden Sea: A time for concerted action. In: *Present and future conservation of the Wadden Sea*. Netherlands Institute for Sea Research Publications 20, 1992; 79–82.

Barbault R. *Écologie générale: Structure et fonctionnement de la biosphère*. Masson Éd., Paris, 1993.

Barnabé G. L'aspect économique des championnats de chasse sous-marine et le dépeuplement des fond. *Etudes et Sports Sous-Marins* 1974; 25: 9–11.

Barnabé G. *Evaluation de la faune ichtyologique dans la réserve sous-marine de Monaco*. C.R. à l'Assoc. Monég. Prot. Nat. 1995 (unpublished).

Boudouresque C.F. *Impact de l'homme et conservation du milieu marin en Méditerranée*. Univ. Merc., Gis Posidonie Éd., Marseilles, 1995.

Carlton L. Espèces exotiques dans la mer. *Bull. Int. Sc. Mer*, UNESCO 1995; 75–76: 11–14.

Castell J.D. Fish as brain food. *World Aquaculture*, September 1988; 21–22.

Constant P. Massacre de Requins marteaux aux Galapagos. *Océanorama* 1994; 23: 33–35.

De Vlas J. The Wadden in maximum, sustainable use. In: *Present and future conservation of the Wadden Sea*. Netherlands Institute for Sea Research Publications 20, 1992; 55–66.

Doumenge R. Aquaculture in Japan. In: Barnabé G. (Ed) *Aquaculture Vol 2*. Ellis Horwood Ltd, Chichester, 1990; 849–944.

Essink K. Multifunctional use of the Wadden Sea aiming for high natural value. In: *Present and future conservation of the Wadden Sea*. Netherlands Institute for Sea Research Publications 20, 1992; 35–43.

FAO. *L'approche de précaution appliquée aux pêches de capture et aux introductions d'espèces*. FAO, Directives techniques pour une pêche responsable, No.2, FAO, Rome, 1997.

FAO/Scialabba N. (Ed). *Integrated coastal area management and agriculture forestry and fisheries*. Environment and Natural Resources Service, FAO, Rome, 1998.

Heip C. Diversité biologique dans la zone côtière. *Bull. Int. Sc. Mer*, UNESCO 1995; 75–76.

Lacaze J.C. *La dégradation de l'environnement côtier*. Masson Éd., Paris, 1993.

Lasserre O. Marine biodiversity, sustainable development and global change. In: *Coastal studies and sustainable development*. Proceedings of the COMAR. UNESCO Technical Papers in Marine Science 64. UNESCO, Paris, 1992; 38–55.

Levesque C. *Environnement et diversité du vivant*. Coll. Explore, Pocket Ed., 1994.

Mauvais J.-L. *Les ports de plaisance: Impact sur le littoral*. Ifremer Ed., Brest, 1991.

McCoy H.D. Aquaculture parks: Zones come of age. *Fish Farming News* 1994; 2(4).

Meijer E.W. and Holsink MJ. Recreation in the Wadden Sea is a matter of good seamanship. In: *Present and future conservation of the Wadden Sea*. Netherlands Institute for Sea Research Publications 20, 1992; 91–95.

Netherlands Institute for Sea Research. *Present and future conservation of the Wadden Sea*. Netherlands Institute for Sea Research Publications 20, 1992.

OECD. *Gestion des zones côtières. Politiques intégrées*. OECD, Paris, 1993a.

OECD. *Gestion des zones côtières. Quelques études de cas*. OECD, Paris, 1993b.

Pirazzoli P.A. *Les littoraux, leur évolution*. Nathan Université Éd., Paris, 1993.
Purves, W.K., Orians G.H. and Heller H.C. *Le monde du vivant: Traité de biologie*. Flammarion Éd., Paris, 1994.
Steyaert M. ad Suzyumov A. The Coastal Marine Programme (COMAR). In: Coastal studies and sustainable development. Proceedings of the COMAR. UNESCO Technical Papers in Marine Science 64. UNESCO, Paris, 1992; 251-266.
UICN. *Stratégie mondiale de la conservation. La conservation des ressources vivantes au service du développment durable*. UICN, Gland, Switzerland, 1980.
UNESCO. *Island agenda*. UNESCO, Paris, 1994.
Weber P. Sauver les océans. In: *L'état de la planète 1994*. Worldwatch Institute, Édition la Découverte, Paris 1994: 65–92.
Wilson E.O. *Biodiversity*. National Academic Press, Washington, 1988.
Wolf W.J. Ecological developments in the Wadden Sea until 1990. In: *Present and future conservation of the Wadden Sea*. Netherlands Institute for Sea Research Publications 20, 1992; 23–32.

10

Managed ecosystems in sheltered waters

10.1 MANAGED ECOSYSTEMS AND SHELTERED WATERS

10.1.1 Aquatic production

When the locations of the most productive marine aquaculture zones are examined, whether they are situated in Northern Europe, on the Atlantic coast of France, in the Adriatic, in central America or the whole of south-east Asia, it can be seen immediately that these are shallow, sandy coasts or coasts with jagged contours, or areas protected by coral reefs, natural banks of rock, or islands; in every case, they are sheltered waters. These virgin areas which, not so very long ago, formed the frontier between land and sea, are no longer a barrier to mankind; their development and management contribute significantly to current global aquaculture production (28 million tonnes in 1998).

The separation we use between sheltered waters and the open sea is not arbitrary: on the one hand, there is no place within the classification system for lagoon waters and, on the other, the functioning of coastal ecosystems is very dependent on storms which constitute major events, as do floods in rivers. Cyclones and storms disrupt the profile of the coastline and also all the benthic flora and fauna, destroying tropical corals, Mediterranean weed beds and populations of echinoderms and molluscs. The impact on pelagic life is less well known, but we have seen that hydrological fluctuations have an impact on the planktonic stages of exploited species (Chapter 4). This separation also underlines the significance of hydrological aspects on ecological functioning and takes into account the historical context—the installation of traditional aquaculture in the sheltered zone.

This form of management in coastal waters has been described in specialised publications (Barnabé, 1990; Nash and Novotny, 1995). We shall only refer to the major types of managed aquatic ecosystems, since it is impossible to cite them all. Sometimes they involve very ancient and extensive activities and their survival proves their ecotechnological and economic validity. Broadly speaking, three management methods can be distinguished:

- The first consists of culturing species, which will exploit natural production but without intervening in the production process, for example, traditional shellfish culture.
- The second management method consists of culturing chosen species, but attempting to control them, to improve natural production by fertilisation, as in agriculture.
- More recently (since the 1970s), intensive rearing of fish in cages in protected coastal waters, but connected with the open sea, has allowed the colonisation of new sites.

10.1.2 Categories of sheltered coastal waters

Sheltered waters are those which, because of their particular geographical position, are not subjected to the extreme effects of storms, strong currents or, in short, to all the irregular, but devastating, physical perturbations of the open sea. Sheltered waters are also defined by physical barriers (coast, offshore bar, island, barrier reef, etc.). They include very varied types of ecosystem that can vary greatly in size; some can be completely separated from the marine environment by an offshore bar (enclosed lagoons), others maintain one or more connections with this environment (bays, open lagoons and inland seas).

The concept of sheltered waters remains fairly subjective. An area of water can be sheltered under certain wind conditions, but become very exposed in a wind from the opposite direction (lee shores, for example). To simplify matters with reference to the marine environment, it is assumed that sheltered waters have waves which are no greater than 1 m in height. In fact, access to a site or installation at a very rough marine location always poses problems and it is convenient to distinguish sheltered sites and the open sea, as the sailor does between harbour and the open sea. Despite this practical criterion, there are all the intermediates between sheltered waters and the open sea.

- Lagoons correspond to a section of water separated from the sea by a sandy bar (coastal bar). Lagoon environments originate from the import of marine sediments which create the littoral bars. They are shallow, often less than 2 m deep. Tides create water movements which are capable of totally or partially renewing the volume of water in the lagoons, thus ensuring their productivity. The absence of tides in the Mediterranean distinguishes the lagoons associated with this sea; communication with the sea is permanent or temporary, via channels or pipes. Sometimes, the littoral bar completely isolates the lagoon, which becomes a portion of sea surrounded by land—an expanse of stagnant water. Shallow lagoons (less than 2 m deep) are often distinguished from deep lagoons (2–10 m deep). Inland seas are very large. There are artificial structures (man-made harbours and outer harbours and lakes) which can be compared with lagoons.
- Although estuaries do not themselves constitute a particular type of site, except from the point of view of shelter, the alluvium carried by rivers into

these areas often creates deltas, marshy areas and alluvial plains which are favourable for the development of culture ponds. The meeting of fresh and salt waters leads to the development of a characteristic and abundant phytoplankton and, as a result, estuaries have often been the first aquatic ecosystems to be developed.

- The mangroves of the tropical belt are shallow, sandy or muddy marine areas, invaded by mangrove trees. This vegetation anchors the seabed and limits erosion during storms. As a specific aquatic ecosystem, it also plays the role of nursery for many species exploited by fisheries.
- A different type of lagoon consists of an expanse of marine water situated at the heart of an atoll and connected with the sea; there is no continental influence.
- In the open sea, strings of islands, archipelagos, islands close to the coast, the constant direction of winds such as the trade winds, and the raggedness of the coast in fjords or rias, create sheltered zones which still connect with the open sea, or are not lee shores. Often deep, these coastal waters are favourable for several types of rearing.
- Over the centuries, salt marshes have often been managed in mangroves or lagoon areas. Salt production amounts to millions of tonnes throughout the world, since the salting of icy roads and the chemical industry are large consumers of this product. This exploitation is based on the hydraulic control of billions of cubic metres of water across surfaces extending for thousands of hectares; this is an example of productive artificial management, integrated through time with the coastal ecosystem. In the Mediterranean, a sea with small tides, water movement in the salt-pans is totally artificial. The increase in salt yields, technological developments or the lack of profitability have rendered many salt-works redundant and these constitute new potential sites for aquaculture throughout the world. This is also true for ancient dyke areas and polders returned to the sea and certain old outer harbours and harbours which are now too shallow.

10.1.3 Brackish coastal waters

Brackish waters are those where the salinity, while greater than 0.5‰, remains less than that of sea water (35–37‰).

In ecological terms, the marine environment towards the sea and the freshwater environment towards land constitute very uncertain and arbitrary frontiers, and brackish waters form a transition zone. These are estuarine waters, as well as those of mangrove swamps, marine bays which receive freshwater and coastal lagoons which are influenced by the tide. In simple terms, brackish waters are those which are influenced both by fresh and salt waters. Numerous classification methods exist for attempting to separate categories of brackish waters; some are based on salinity or ecological zonation while others take climate into account, but none explains the biological or ecological processes involved. As they are of little use, we shall not describe any of them.

In biological terms, the species frequenting these environments all show adaptation to living in waters of variable salinity (euryhaline). In physiological terms, the effects of salinity interact with those of temperature and dissolved oxygen levels and thus cannot be considered independently. Thus, at equal salinities, penetration into warm waters would be easier since the increase in temperature improves the potential for osmoregulation. There are many interactions between parameters: the calcium ion, for example, increases resistance to low salinities since it decreases membrane permeability. Brackish water species are also characterised by a certain degree of polymorphism (morphological or anatomical variations, dwarfism or gigantism).

In ecological terms, these waters, which are in contact with land and which benefit from its input, are rich in nutrients and are termed "eutrophic waters". Without doubt, for this reason, these waters are characterised by populations with little species diversity. In contrast, these species are very abundant in numbers of individuals and their biomasses are very high. These are the characteristics of a young, unstable eutrophic ecosystem.

Numerous species are only temporarily adapted to these waters. They spend part of their lives there, then leave to reproduce in the sea or freshwater. When the salinity exceeds 5–8‰, because of their richness in planktonic (diatoms, rotifers and copepods) or benthic (worms, oysters, prawns and fish) biomass, the waters constitute a sort of larder accessible mainly to marine species, which are more able to tolerate a decrease in salinity than freshwater species can tolerate an increase in salinity.

Brackish waters are very turbid owing to the many particles in suspension. These particles are colonised by populations of bacteria and protozoa and even rotifers (see Section 3.6). With such nutrient richness, it can be understood why primary and paraprimary production should be high (see Sections 3.5 and 3.6.1) and this provides the explanation for the high biomasses which characterise all food chains in brackish waters. The activity of bacteria and protozoa and the degradation of organic matter is comparable to that going on in water treatment stations by activated sludge or settling ponds: it is an effective purification process which takes place with the degradation of organic matter into inorganic matter, etc. (see also Section 11.2).

For man, brackish waters often constitute sheltered waters, but also easily accessible ones, because they are close to land. Thus it is understandable that natural brackish waters have been the first places for development. Their spectacular richness, proximity, the ease with which ponds can be hollowed out or dykes constructed on alluvial plains or littoral bars, have made them the cradle of traditional aquaculture. Fisheries have provided the first subjects for rearing, as shown by the example of "valliculture" in Europe (Section 10.3.1) or "tambaks" in Asia (Section 10.3.3. below). As a result of human requirements, but also taking into account climate and the species adapted to these ecosystems, a diversification of rearing methods has taken place. The main types will be described later.

10.2 TRADITIONAL MOLLUSC CULTURE

As we shall see in the following chapter, primary production of waters which can be used for the purification of waste water or obtaining microalgae often contributes to the culture of edible molluscs.

10.2.1 Mussel rearing on the seabed

In a sheltered area, at the end of their larval life, planktonic mussel larvae attach to any virgin substrate which they encounter (Chapter 4). The low-lying coasts of Holland and the Wadden Sea are rich in gravel beds where mussels are naturally abundant. Access to plots has been allowed to concession holders, but it is not simply collection which is involved. Young mussels (spat) are taken from areas where they are naturally abundant and transferred to beds set aside for their growth where they stay for more than a year before being collected and treated prior to sale. Collection operations for spat or full-grown mussels are carried out using large dredges from a boat 20 m long. This boat is also used to scatter the juveniles in the growing areas: its bottom is pierced with vertical holes through which the little mussels are released, while the boat moves slowly along. More details can be found in Korringa (1976).

In this case, human intervention is limited to favouring growth by placing the juveniles in better ongrowing conditions. Recruitment and growth are left to nature. Holland produces more than 100,000 tonnes of mussels reared on the seabed annually. The low rearing costs allow their sale throughout Europe at less than €0.76/kg.

10.2.2 Mussel rearing on bouchots

The rearing of mussels on bouchots* on the Atlantic coast of France also involves the attachment of young mussels or spat to a submerged solid substrate. Rows of posts or stakes are placed in the intertidal zone, allowing the capture of young mussels which attach to these posts and which then feed solely by water filtration. Nowadays, capture is carried out on ropes which have been wound around the posts, where the spat grow (for more details see Dardignac-Corbeil, 1990). Other operations, such as the thinning out of mussels, their maintenance on the bouchots and their mechanised collection, also improve production. Since human intervention is more frequent and less mechanised, the mussel price is higher but, because they are reared above the seabed, they are of higher quality.

10.2.3 Suspended mussel culture

In deeper, but sheltered sites, such as the Galician Rias (in north-west Spain), bays in New Zealand or the Étang de Thau in the Mediterranean, mussels are reared on ropes running from the surface towards the seabed. Spat is collected from rocks on the shore, where it abounds. Clusters of little mussels attached by their byssus are placed in a tubular net, like a large-meshed sock; this holds them in place while the byssus of the young mussels develop, and holds them together with each other and to the rope which runs through the middle of the tubular net. Thus ropes are formed, in the form of "sausages", about 5 cm in diameter, the length of which (most often 3–8 m) depends upon the water's depth. Suspended from rafts or supports fixed to the bottom, growth is favoured by the permanent immersion in water and the easy access for the mussels to plankton in the open water. Currents ensure the turnover of the rearing water. In favourable conditions, production can reach 250 tonnes (fresh weight) of mussels/ha/year. This happens in the Galician rias where tides are large (3.5 m) and the current reaches 1 m/s. However, the water is clear, with a Secchi disc disappearing at between 5 and 6 m. The mussels grow to harvest size in 6–9 months on these ropes, reaching weights of 8–10 kg/m at harvesting. Studies have shown that 35–40% of phytoplankton and fine particulate material is retained during one single passage of seawater under the raft. Spain expanded this rearing method in the 1970s and 1980s which has resulted in it being the main European producer with more than 200,000 tonnes per annum.

All these rearing methods are very similar and have been described by many authors (see for example: Dardignac-Corbeil, 1990; Barnabé, 1990; Bompais, 1991).

All these rearing units are set up in sheltered zones, but mussel-rearing units have recently been established in Languedoc (France) and in the Sagres region of Portugal, although the Japanese have been cultivating other species in exposed areas for a long time. Mussel production in Languedoc has come up against problems with profitability (competition with Spanish mussels), but many methods have been shown to be viable.

10.2.4 Oyster farming

The example of the mussel is not unique. French oyster culture (150,000 tonnes) also results from the adaptation of a natural process to man's requirements; here again, a bare substrate is provided for oyster larvae ready to attach (there are few virgin substrates in coastal waters) and, after capture, ongrowing is carried out either directly on the seabed or up on trestles on the shore in the Atlantic or suspended in the Mediterranean or Japan, for example. Korringa (1976b, c), Héral (1990) and Doumenge (1990) have described in detail how oyster farming is carried out in various parts of the world.

The coastal landscape is often profoundly moulded by such developments; in the Marennes-Oléron, small developed portions of water constitute oyster beds or "claires". The distinctive management of the waters in these oyster beds is at the root of the "finishing" of these famous oysters. More details can be found in Clément (1990).

10.2.5 Scallop culture

The capture of juvenile scallops (*Pecten yesoensis*) is carried out on various substrates. Once the branches of fir trees were used but these have given way to old nets or fishing lines placed, jumbled up, in sacks made of net of several millimetres mesh size, themselves suspended in strings on submerged ropes (culture in suspension).

The larvae attach to this substrate by their byssus and live there for more than a year. The byssus then disappears, but the young scallops are trapped in the bag. Their growth in suspension continues in these bags, or a hole is bored in the scallop, in order to suspend it (rearing by ear hanging method).

Floating lines are installed to support the submerged ropes in sheltered coastal waters. Little by little they have been modified and adapted to go deeper and deeper and further and further out into the sea, resulting in vast "culture fields" made up of long lines which can reach lengths of 1800 m. Japan produces more than 200,000 tonnes of scallops mainly using this method (see Doumenge, 1990).

10.3 TRADITIONAL FISH CULTURE IN COASTAL PONDS AND LAGOONS

Many lagoon or marshy areas have been developed into coastal ponds for the rearing of fish, molluscs and crustaceans. Except in certain closed seas such as the Mediterranean, the tides renew and fertilise these ponds, which have different names depending on where they are—"fish reservoirs" in Arcachon, "valli" in Italy, "tambaks" in south-east Asia, "claires" in the Marennes. Fish or shrimp culture basins or ponds belong to this category of water bodies developed for the biological production of species which are useful to man.

10.3.1 Valliculture

Valli are natural lagoons whose outlets to the Adriatic Sea (Italy) are equipped with grills located in dams which allow the entry and exit of fish during immigration between sea and lagoon. Juvenile sea bream and bass, previously caught in the sea and now produced in intensive hatcheries, are used to repopulate these managed lagoons; survival during growing on is of the order of 30% and the sea bass reaches 350 g in 30 months. Stocking levels are limited, since the sea bass is piscivorous. Gilthead sea bream survival varies between 50 and 90%. They reach 300 g in their second winter in northern Italy and 500–600 g in the south, which is warmer. Production can reach 100 kg/ha/year but this species dies when the temperature drops to 5°C (the bass is resistant to 1°C).

This extensive system for exploitation, or valliculture, results in the production of several thousand tonnes of fish per year (Chauvet, 1989). The explanation for this success is the relatively large tides of the Adriatic, combined with management ranging from the control of water movements to the sorting of catches and restocking to over-wintering installations with oxygenation of the pond beds. It is also due to the capacity for group management and to the particular legal status of these lagoons (private or collective ownership). A book devoted to valliculture has been written by Ravagnan (1978).

The public financing of this type of development in the Étang de Salses-Leucate (French Mediterranean) has never allowed its exploitation according to the valli model, because of the individualism of the fishermen who have never maintained the fishery (Bourquard and Quignard, 1984).

10.3.2 Integrated pond fish culture

In valliculture, the only fertilisation is the introduction of coastal water or the treatment of the seabed; there is no direct form of fertilisation. In contrast, pond fish culture is based on fertilisation which allows natural production to be multiplied by 10–15 times.

We shall only deal with the main principles since they are identical in fresh, brackish and sea water. In freshwater, carp production is about 13 million tonnes, with the Chinese responsible for about half of these; the type of fish culture practised in the Orient is spread throughout eastern Europe and also France, Israel, etc. The methods and techniques specific to this fish culture have been described elsewhere (Michael, 1987; Marcel, 1990; Barnabé, 1990; Billard, 1995).

Fish culture ponds are bodies of water between 1.5 and a maximum of 3–4 m in depth; their surface area can be less than a hectare when they are used for a specific purpose (rearing of fry, holding of broodstock, spawning), but most often their surface areas vary from several hectares to a hundred hectares. In most cases, movements of freshwater into and out of the ponds can be controlled, which differentiates these ponds from salt water ones.

The other feature of fish culture ponds is that fertiliser is applied either in mineral form, or as waste products (manure, fresh droppings, dried dung). In the latter, as in a purification lagoon or in the sea, the mineralisation of organic matter by bacteria plays both a purifying role and forms the starting point for new primary production.

The pond constitutes a developed ecosystem which is managed in several ways: by the amount of waste, fertiliser or food added; by the number of individuals being reared, which will utilise the biomass produced; lastly, regulation of the characteristics of the environment (oxygen concentration, acidity, etc.) is carried out by emptying or diluting the water, liming or adding ammonium or alum to regulate pH. More recently, forced water circulation and aeration have allowed increased tilapia production (Costa Pierce, 1989).

Fish in ponds are very often fed by-products from agriculture, which are consumed directly: these can include rice flour, soya cake, grass, etc. According to Marcel (1990), 6.8 kg of rice flour or 4 kg of soya cake allows the production of 1 kg of pond fish. A similar result can be obtained from 15 kg of terrestrial plants or 45 kg of aquatic plants. This type of exploitation constitutes a true combined production system. The situation is similar for carp culture in eastern European countries, under temperate climate conditions.

Primary plankton production induced by fertilisation can be used directly by phytoplanktonivorous fish or by zooplankton which are then consumed by zooplanktonivorous species. Waste products from the fish (faeces, urine, waste products excreted by the gills) in turn contribute to the enrichment of the environment in organic matter. Particulate or solid materials which sediment out are used by detritivorous fish that feed by filtering the mud in which a characteristic fauna develops.

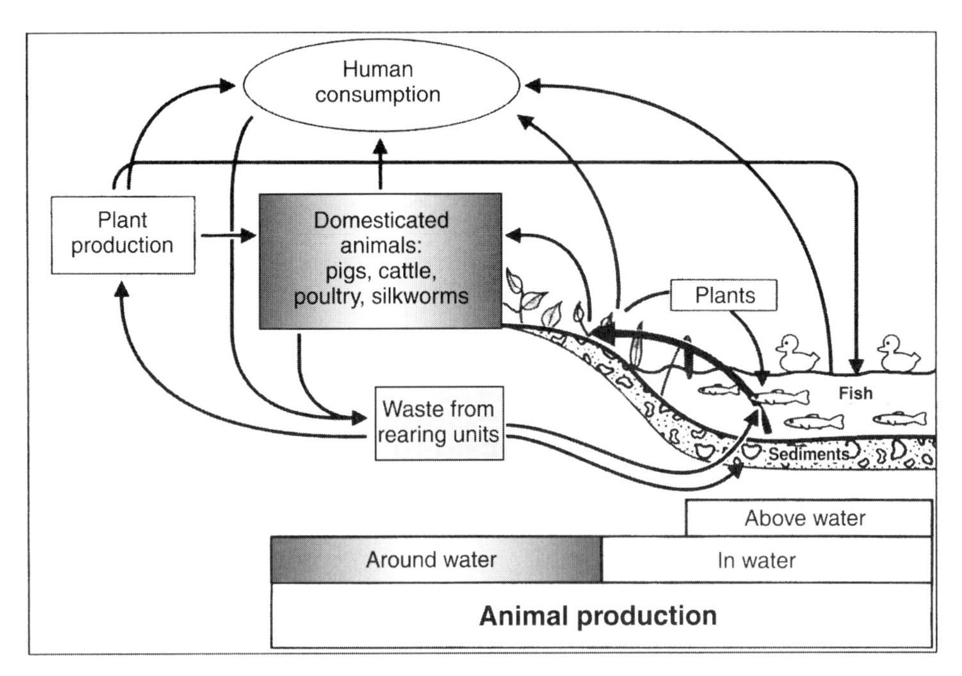

Fig. 10.1. Integrated fish culture and agriculture in China. Waste products from the rearing of various species along the edge of the pond are distributed as high-dose fertiliser (up to 100 kgDM/day/ha) into the fish culture ponds. These stimulate all levels of the food web, different levels of which are exploited by the various fish speces: phytoplanktonivores, benthophages and herbivores. The sediments on the pond bed are rich in organic matter and are periodically transferred onto the banks where plants used to feed animals, including herbivorous carp, are cultivated. The fish also receive plant products (cereals, soya). It should be noted that, in eastern China, animals are concentrated around, above and in the water.

The carp or gudgeon in freshwater and the grey mullet in salt water, stir up the bottom: nutrients are thus resuspended and recycled.

Pushed to its limits in freshwater in Asia, this system constitutes an integrated aquaculture system in which waste products from domestic animals are recycled by fish. Castell (1989) reported the example of this type of farm (Wuxi Integrated Fish Farm, near Shanghai). There, in an area of 70 ha, eight species of fish are reared, including several carps, bream and tilapia. On the banks between the rows of ponds, buildings house ducks, pigs or cows. Their waste products are sent to the ponds (2 ha across and 3 m deep). Chicken droppings produced at another site are also used. Fish production in the ponds is around 800 tonnes/year while 400,000 ducks, 1060 pigs and 123 cows are produced annually on land, resulting in a total production of 13,500kg/ha/year. Duck livers are even exported to France for foie gras. A restaurant completes the set-up; it serves farm produce. The entrails from the pigs, ducks and chickens are used to feed carnivorous carp (black carp).

Detailed descriptions of rearing procedures can be found in aquaculture books (Barnabé, 1991; Nash and Novotny, 1995) and Fig. 10.1 (Billard, 1995) shows a diagram of the integration of fish culture and agriculture in China. Jingsong (1989) gives some interesting data on this fish culture: oxygen production in carp culture ponds is between 26 and 36 tonnes/ha/year and phytoplankton production (wet weight) between 127 and 155 tonnes/ha/year, similar to production levels in purification lagoons.

10.3.3 Rearing shrimp and fish in coastal lagoons

Almost everywhere in the world, lagoon areas, low-lying coasts and sometimes mangrove swamps, have been developed for the rearing of fish or crustaceans. In the Far East, lagoon areas developed into private ponds ("tambaks") are used for extensive growing on of milk-fish (with fertiliser applied); Indonesia and the Philippines produce 167,000 and 161,000 tonnes, respectively in 1999 (FAO, 1994). Spawning in captivity has recently been controlled and many of the juveniles produced in the sea are caught during their migration to the shallow coastal waters. These tambaks are also used for the growing on of various penaeid shrimps (79,000 and 53,000 tonnes in 1991 in the two countries cited above).

Rice growing is sometimes combined with shrimp or fish culture in tambaks and rice flour is used for fertilisation. Vu Do Quinh (1996) demonstrated the different methods for the integration of prawn culture with natural or developed coastal ecosystems (mangroves, salt-works and paddy-fields) in Vietnam.

This is another example of a combined production system, this time in coastal ponds/lagoons. Many other examples can be found in articles by Marcel (1990), Doumenge and Marcel (1990), Jhingram (1990) and in a book by Nash and Novotny (1995). Sometimes, an industrially-made artificial food can complete the food which shrimp, prawns or fish obtain from the natural environment; this is termed semi-extensive rearing. The robustness of such operations and the limited investment they require, explain why China, which produced 19,000 tonnes of shrimp in 1984, and announced a production of 186,000 tonnes (largest producer in the world) in 1990 (FAO, 1992) and more than 500,000 tonnes now.

Tambaks are also used for the rearing of *Lates calcarifer,* the Asiatic bass or baramundi. This predatory marine species requires complementary feeding in the form of dried food compound for its growth, but is reared in the tambak from the larval stage until it is sold.

In central America, this type of lagoon area or coastal marsh has been developed into large basins, several tens of hectares in size, for the quasi-exclusive rearing of penaeid shrimps. Ecuador produced 150,000 tonnes in 1996. Taiwan and also Indonesia, Malaysia and India now produce shrimp since rearing has evolved from fish towards penaeids or vice versa, according to profitability, disease, etc. This demonstrates the adaptability of these ponds developed in the coastal zone. The production of shrimp by aquaculture now exceeds that from fishing. The organic residues, the mud accumulated on the bed of these rearing basins and machines capable of extracting them have been studied to this end (Williamson, 1990).

The plan of an integrated farm tested by ICLARM (International Center for Living Aquatic Resources Management in Manilla) is reproduced in Fig. 10.2 (Hopkins and Cruz, 1982). These authors provide more precise data concerning

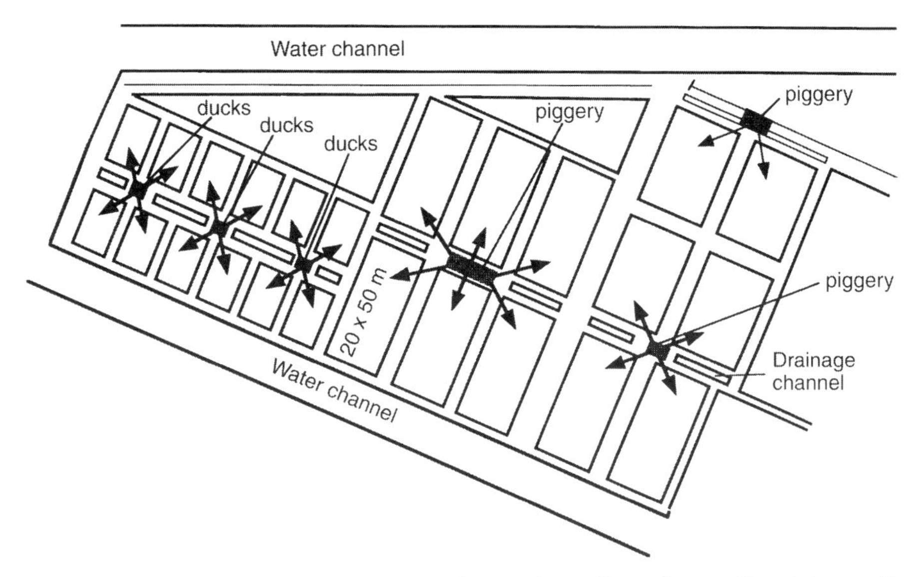

Fig. 10.2. Plan of experimental site at the Freshwater Aquaculture Center, Central Luzon State University, Munoz, Nueva Ecija (Philippines). (Adapted from Hopkins and Cruz, 1982.)

the influence of various types of wastes on production but especially noted the drop in oxygen at the end of the night (1 mg/l) in the most enriched ponds. No parasites common to fish and man were found and the system is very profitable in economic terms. It allows better use of proteins than a simple livestock rearing unit.

A marine fish culture pond which recycled waste from a school canteen functioned for several years in Monastir (Tunisia), but the fish were difficult to recapture since the walls of the ponds were made of blocks of rock, the crevices of which provided refuge for the reared individuals when harvest was attempted.

10.3.4 Rearing marine plankton in earth ponds

Food chains are not exactly the same in freshwater and salt water, but function in the same way, as we shall see: thus, in Martinique, rearing of marine plankton has been successfully achieved in tropical conditions, to provide food for fry of the sea bass, *Dicentrarchus labrax*. Earth ponds 100 m^2 in area, 0.8 m deep are filled with sea water and enriched with dried poultry waste, which is placed in jute sacks, 10 kg every 15 days. Intense development of phytoplanktonic algae precedes the development of copepods which are collected by the million using a filter bag into which the water from the pond is pumped. In this case, waste water is replaced by poultry waste. The system is not without problems: the high temperatures and strong sunlight stimulate photosynthesis so much that carbon dioxide is used up, which in turn, through the carbonate-carbon dioxide buffering effect (Chapter 1) sometimes increases the pH to above 9, the lethal value for zooplankton. This example demonstrates both the potential intensity of natural processes in extreme conditions and also the limits to their utilisation.

10.3.5 Rearing aquarium fish in ponds

A specialised type of rearing in small ponds involves ornamental or aquarium fish: one of the most active regions for this type of rearing is Florida (USA). Farms consist of hollowed basins dug in the ground, similar to small ponds. They are around 10 ha in overall area, divided up into small ponds approximately 200 m^2 and 2 m in depth. Some are covered with glass to overcome winter mortalities owing to bad weather and low temperatures. Management is identical to that for ponds: reared species (guppies, platies, mollies, etc.) are stocked in fertilised earth ponds. Spawning is not controlled and occurs naturally in the ponds, even in nesting species (*Cichlidae, Centropomidae*), but food is added and macrophytes and predators are controlled. As emptying is not possible, anaerobic sediments are eliminated by pumping. There are very few installations of this type for marine fish.

10.3.6 Common features of the operation of managed ponds

Managed ponds function like a natural ecosystem (Part A), but the higher nutrient richness and biological manipulations speed up the processes, resulting in specific management problems. We shall only summarise these, since there are numerous publications about the subject.

The control of biological production in ponds, whether in fresh or salt water, rests on two, often contradictory, criteria: water of a quality compatible with the requirements of the species living in it, and sufficient fertilisation to obtain optimal production of cultivated species through that of prey species.

In freshwater, the maintenance of pH above 7 in acid soils is sometimes achieved by adding lime, but in the saltwater environment, a natural buffering effect regulates the pH, as long as the capacity of the ecosystem is not exceeded (Chapter 1). We have seen above (Sections 10.3–10.4) how intense photosynthetic activity can increase the pH in shallow salt water ponds where, because of the harvest of zooplankton, natural equilibria are no longer maintained.

When there is a significant animal biomass present, it produces N and P; the phytoplankton and benthic algae can possibly reintegrate these elements into the food chain, but ammonia, one of the forms of nitrogenous waste, is toxic to many reared species at levels of less than 1 mg/l . This ammonia is normally degraded into nitrite, then nitrate, by bacteria (nitrates are a non-toxic, inorganic form of nitrogen, which can be assimilated by plants). This is another equilibrium which must be respected, so that the recycling capacity by plants of animal waste allows the viable functioning of the ecosystem.

The most sensitive parameter in the operation of managed ponds is oxygen concentration. Plants produce oxygen during the day, but consume it at night; with oxygen constantly being consumed by the biomass of animals, this often results in too low concentrations for the animals at the end of the night (1 mg/l). This is one

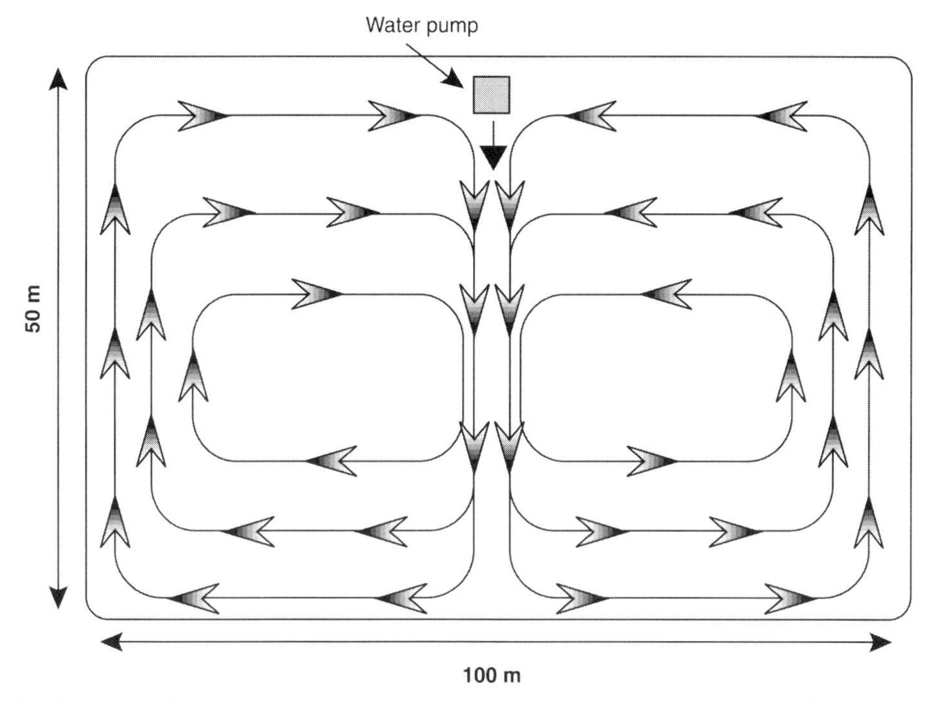

Fig. 10.3. Induced water movement in a half-hectare lagoon, using a water pump (adapted from Faivre, SARL).

of the main limitations to the capacity of managed ponds. The current trend is thus towards artificial aeration using equipment with electric or fuel-driven motors, with machines becoming more and more efficient. Aeration and water circulation are carried out simultaneously by various aerators, even in the largest ponds (usually a power of 1–2 kW/ha is used). This type of equipment is found in Chinese fish ponds and in shrimp ponds in Ecuador or Malaysia. Figure 10.3 shows the type of currents created by an aerator/water circulator marketed in Europe.

The manipulation of the aquatic ecosystem by man in this situation is complex and depends on many factors: the number of individuals being reared, their position in the food chain (planktivore, herbivore, detritivore, carnivore), turnover rate of the water, aeration, fertilisation or adding of food. If one adds that the different stages of rearing require different ponds (spawning, fry rearing, different ongrowing stages), management is not a meaningless word. Depending on the country, site and whether one or more species is being cultivated, traditional methods differ.

Fertilisation of the water, for example, puts into practice three main techniques:

- The adding of mineral fertilisers (nitrogenous and phosphoric fertilisers); this happens in fish culture ponds in Europe;
- The adding of wastes to the ecosystem (organic fertiliser such as manure, droppings, etc.): the natural processes of waste recycling ensure the

fertilisation of the ecosystem and the reared species consume the natural production produced by this fertilisation; this happens as we have seen, in fish culture throughout Asia and the far East;

- Feeding reared individuals: the food is used directly, but that which is not consumed is degraded and also recycled. The waste from fed animals also contributes to fertilisation. Feeding of fish in ponds is practised in summer in Europe but is expanding throughout the world.

10.4 INTENSIVE CAGE FARMING

Intensive rearing in cages or pens is not the oldest or most traditional of the forms of management; it is a recent method of rearing, made possible by the feeding of fish using a dried food made since the 1960s and by the control of reproduction of marine species in hatcheries in the 1980s. It is the aquaculture of transformation, which differs from the production aquaculture represented in sheltered waters by mollusc culture and rearing in ponds.

In contrast to shellfish culture or pond culture, which draw part of the production directly from the rearing environment to nourish the species involved, food in cages comes from elsewhere: it is exogenous in origin. It takes the form either of low market-value fish distributed in frozen form (yellowtail rearing in Japan) or dry compound foods, made largely of fish meal. It is thus an activity which is dependent on the fish meal industry, that is 15-20 million tonnes of fish, representing nearly a quarter of the world's fish catches (mainly Peruvian anchovy and Norwegian herring). The ecological validity of this use could be discussed, but these millions of tonnes of fish have no chance of being marketable in their fresh state.

Cage farming uses the physical shelter of coastal waters and these waters also import oxygen and absorb the waste products of the reared individuals. As a result, cage farming is often accused of polluting coastal waters. We shall return to this problem in the following chapter. Detailed accounts of this technology can be found in Barnabé (1991) and Beveridge (1989).

10.5 THE ARRIVAL OF A MANAGED ECOSYSTEM IN COASTAL WATERS

10.5.1 Preservation of traditional managed ecosystems

If we exclude the recent use of cage farming, the methods used in these managed areas, however empirical or rustic, allow excellent control of pond ecosystems with the aim of producing aquatic species and recycling waste products from terrestrial rearing operations. These developments and this management form the basis of three-quarters of aquaculture production (36 million tonnes, the majority of which is animal protein). This biomanipulation of production processes by man can involve aquatic ecosystems up to the size of the great American lakes (Barraclough and Robinson, 1972).

These examples should be carefully considered, since they involve productive and profitable ecosystems, in the economic sense of the term. They create employment in areas distant from large cities and contribute to the maintenance of the land and its aquatic margins. Man's intervention, often going back hundreds of years, has led to true managed ecosystems, the functioning of which is identical to those of natural coastal ecosystems (Chapters 1 and 4).

The great age of these installations and their traditional character mean that they are now considered as natural zones, part of the countryside and some of these areas are protected, like a wild place. Even the purest ecologists take part in the conservation of ecosystems managed by man, which now form part of the littoral scenery.

This is the case, for example, for the "claires" in France where oysters are matured and the "valli" of Italy, which produce eels, grey mullet and sea bream. We are witnessing the constitution of a real landscape. Other uses are found (hunting, walking, tourism, nesting ground for aquatic birds, conservation of protected species), while at the same time the area remains productive.

The salt marsh has also become an integral part of the littoral countryside and some salt marshes are protected and classified as natural heritage under this name, even though it all started as an industry. This example is interesting since ecology rejoins the economy: with time, salt production has become a tradition.

10.5.2 Expansion of traditional activities

In addition to traditional managed ecosystems, such as valliculture and fish culture in tambaks, there have been more recent developments linked to new biotechnology. Throughout south-east Asia and also in China, central America and Africa, there has been widespread and sometimes excessive colonisation of low-lying coasts or sheltered zones (lagoons, bays, etc.): littoral waters, lagoons and mangrove swamps have been developed for the production of thousands of tonnes of molluscs, crustaceans and fish (see Doumenge, 1990; Nash and Novotny, 1995).

The expansion of mollusc culture in as yet unexploited, sheltered sites, or the explosion of penaeid shrimp culture in tropical coastal waters, are certainly the best well-known examples of recent exploitation (Doumenge, 1990). We have qualified these new aquaculture activities in order to distinguish them from "traditional" aquaculture. Cage rearing of fish belongs to this new wave.

There are at least 200 million hectares of earth ponds and 0.5 million hectares of coastal ponds devoted to fish culture (Nash, 1995), but the areas of coastal water developed for mollusc production and marine sites used for restocking purposes are "immeasurable". The volume of pens used for intensive fish farming is greater than 10 million m^3.

10.5.3 Criteria for the development of management of coastal waters

The type of biological production can vary in its form depending on the species being reared and the nature of the site, but examination of the examples given above shows that there are several constraints:

- ensuring easy access for man to the managed waters, which explains the choice of sheltered waters;
- culturing species which can exploit aquatic production which is inaccessible to man (microalgae, zooplankton, benthic plants and animals);
- ensuring an economic outlet for the aquatic product;
- improving natural production of ecosystems through development and management of coastal environments and continental waters; the complexity of the manipulations to be controlled is described above (Section 10.3.6).

Despite their traditional character, managed ecosystems which respond to these constraints evolve and continue to evolve under pressure from many scientific and technical, and also socio-economic factors:

- Human overpopulation leads to increased competition for coastal space and an increased requirement for animal proteins, leading to intensification of production and the development of new areas. Coastal waters are not inexhaustible and development extends to more distant sites, in countries where there is no tradition of this type. There is also expansion towards the open sea, away from shelter, using new techniques; this expansion has already been broached for mollusc rearing and fish in cages and we shall return to this in the following chapters.
- The globalisation of the economy profoundly affects the development of coastal waters: fishponds in the Arcachon region have disappeared due to lack of sufficient profitability, as has the rearing of the Manila clam in France, which could not compete with Italian production. To remain profitable, aquatic production must adapt to match the characteristics of the environment available, and the requirements of the species.
- The site–species relationship becomes one of the determining factors in management. Each aquaculture site can differ greatly from neighbouring sites (presence or absence of shelter, currents, depth, concentration of dissolved gases, transparency, etc.). Each site is a separate entity. The environmental conditions offered by a natural aquatic site must correspond as closely as possible to the biological requirements of the reared species. The system of relationships existing between a species and its environment, or its "ecological niche" is very varied in the aquatic environment and it integrates very many environmental factors (food, temperature, salinity, pH, oxygen, etc.), while birds and mammals, the main animals reared in agriculture, have a constant internal body temperature, live on one surface (the ground) and breathe the same air.
- The concept of sustainable development must develop in relation to both the site–species constraints, and also market preferences, which determine the

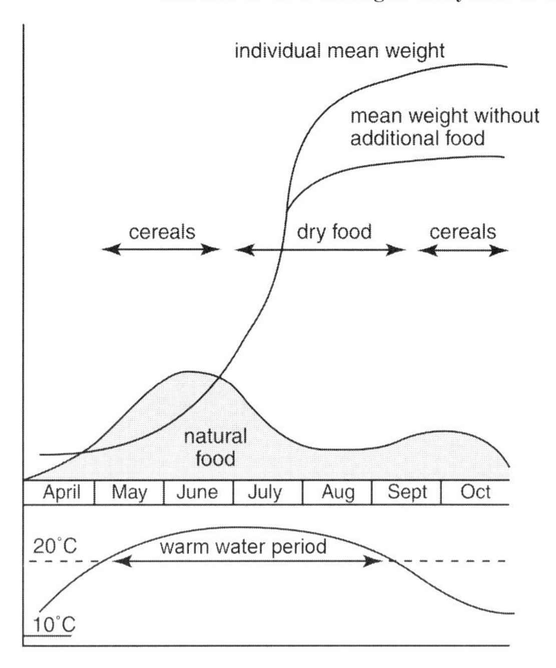

Fig. 10.4. Diagram showing changes in temperature, natural production and fish weight in a pond with and without complementary feeding. (Adapted from Matena and Berka, 1987).

price and thus the profitability of the culture. In addition to speaking of sustainable development, we should accept the notion of sustainable and opportunistic development—bear in mind the milk-fish tambaks, adapted to prawn rearing in Indonesia, or grouper rearing in Taiwan, to solve profitability problems. Sustainable and opportunistic development will bring together ecology and the economy.

- The transfer of species will become widespread in order to respond to these criteria. It has already started with the European production of oyster culture and clam culture resulting from the rearing of introduced species (Japanese oyster and Philippine clam). The rainbow trout, which occupies the majority of intensive rearing units in Europe, is also an introduced species. The potential of many marine or brackish species for culture is still unknown and may develop in the future. We shall return to the problems with these introductions in health terms in the following chapter, but many ecologists have proposed such transfers and true agriculture rests on their practise.
- The widespread use of semi-extensive rearing in which part of the food is provided by the working of the ecosystem and the other part is provided by man, will continue. This leads to a very slight improvement in production (Fig. 10.4). This technique is already used in China and Europe and is spreading through tropical waters within the framework of semi-extensive

rearing of prawns and marine fish. Complementary feeding can be very diverse in nature, but the aim remains the same—to intensify production. In ecological terms, adding food corresponds to an auxiliary input of energy to the ecosystem (see Chapter 8). This type of rearing often renders necessary the artificial oxygenation of the ponds.

The situation is far less developed in the open environment where the absence of physical barriers makes control of the environment more difficult. The Chinese, however, fertilise algal and shellfish culture areas by submerging porous earthenware jars filled with manure in the rearing area; other experiments with fertilisation or adding food in the open sea have also taken place and we shall return to these.

Technological progress constitutes the main factor in development, not only in the management of sheltered waters, but also that of all coastal waters. We shall study these in specific chapters. There are four main categories:

- the creation of water movement and oxygenation of water using special equipment;
- the availability of juveniles of marine species at a low price for culture, or restocking from the potential of hatchery-nursery set-ups;
- the utilisation of substrates and artificial reefs and various concentrating devices to increase production at interfaces in the aquatic environment;
- the synergistic use of these innovations for the complete management of coastal waters.

There are many other possible management prospects, thanks to technology: nowadays there is the possibility of transforming cereal-growing areas into fisheries (this has been envisaged in the Pô delta); in the Camargue, carp ponds have been transformed into wheat and rice fields, since the value of carp is low. On the other hand, 15,000 hectares of land in Holland have been returned to the sea: all this shows that choice is possible.

10.5.4 Imminent threats

Despite these prospects, the degradation of water quality in coastal waters (red tides, bacterial pollution, eutrophication and especially chemical pollution) remains a deadly threat to development. Many interactions are involved and the problem is immense; we shall describe these in the following chapter.

In simple terms, the solution is to use the management of aquatic ecosystems for getting rid of pollution and not only for production. Already, in Europe, reared molluscs must pass through a purification plant, because of the problems they have caused in the past and continue to cause. This still only concerns the health aspect—it is not known how to treat chemical pollution (pesticides, heavy metals, toxins, etc.).

10.6 CONCLUSIONS

In addition to their role in aquaculture production, sheltered and managed aquatic zones play a significant ecological role in fauna protection and the conservation of particular biotopes; these are often a major tourist attraction (either in aesthetic terms, for bathing, for fishing or sailing). While in food-producing agriculture, the monotony of monocultures constitutes a handicap, management of coastal waters can integrate the ecology of the countryside with the particular perception of the aquatic environment (Chapter 8).

Given that there is an area of encounter between man and nature (ecosystems), there are inevitable conflicts of usage created by this profusion of activities, which themselves result from the development. Without management, there would be less production and, hence, fewer or no aquatic birds or rare species, fewer accessible sites and thus fewer tourists or walkers. By exacerbating biological production, other uses are also exacerbated. Thus a compromise must be reached, which is often a form of success.

Although science is directed towards these problems and ecologists have cogitated over them for the past 50 years, it has not been ecological science which has defined these developments and their management; ecology has not added any new forms of exploitation to these traditional activities, despite some major inputs to the understanding the ecosystem function. On the basis of empirism, tradition and technology now leads us towards sustainable, yet flexible, development

Development stimulates ecology and reinforces the economy.

10.7 BIBLIOGRAPHY

Barnabé G. (Ed). *Aquaculture, Vols 1 and 2*. Ellis Horwood, Chichester, 1990.

Barnabé G. (Ed). *Aquaculture: Biology and ecology of cultured species*. Ellis Horwood, Chichester, 1994.

Barnabé G. Open sea culture of molluscs in the Mediterranean. In: Barnabé G. (Ed). *Aquaculture, Vol 1*. Ellis Horwood Ltd, Chichester, 1990; 429–442.

Barraclough W.E. and Robinson D. The fertilization of Great Central Lake. II: Effect on juvenile sockeye salmon. *Fish. Bull.* 1972; 70: 37–48.

Beveridge M.C.M. *Cage aquaculture*. Fishing News Books, Oxford, 1989.

Billard R. The major carps and other cyprinids. In: Nash C. and Novotny A.J. (Eds) *Production of aquatic animals: Fishes*. Elsevier Science, Amsterdam, 1995: 21–55.

Bompais X. *Les filières pour l'élevage des moules*. Ifremer Éd., Brest, 1991.

Bourquard C. and Quignard J.P. Le complexe de pêche de Salse-Leucate: Bordigue et barrages de poissons. *La Pêche Maritime*, 1984; 1272: 3–11.

Castell J.D. An integrated fish farm. *World Aquaculture*, 1989; 20(3): 20–23.

Chauvet C. L'aménagement des milieux lagunaires méditerranéens. In: Barnabe G. (Ed) *Aquaculture*. Tec and Doc Lavoisier, Paris, 1989; 857–888.

Clément O. Aquaculture in marshes: The salt marshes of the French Atlantic coast. In: Barnabé G. (Ed). *Aquaculture, Vol 2*. Ellis Horwood Ltd, Chichester, 1990; 786–800.

Costa-Pierce B. Stirring ponds as a possible means of increasing aquaculture production. *Aquabyte* 1989; 2(3): 5–7.

Dardignac-Corbeil M.J. Traditional mussel culture. In: Barnabé G. (Ed). *Aquaculture, Vol 1.* Ellis Horwood Ltd, Chichester, 1990; 285–338.

Doumenge F. Aquaculture in Japan. In: Barnabé G. (Ed). *Aquaculture, Vol 1.* Ellis Horwood Ltd, Chichester, 1990; 849–944.

Doumenge. F. and Marcel J. Aquaculture in China. In: Barnabé G. (Ed). *Aquaculture, Vol 2.* Ellis Horwood Ltd, Chichester, 1990; 946–963.

FAO. *Production de l'aquaculture 1984–1990.* FAO Fisheries Circular No 815, Revision 4. FAO Rome, 1992.

FAO. *Review of the state of world marine fishery resources.* FAO Fisheries Technical Paper No 335. FAO, Rome 1994.

Héral M. Traditional oyster culture in France. In: Barnabé G. (Ed). *Aquaculture, Vol 1.* Ellis Horwood Ltd, Chichester, 1990; 342–380.

Hopkins K.D. and Cruz E.M. *The ICLARM-CLSU integrated animal-fish farming project: Final report.* ICLARM Technical Report 5, Metro Manilla, Philippines, 1982.

Jhingram V.G. L'aquaculture en Inde. In: Barnabé G. (Ed) *Aquaculture.* Tec and Doc Lavoisier, Paris, 1989; 1117–1152.

Jingsong Y. Integrated fish culture management in China. In: Mitsch W.J. and Jorgensen S.E. (Eds) *Ecological engineering: An introduction to ecotechnology.* John Wiley and Sons, New York, 1989; 375–408.

Korringa P. (Ed). *Farming marine organisms low in the food chain.* Elsevier Science, Amsterdam, 1976a.

Korringa P. (Ed). *Farming the flat oyster of the genus Ostrea.* Elsevier Science, Amsterdam, 1976b.

Korringa P. (Ed). *Farming the cupped oyster of the genus Crassostrea.* Elsevier Science, Amsterdam, 1976c.

Korringa P. (Ed). Farming the European flat oyster (*Ostrea edulis*) in a Norwegian poll. In: Korringa P. (Ed) *Farming the flat oyster of the genus Ostrea.* Elsevier Science, Amsterdam, 1976d; 187–204.

Marcel J. Fish culture in ponds. In: Barnabé G. (Ed). *Aquaculture, Vol 2.* Ellis Horwood Ltd, Chichester, 1990; 593–626.

Michael R.G. (Ed). *Managed aquatic ecosystems.* Ecosystems of the World 29. Elsevier Science, Amsterdam, 1987.

Nash C. Introduction to the production of fishes. In: Nash C. and Novotny A.J. (Eds) *Production of aquatic animals: Fishes.* Elsevier Science, Amsterdam, 1995; 1–20.

Nash C. and Novotny A.J. (Eds) *Production of aquatic animals: Fishes.* Elsevier Science, Amsterdam, 1995; 1–20.

Ravagnan G. *Vallicoltura moderna.* Edagricole Ed., Bologna, 1978.

Vu Do Quinh. La culture de crevettes marines au Vietnam. Carnets de la SFJO; *Soc. Fr. Jap. Oceangr.* 1996; 7.

Williamson M.R. Development of a silt pump for aquacultural ponds. *Aquacultural Engineering* 1989; 8: 95–108.

11

Freshwater, waste water, coastal water

We have shown (Chapter 6) the real, often negative, impacts of inputs of freshwater on coastal seas. However, freshwater and waste water have an immense role to play in terms of ecological management of coastal areas. There are many interfaces between freshwater, waste water and the managed ecosystems studied in the preceding chapter. The separation made is often arbitrary.

11.1 MANAGEMENT OF COASTAL WATERS AND THE REQUIREMENT FOR PURE FRESHWATER

Freshwater, produced by rain on land, is essential to man and requirements for water continue to increase, both because the human population is increasing and because the mean quantity of water used per inhabitant increases with their standard of living. Flushing toilets, bathrooms and washing machines consume a great deal, and industry even more: thus, public authorities have evaluated the mean water consumption per inhabitant at around 100 l/day for a camper, but 200–300 l/day for a city dweller. This is the "inhabitant equivalent" which, after use, contains an average of 90 g of suspended matter, 57 g of oxidisable matter, 15 g of organic nitrogen and ammonia and 4 g of phosphorus.

Agriculture's requirements for water are also very large; to produce 1 kg of food, the following estimated quantities of water are required, depending on the nature of the food being produced:

500 l:	potatoes
900 l:	wheat
1100 l:	sorghum
1400 l:	maize
1900 l:	rice
2000 l:	soy

3500 l:	poultry
100,000 l:	beef

Agriculture uses excesses of fertilisers which often pollute ground waters and many of these have nitrate levels close to those declared as unfit for human consumption (50 mg/l). Agriculture also uses 90% of the pesticides used and 7700 tonnes of insecticides are utilised each year on maize, beet and oilseed rape fields. Herbicides used on cereal fields represent 40% of the pesticides used (Bérard, 1994). Mineral fungicides (copper and sulphur) represent 18,000 tonnes/year in contrast to 23,000 tonnes/year for synthetic fungicides. It is estimated that there are about 450 types of pesticide molecule and treatment products currently being used. These active compounds degrade very slowly. Their concentration in drinking water, according to European standards (80/778/CEE), must be less than 0.1 μg/l and the total concentration of the whole, less than 0.5 μg/l. These products turn up again firstly in freshwater: in Lake Geneva, for example, concentrations of triazines vary between 0.05 and 0.09 μg/l. The majority of mammals, including humans, contain DDT, although this product is no longer used.

The lack of pure freshwater is no longer a distant threat, but a reality which can be seen in the abundance of waste freshwater or water laden with nutrients and pesticides.

Once more, the management of coastal waters may provide the solution to such a problem. A large project is planned in northern Europe on the Gulf of Bothnia, situated between Finland and Sweden, at the northern extremity of the Baltic Sea. This Gulf receives large amounts of freshwater and its salinity never exceeds 4‰. The Northern Baltic Water Exploration Council (Stockholm) proposes the construction of a dam between the towns of Umea in Sweden and Vassa in Finland (Fig. 11.1): this 20 m high and 60 km long dam would allow the storage of 1500 billion m^3 of water, for less than €3 billion of investment. Numerous unpolluted rivers are situated in the catchment area, constituting an immense funnel. Such a storage system would suffice not only for Europe, but also the other Mediterranean countries.

This example is not restrictive, but its attainment has not been approved unanimously and poses problems. Detailed mathematical and especially physical modelling studies are required to predict the impact. One large-scale, but catastrophic example comes immediately to mind, carried out by the former USSR in the Aral Sea: the diversion of two large rivers to irrigate cotton fields has led to the lowering of the level of the Aral Sea by 15 m since 1960, since the input from these rivers decreased from 50 km^3 in 1960 to nothing in 1985 and 1990 and 5 km^3 in 1989. Its surface area has reduced by half and its volume by two-thirds; salinity has increased from 10 to 30 g/l and was predicted to rise to 50 g/l by the year 2000 (Williams and Aladin, 1991). The fauna, particularly the fish, has almost disappeared (20 species out of 24, according to Swaminathan, 1992). The salinisation of the land has made it unfit for farming. Vegetation has disappeared and winds have eroded the soil to nothing. The human mortality rate has led to the exodus of populations (Levesque, 1994). In another example, the building of the Aswan Dam in Egypt, although far

Fig. 11.1. Location of freshwater reservoir in the Gulf of Bothnia.

from the sea, led to the almost complete disappearance of sardine populations in the eastern Mediterranean by reducing the nutrient input from rivers.

Thus there are developments which have harmful effects on coastal waters or large aquatic ecosystems and so, in order to be a success, a development must be a compromise between all potential partners and the subject of a long study beforehand.

Taking from the sea is not always the most profitable management solution: thus Weber (1994) reported that the Dutch planned to return 150,000 ha of farmed land (15% of its polders) to the sea in order to save money and develop in other ways.

11.2 SELF-PURIFYING CAPACITY AND PRODUCTION OF THE AQUATIC ENVIRONMENT

11.2.1 Recycling matter within ecosystems

Purification in the aquatic environment puts into play the natural processes of the functioning ecosystem, both continental (ponds, lakes, large and small rivers) and marine (the sea and also lagoons, salt basins, salt marshes, etc.).

We have seen (Chapters 3 and 4) that not only the oceans but also rivers and lakes have the capacity to degrade organic matter into inorganic matter and to use the latter in processes such as photosynthesis or pH regulation, while bacterial pollution may be eliminated by the sun's rays or by predation by bacterivores.

- Solar radiation is lethal to the bacteria *Escherichia coli* (pollution indicator) and is effective to depths of 5 m in the sea (Curds and Hawkes, 1975).
- Curds and Hawkes (1975) have shown the importance of the role played by ciliates in the control of bacterial populations and flagellates in the consumption of soluble nutrients in freshwater: the half-life of *Escherichia coli* introduced into a system without protozoans is about 16 h in contrast to 1.8 h in the presence of protozoans.

All this results in natural self-purification in biological terms: the polluters have largely relied on this capacity to justify the absence of purification on their part, but this is no longer a credible scenario. Because of the sheer amount of this pollution, there is no possible compatibility with the direct absorption capacity of the natural environment, especially with regard to chemical pollution which can become concentrated millions of times along a food chain (see below).

All these natural processes are brought into play in waste water treatment; lagunage is a simple method, but the various different eco-biological activities and intervention methods of diverse categories of organisms are clearly described in Curds and Hawkes (1975).

Odum (1969) proposed in his ecosystem development strategy the recycling of waste by using the purifying capacities of the aquatic environment. Within this framework, the waste products from a species (CO_2 or nitrogen, for example), constitute the starting point for new primary production. This technique can be used for the treatment of animal waste products. Recycling thus appears to be a form of management.

It has been shown that there is a capacity for rapid self-organisation in food chains in salt water ponds, and the formation processes of these new ecosystems which receive waste water have been described (Odum, 1989). These chains are made up of a few species and are cheap to put into operation. He draws attention to the fact that these conditions, considered eutrophic in the natural environment (green waters, little diversity, extreme variations in oxygen concentration) and judged to be unstable, unhealthy and undesirable, are in fact the basis of well-organised ecosystems whose performance in terms of production are much greater than those of conventional estuarine ponds. New ecosystems rapidly become installed in reaction to new conditions, using the species present.

We shall not return to this process, but it is an "ecological" purification technique which must be remembered, since it is situated at the interface between purification, aquaculture and management and is a good example of how aquatic ecosystems function—it forms the basis of the lagoon system of water purification, referred to as "lagunage".

11.2.2 Typical purifying ecosystem: sewage treatment and lagunage

Lagoons or oxidation ponds constitute the typical purification system for domestic waste water used by medium-sized towns in France: 200–300 l per citizen per day of waste water is carried to the lagoon site by the sewage system and arrives in a large (often several hectares) shallow (1.2 m deep) pond. The aerobic bacteria present in the water mineralise, at least partly, the organic matter while, on the bottom, heavy sediments are colonised by anaerobic bacteria. After passing through this first pond, where it stays for around 10 days, the water is transferred to a second, identical, pond (Fig. 11.2). In this, active photosynthesis occurs as a result of the richness of mineral elements available to the phytoplankton and natural light, as happens in a fertilised fish culture pond. Throughout the water's journey, bacteria are eliminated by the ultra-violet component of sunlight, which constitutes a purifying agent. A third pond completes the process; Fig. 11.2 shows the main features. On leaving the lagunage, 30–80 days after arrival, the water is pure from a bacteriological point of view. Depending on temperature and amount of sunlight, i.e. season, phytoplankton are found in abundance in the third basin (winter) up to the first (summer). Zooplankton (rotifers, daphnia and sometimes copepods) are found in the third pond, since it receives a continuous flow of microalgae from the second pond. It is profitable to recover these zooplankton. This is achieved using net bags through which water is passed using pumps or simply by gravity (Barnabé, 1980, 1990). The function of lagoons in terms of purification has been described by Edeline (1979) and the experimental use of lagoon-derived algae for rearing filter-feeding prey, by Barnabé and Périgault (1983). In aerated lagoons, which are much smaller in area and 2–3 m deep, water is mixed and aerated by electric turbines to accelerate natural processes, especially oxygenation.

Natural lagoons are 1.2–2 m deep to avoid their invasion by reeds, but ones that are not aerated artificially are more than several hectares in area, since 10 m^2 of lagoon is required per inhabitant equivalent. This system is widespread in France, with 2600 lagunages (Sevrin-Reyssac, 1995). It has mainly been adopted by small communities, but the town of Rochefort (30,000 inhabitants) utilises the largest French lagoon system (30 ha).

A method similar to lagunage has been tested in the recycling of piggery waste (slurry and waste water from washing out pigsties). This is one of the most urgent pollution problems in Europe because of the odours emitted, the difficulty of purification and the large amounts of slurry produced. There are 100 million pigs in Europe (of which the 12 million in France produce 100,000 m^3 of manure each year). Each pig creates twice the amount of pollution produced by a human, in terms of nitrogen, and slightly more in terms of phosphorus. Manure spreading requires 22 ha for 1000 pigs, is very malodorous and pollutes ground water. In Brittany, where most of France's pig-rearing occurs, this results in the pollution of coastal waters by macrophytic algae (ulva or sea lettuce) which accumulates on the beaches (see Chapter 6). De Pauw *et al.* (1984) and more recently Sevrin-Reyssac (1995) have proposed "aquatic" solutions for getting rid of pollution.

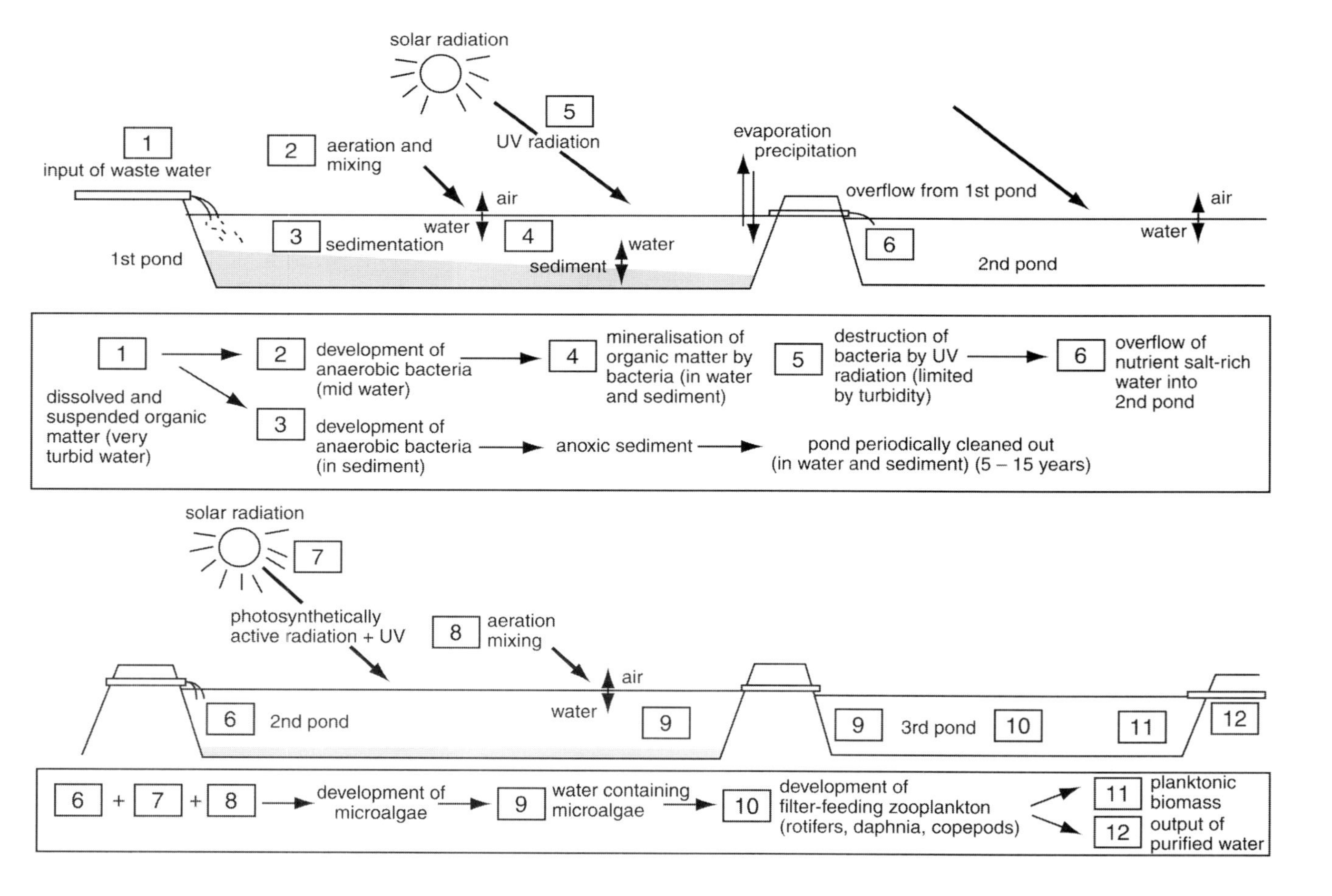

Fig. 11.2. Lagunage system for water purification and aquatic production.

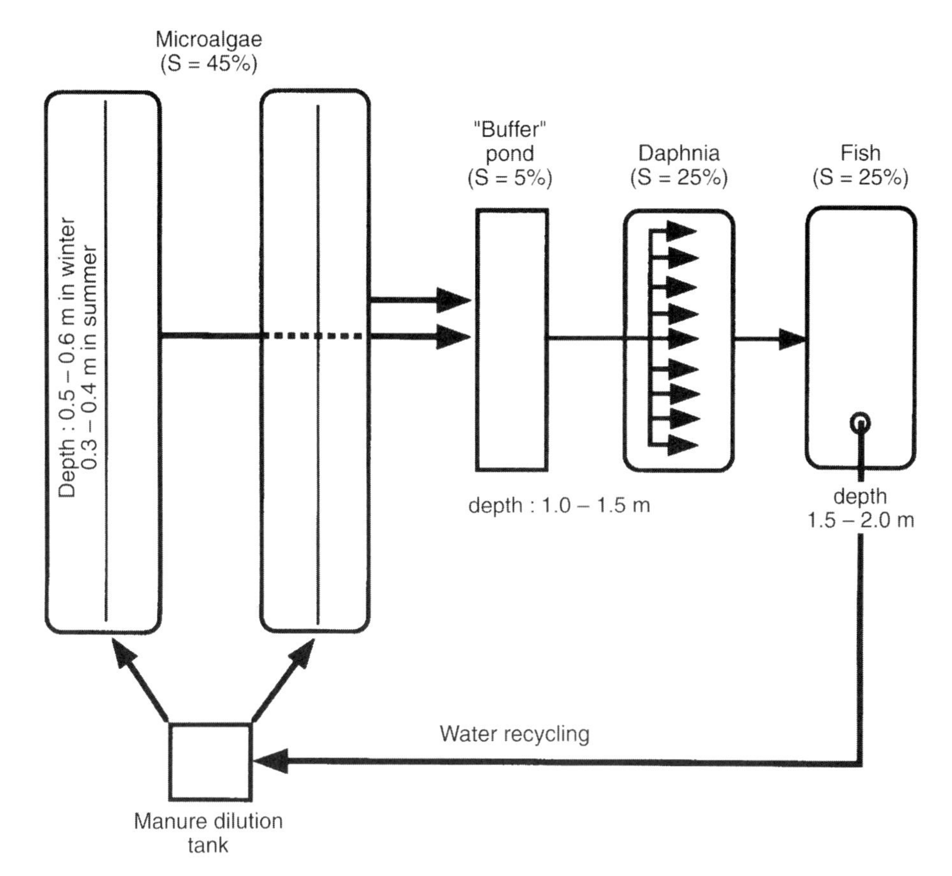

Fig. 11.3. Features of the various ponds in a lagunage system for pig manure recycling. (Adapted from Sevrin-Reyssac, 1995.)

The method proposed by the latter author is similar to lagunage and requires 10 m^2 of pond surface per pig. These 40 cm deep ponds are organised differently (Fig. 11.3). Freshwater is still used , but purified water is reused to dilute the fresh input of slurry. It is vital to inject CO_2 into the algal cultures, but air rich in CO_2 can be taken from the pigsties for this purpose. As clean water is recycled to dilute a new input of manure, the system thus consumes little (evaporation).

Purification of waste water, producing biomasses of algae, is a biological process whose setting up and functioning does not require any human intervention. Of all the microbial processes producing proteins, it is that of algae which requires the least sophisticated methods (Grobbelaar, 1979).

11.2.3 Purification by lagunage to the production of living matter

There are many indirect applications of lagunage to fish culture. In the Mèze lagunage (Hérault, France), ornamental tropical fish are reared and, in addition,

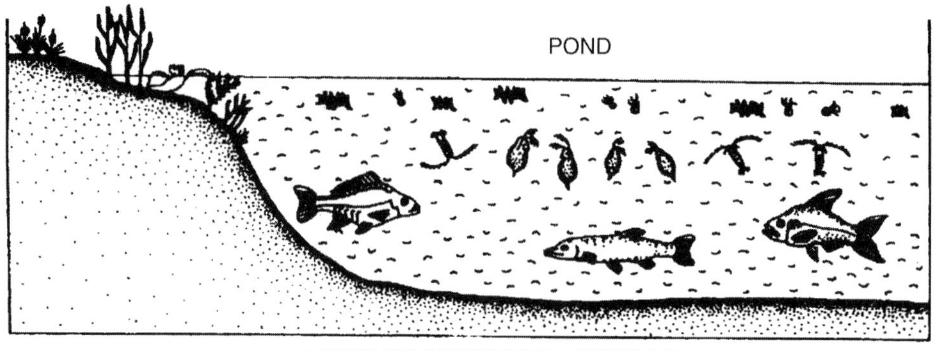

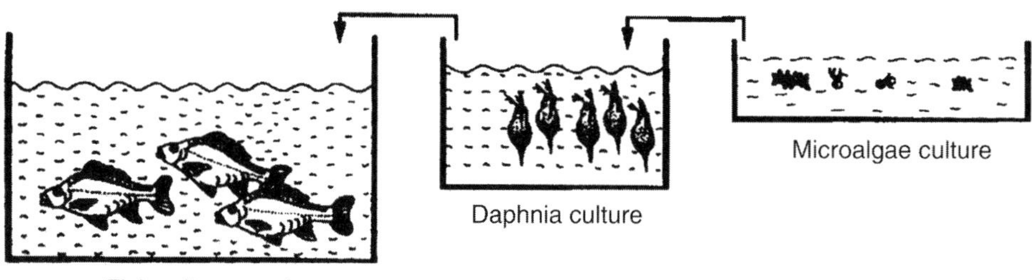

Fig. 11.4. In a fish culture pond (top), all organisms in the food chain are in the same environment. In compartmentalised lagunage (bottom), the different elements in this chain (algae, zooplankton and fish) are in separate environments. (From Sevrin-Reyssac *et al.*, 1995.)

plankton is collected for sale to aquarists or aquaculturists. Elsewhere, plankton is harvested for fish culture for restocking purposes (trout, carp, pike). The methods which have been put to use for harvesting plankton in freshwater (Barnabé, 1981) have also been used on a commercial scale for collecting marine copepods or Artemia from the ancient saltworks by the Cie des Salins of Midi on the French Mediterranean coast.

In theoretical terms, the productive capacity of lagunage in tropical areas reaches 170 tonnes of microalgae (dry weight) per ha/year. This is the record for plant production (a few tonnes for beet or maize and several tens of tonnes for sugar cane).

Much detail about the recycling of waste water and microalgae production (ecological, food and health aspects, etc.) can be found in publications edited by Shelef and Soeder (1980), Grobbelaar, Soeder and Toerien (1981) and Becker (1985). The French periodical *Aqua-Revue* has devoted an edition to reports on purification by lagunage and evaluation of biomasses produced (1994).

In relation to traditional culture of fish in ponds, lagunage is distinguished by the separation of different organisms into separate ponds, as shown in Fig. 11.4.

Rearing of freshwater prawns (*Macrobrachium*) in the end tanks of tropical lagoon systems has been successfully tested in several sites in Bengal, India (Ali *et al.*, 1993): growth is increased four-fold without any additional food. This "symbiotic" use of lagoon food chains is thus effective.

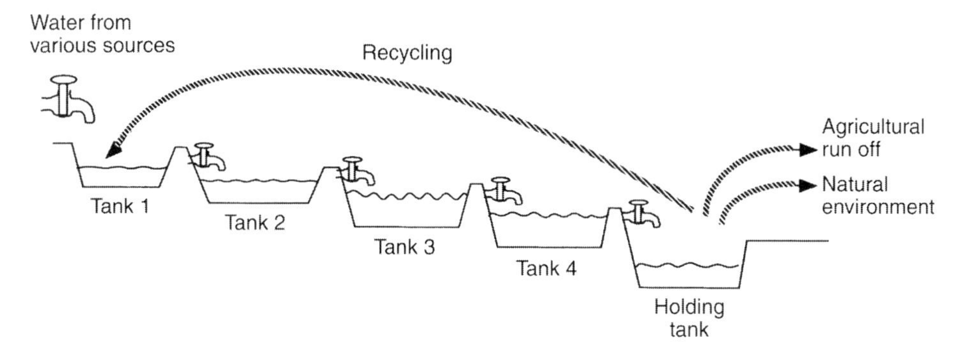

Fig. 11.5. New production methods in freshwater (adapted from various sources).

In France, especially in the Central and Champagne/Ardennes regions, new fish culture ponds no longer function according to the model of an individual pond considered as a closed ecosystem, but as a succession of semi-intensive rearing ponds with a blocked-off terminal basin used as a purifying lagoon before the water is recirculated or reused for agriculture (Fig. 11.5). The Central region has become the leader in French freshwater aquaculture, from almost zero production 10 years ago to more than 2000 tonnes today, with the accompanying creation of jobs.

We have described (Fostier *et al.*, 1980) the importance of the physical separation of the different components of an aquatic food web. In waste water lagunage, as for the purification of pig slurry or for new fish culture ponds, the different components of the food web are separated physically, but in a land-based environment using excavated earth ponds, this separation is easily achievable. This is not always possible in salt water ponds and, by definition, not in coastal waters. The problem is also one of scale. All this makes the control of food webs more uncertain.

11.2.4 Spirulina, water recycling and human health

Spirulina is a blue-green planktonic alga (*Cyanophycae*). It occurs in spiral filaments, several hundreds of μm long, made up of cylindrical cells approximately 10 μm in diameter, joined end to end. It is a warm water alga, growing fastest at 35°–38°C. It is relatively tolerant to high salinity (20–70‰) and can be cultured in sea water, requiring alkaline conditions with high pH (>9). Often it is the only entity to colonise extreme environments (pH 10.5). It is very digestible, containing 60–70% protein, one of the best polyunsaturated fatty acids (linoleic acid), a lot of beta-carotene, vitamins including vitamin B12, and the only naturally-occurring blue pigment, made up of phycobiliproteins. As a result, it is a useful food supplement for populations suffering from malnutrition, but also for westerners with poor diets containing few natural substances.

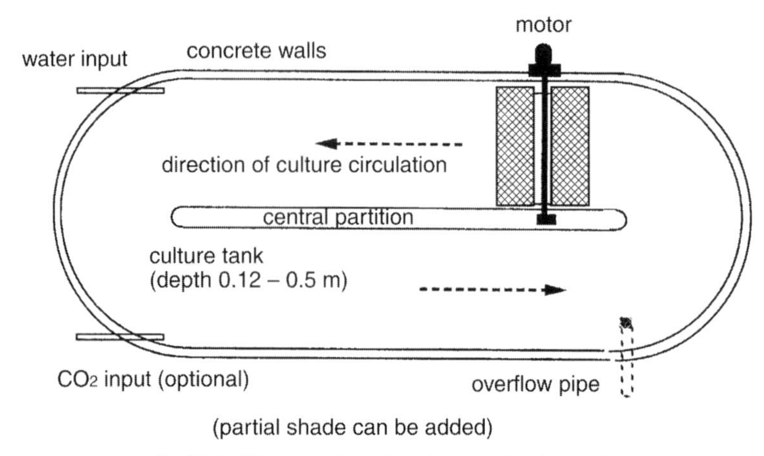

Fig. 11.6. Diagram of a microalgae production tank.

Numerous articles and publications have been devoted to spirulina. The Museum of Monaco has published the proceedings of a colloquium (Doumenge *et al.*, 1993) which gives a detailed account of the subject.

Spirulina is cultivated for food in Mexico, India, Belgium and in south-east Asia (Shelef and Soeder, 1980). Production is of the order of 1000 tonnes/year. The features of the culture ponds are summarised in Fig. 11.6. The culture ponds are most often 15–30 cm deep and their surface area varies from a few m^2 to 5000 m^2. Water is circulated (at 15–60 cm/s) and aerated by paddle wheels or, more rarely, by air lift systems. Aeration results in photosynthetic activity while reducing oxygen requirements for respiration. Alkalinity is most often achieved by adding calcium carbonate (between 0.5 g/l and 4.5 g/l). This salt constitutes the biggest running expense for this type of culture. A description of this type of system is given by Kawaguchi (1980), including the economics of operation. Production varies depending on site—under tropical conditions, it is most often between 8 and 30 g of spirulina/m^2/day (dry weight).

One of the largest algal production systems is represented by a giant solar evaporator (3200 m in diameter, 940 ha), used for the culture of spirulina in a long helical canal (the "snail"), installed by the French Petroleum Institute in Mexico (Durand-Chastel, 1980). It produces about 600 tonnes of dry spirulina/year (14 tonnes/ha/year) and constitutes the largest man-made bioreactor in the world.

Spirulina can also be used to purify domestic waste water in basins which function like lagoon systems and produce phytoplanktonic biomass from waste. Because of their organisation into spirals, spirulina can be collected by sieving, a method which is also possible for other microalgae which are several μm in size.

This type of integrated system—used for the recycling of human and animal waste in small, third-world villages and transforming it into spirulina, which can be used by man—has been proposed by the Association for the Combating of Malnutrition by Algoculture (ACMA), led by Fox (1993). Effluent from latrines and animal waste are treated in a digester and then transferred to algal ponds and stirred periodically either manually or by machines run on solar energy. Bicarbonate of soda is used to make the water alkaline. The algae are drained on a filtration cloth, then dried in the sun before consumption. There are some religious or psychological barriers to this form of recycling, but this is not the case in India or China.

Spirulina is cultivated commercially in sea water in China and Hawaii (Cyanotech commercial farm).

11.2.5 Utilisation of microalgae in coastal waters

Since the installation of a lagoon system for new tourist areas in Languedoc in the 1970s, it has proved to be a reliable source of clean seawater. The services concerned have noted very good purification of salt water, with significant production of phytoplankton.

Marine microalgae production farms exist: in Australia, *Dunaliella salina* is cultivated in 70 ha ponds. The high salinity accelerates the production of beta-carotene by this alga.

At least one marine fish farm functions with a semi-closed circuit, similar to the recycling ponds of the Central region and that is the Poissons d'Argent farm close to Grau du Roi. Salt water pools, created by the extraction of sediment for hard core for road construction, have been developed into extensive rearing ponds for sea bream, rearing water being recycled in a large, natural lagunage pond. An example of marine plankton rearing in tropical conditions in Martinique was given in Chapter 10.

In some small salt water ponds like the "claires" of the shellfish culture basin in Marennes Oléron, water management is identical in nature to that described above. By playing on the nature and origin of the waters (fresh or marine), which contain different concentrations of fertilisers, the farmer influences the composition of the phytoplankton and its dilution.

11.3 PURIFICATION AND BIOLOGICAL PRODUCTION IN COASTAL WATERS

11.3.1 A continuous process of breakdown and dilution

We have seen (Chapter 6) that waste water used by man in industrialised countries is purified before discharge, at least in theory, but that the various methods of purification used are never 100% effective. At least a portion of the polluted water returns to the ground water, rivers and then the sea. Along the way, some types of pollution evolve (by bacterial breakdown, for example), or disappear (certain minerals are trapped in the sediment or self-purified and taken up into the food chain), while other pollutants are simply diluted in the water (non-biodegradable chemicals). Examples of such pollutants are heavy metals and synthetic products such as organochlorides or polychlorobiphenyls (PCB). For a long time, the philosophy of the polluters has been that the solution is dilution.

Throughout the discharge of these products into streams, rivers, large rivers and especially coastal waters, these different aquatic ecosystems contribute to the purification of biodegradable products, as happens in lagunage.

The end products of the degradation of domestic waste water (to distinguish them from industrial water) are mainly CO_2, nitrogen in the form of nitrates and phosphorus in the form of phosphate. These are secondary constituents of sea water which are most often the fertilising substances limiting primary production in coastal waters (see Chapter 1).

According to Aubert (1965), it was Harvey in 1933 who was the first to show, in *Nitzschia,* that photosynthesis is reduced when the concentration of phosphates falls lower than 10 mg/m^3 and even lower at less than 5 mg/m^3. The development rate is independent of phosphate concentration as long as it stays above 17 mg/m^3. The same is true for nitrates, as long as their concentration does not fall below 47 mg/m^3, the lowest concentration tested.

Other substances, such as metals or silicon, are vital at very low concentrations and are sometimes factors limiting biological production. Domestic waste water contains all these substances and thus constitutes a form of fertilisation for the environment in which it ends up. The sun and osmotic shocks can get rid of residual viral or bacterial pollution.

11.3.2 Huge purifying and production capacity

In gross terms, domestic pollution can be treated and digested in some way in coastal waters and can constitute the starting point for new biological production, rather like a "giant" lagunage system.

The preceding examples demonstrate this point—they prove that food chains functioning naturally in extremely oligotrophic* coastal waters slow down well below true potential production levels due to a lack of nutrient salts (see also Chapter 4). Thus coastal waters are not sufficiently enriched to be very productive.

It can be said that the increase in fisheries catches in some oligotrophic seas can be attributed to the increased input by rivers of nutrient salts (Chapter 7). Thus not everything in waste water of human origin is harmful. In the Mediterranean, sediments from the Rhone are deposited in the Golfe du Lion (Fig. 6.2) (Ekman transport, see pp. 59 and 138); as a result, this gulf is one of the most abundant areas for fish in this sea. These sediments also contain many toxic substances; we shall return to this. In contrast, the construction of the Aswan Dam, depriving the Mediterranean of Nile spates, led to a catastrophic decrease in production of small pelagic fish in the coastal waters fed by the river.

In coastal waters, even if the actual separation of material at different trophic levels no longer exists and if biological productivity is lower than in lagunage, there is a change in dimensional scale. These waters must now be seen as an immense solar photoreactor with the same transforming and production capacities as lagunages, but with surface areas of hundreds of km^2 and volumes measured in km^3.

The consequences are essential in terms of ecotechnological development. Without being infinite, the capacities of aquatic environments for production and recycling are immense, in relation to their current natural production (exploited by fisheries) which is minuscule. However, it is not proposed that more waste should be discharged into the coastal zone, but that its potential should be considered.

The management of these immense coastal ecosystems, often limited by a single physical barrier in the form of the coast, is however as simple as that of continental (fresh) waters, and biological production is also affected by physical and climatic factors (Chapter 4).

Water carried by rivers (and thus waste water), ensures the fertilisation of coastal waters, but also carries chemical pollution and undoubtedly contributes to the increased frequency of coloured waters in coastal seas (Chapter 6).

11.3.3 Pollutants and fertilisers: quality and quantity

We have seen (above and Chapter 1) what the optimal levels are for secondary constituents for good primary production. The problem is that there is rarely sufficient arriving in coastal waters to allow this optimal production. In terms of what is required for good primary production, we are considering oligotrophic waters, whereas beyond that, we are speaking of eutrophication (Section 6.4).

Nitrogen is often limiting in the marine environment, but one of its pre-breakdown forms, ammonia, although capable of being assimilated by some algae (as fertiliser), is toxic to reared animals above 0.1 mg/l and must be considered as a pollutant. Iron is considered as a pollutant by the FAO (see Chapter 6), but limits production in some large oceanic ecosystems. Copper is essential at low concentrations (a few μg), but at slightly higher levels (50 μg), leads to mortality in fish larvae. Distinguishing between pollutant and fertiliser is thus a matter of quantity for these substances. We have already reported some figures concerning the significance of marine pollution (Chapter 6).

Alongside these naturally-occurring compounds, which take part in ecosystem function, there are around a million types of man-made molecules. Their destiny is not always known, but some of them are highly toxic (pesticides, detergents, chloride products, phenol derivatives, etc.). This is one of the main problems in environmental protection, but there are others which relate to aquatic food chains.

11.3.4 The impasse of chemical pollution

Phytoplankton is sensitive to pollutants such as acids or oils, but DDT or PCBs are extremely toxic. Two parts per billion of a DDT derivative, DDE, inhibits photosynthesis and cell division in at least one species. Examples of chemical pollution in coastal waters are numerous, and their quantitative and qualitative significance should be reviewed.

- PCBs change the composition of planktonic communities, favouring forms which are small in size, which has repercussions for the upper levels of the food chain structure with regard to the relationship between the size of the eater (predator) and the eaten (prey). Plankton in the open sea, which are

adapted to life in an environment characterised by its stability and water "purity", are more sensitive than coastal plankton—thus discharging more and more waste into the open sea is not without consequence. In addition, phytoplankton largely determine the oxygen concentration of the atmosphere.
- Hydrocarbons are one of the most abundant categories of pollutants in the oceans. According to the American Academy of Sciences and the United Nations, between 1975 and 1982, 6.5 million tonnes were discharged into the sea every year. Hydrocarbons can asphyxiate, poison and deform marine organisms and are toxic to eggs and larvae at levels of less than 10 parts per billion (ppb).

The most abundant waste product in the oceans is without doubt human waste which amounts to hundreds of millions of tonnes.

- The USA discharges 100 million tonnes of waste annually into coastal waters, and 30 million m^3 of waste water. Part of this waste is biodegradable and so is thought to be absorbed by the assimilation capacity of the oceans, but such large figures are not on the same scale as this capacity. An area of the Atlantic, 12 miles out from New York and New Jersey is the dumping ground chosen by these conurbations which deposit solid waste or sludge from barges at a depth of 30 m. In 1987, it was estimated that the amount dumped was 9 million tonnes of sludge waste, part of which originated from New Jersey where 15,000 enterprises produced more than 4000 m^3 of liquid chemical residue per annum, in addition to 3 million tonnes of industrial waste. These sludges are more toxic than the waste from New York because of the industrial content.

These waste products are very dangerous since they contain solid human waste, pathogenic viruses and bacteria, toxic waste (PCB, DDT) and heavy metals. The mud comes from rain, city streets, some from treatment stations and some from WCs (half of the waste water from New York is not treated).

On this scale, dilution, dispersion, sedimentation and the assimilation capacity have little effect. Such an assimilation capacity does not exist for chemical pollutants which, in contrast, accumulate in food chains (see section below) or intervene in the delicate interplay of telemediators*.

In France, pesticides (insecticides, herbicides and fungicides) used per marine catchment area have been estimated as follows (in tonnes/year) (source: *Science & Vie* 1996; 945; 117):

North	3,141
Brittany/Normandy	3,766
Poitou-Charentes	5,597
Aquitaine	1,458
Languedoc-Roussillon	11,170
Provence	2,245

Figures cited (in tonnes/year) for inputs of heavy metals to littoral areas by French rivers are:

Mercury	from 2.6–5.6
Cadmium	35.5
Lead	500
Copper	60
Zinc	230

According to the same source, concentrations of mercury in molluscs vary from less than 0.2 to more than 0.4 mg/kg of dry weight in relation to rearing zone.

Chemical pollution not only constitutes a threat to life in the oceans, but also the survival of the human species, as incident prevention and control are poorly understood.

11.3.5 Biological concentration of pollutants or bioaccumulation

Substances such as mercury, other heavy metals and numerous synthetic products (pesticides) are not only not broken down in the aquatic environment, but are concentrated by factors of millions by aquatic organisms all along the food chains. Beluga whales (Cetacea) in the Gulf of St. Lawrence have, for example, been rendered sterile by organochloride pesticides over a period of several years. This type of concentration also implies that a mother who eats tuna and then breastfeeds her child transmits more poison to it than she absorbs herself.

The terrible Minamata disease caused the death of many people who had eaten fish which had concentrated the mercury discharged in water and recycled by food chains. It is estimated that 5% of the mercury discharged into the Mediterranean turns up again in the fish in this sea, which could render some pelagic species unfit for human consumption (Cossa, 1995).

Throughout the world, the consumption of mussels and oysters is prohibited in various production basins from time to time, as they have concentrated filtered phytoplankton toxins or pathogenic bacteria in their flesh. This is thus a widespread threat linked to either a direct or indirect process of pollution.

The solution to this type of problem is easier when it concerns bacteriological than chemical pollution. The purification of waste water, or the obligatory passage through a purifying system of shellfish destined for human consumption (as has been established in the EU) could more or less resolve the problem. In contrast, chemical pollution must be treated at source according to the principle of polluter pays, but this is a long way from being put into practice.

We have seen how bioaccumulation can also be used to get rid of pollution (e.g. in mussels or water hyacinths). Algine, extracted from marine algae, has the capacity to concentrate strontium 90.

In anecdotal terms, we can add here types of pollution which are even less known than the hydrocarbons: the bed of the Mediterranean is covered by about 200 million objects and along the beaches of the coast of the United States, around 1 million pairs of old shoes are found every year.

11.4 NEW IDEAS RELATING TO PURIFICATION IN COASTAL WATERS

11.4.1 Direct dilution of waste water in coastal waters

Waste water, whether purified or not, is most often discharged into the sea at varying distances from the coast, through an outlet pipe anchored to the seabed. The low density of the freshwater reaching the end of the pipe means that it ascends to the surface, forming a cone-shaped "plume" which results in an often circular area of freshwater at the surface, which has not mixed with the sea water. This mixing occurs over several hundred metres around the discharge point, as a function of the physicochemical parameters regulating this difficult process (see Chapter 2). Dilution from freshwater to salt water is very progressive, and allows many bacteria to adapt and survive, whereas a severe osmotic shock would prove fatal.

Mathematical modelling of these wastes shows that, in terms of hydrology (which is itself driven by winds and currents, i.e. the climate), the mixing and dilution of effluents in sea water are not inevitable. The drift back towards the coast and the trapping under the thermocline are still threats, both for water quality on beaches for bathers and for mussel-rearing operations. The physical problems described in the Chapter 2 are found again here.

It can be noted that the only solution currently envisaged in all of these studies consists of leaving the waste to spread passively from the end of its outflow pipe, despite all these drawbacks. The sophistication of the theoretical study is in stark contrast to the passivity with which man releases waste water into the sea. This is a new example of the importance that pure research has taken on, over that of research into practical technological solutions (see Chapter 8).

However, various methods can contribute to the rapid mixing of waste water with sea water.

- The first consists of injecting waste water under pressure in such a way as to use the flow of waste water to initiate mixing on the sea bed, using a Venturi type mixer, for example. The kinetic energy of flow and the differences in density mix the fresh and sea water. Dilution is rapid and avoids the formation of a plume at the surface. The toxicity caused by high concentrations of possible pollutants is diminished and the osmotic shock destroys the majority of bacteria and viruses almost instantaneously.
- Mixing can also be achieved using submerged circulators which are already used in many areas (see Chapter 12). This equipment is capable of mixing waste and sea water very quickly, which also contributes to bacterial purification.
- The injection of compressed air to achieve this dilution or to move effluent to the surface is also possible, without the need to install submerged equipment; devices already exist which use this principle of gas injection. They are used in freshwater lakes to mix oxygen-poor water near the bed with surface water, which is stratified into layers which do not mix naturally (see Chapter 12).

- The splitting up/division of the waste at several points or several tens of points over a hundred metres from the end of the outflow favours the rapid dilution of freshwater in salt water.

These solutions favouring mixing would speed up the functioning of a typical marine water purification plant, whereas current methods allow bacteria—indicators of pollution—to survive, because of the slowness or absence of mixing of fresh with salt water.

11.4.2 Recycling of waste water and microalgae aquaculture on land

In addition to the potential for lagunage in sea water as described above, Fox (1993) proposed various systems which may or may not involve the recycling of waste water in spirulina production; we shall only refer to those systems using sea water.

This author noted that 20,000 km of arid marine coastline could be used for this culture. The installation of 5000 ha of culture basins alone could produce enough spirulina to eradicate the malnutrition which affects 50 million children in the world. To achieve this, it would suffice to provide 15 g of spirulina per day. Fox calculated that 750 tonnes of spirulina/day would have to be produced. It would also require 500,000 m^3 of water/day and 100,000 tonnes of bicarbonate.

He proposed the digging of ponds, 1000 m long, 50 m wide and 2 m deep at sea level, in coastal sand, with a floating excavator. The extracted sand would be used to create a dyke isolating these ponds from the sea; this dyke would serve as a breakwater and a filter for sea water. It would be covered in halophilic vegetation and watered by salt water. Floating tanks, 15–20 cm deep, made of PVC film, would be installed on these "ponds". What is feared is that, over the surface of the farm (120 ha), ground movements would be likely to change the wave height in very shallow water, if digging was carried out directly on the littoral sand. Evaporation would be used to obtain salinities of 60‰. The author gave figures for his project: 40 farms would require an investment of about €27.5 million each, with an annual running cost of €11 million each. He comments that this price corresponds to that of 1.33 days of the Gulf War.

The second concept proposed by the same author is an analogous system in which urban waste water is transported by tanker, the cost of treatment being covered by the town. Other methods of finance (personal subscription, tax or surcharge on military expenditure) are proposed.

The economic validity of such concepts can be contested, but their technical validity is accepted. Methods for rearing spirulina are well established, even on a large scale (900 ha in the case of the Sosa Texaco bioreactor). Spirulina is a promising alga whose production reaches 15–20 tonnes of dry flour/ha/year (as opposed to 4 tonnes of wheat flour, which is 20 times less rich in proteins) on agricultural land. It is easy to harvest, can be grown in waste water and also in marine waters and is, in fact, an ubiquitous species in tropical waters.

The only element which appears to pose a problem is that of land use: the shoreline, even when arid, is coveted space and experience has shown that, in low-lying tropical areas, the rearing of penaeid shrimps takes precedence, as it is more profitable.

11.4.3 Floating purification plant/station

A prototype floating purification station was constructed in Japan for the municipality of Nagasaki. With a 37,000 m^3 treatment capacity (activated sludge, oxygenation tank), the plant measures 80 m by 40 m, with a depth of 8 m. It can be moored at a quay or at moorings. Wave height must not exceed 3 m for it to function effectively. At a cost of €70 million, it costs €15 million less than a land-based purification station.

A floating purification station can also be developed on board a boat, like the aquaculture hatchery installed on an old cargo ship in Monaco, by the P2M company (cited in Barnabé, 1995). This 180 m long ship, anchored in 250 m of deep open water not only functioned very well as a hatchery, but also completely resisted the forces of the sea. It was big enough to have installed a purification plant for a large city, using the aerated lagunage or activated sludge method. Recycling waste water thus allows the redeployment of old ships which are no longer profitable for use in marine transport. This is therefore a very economical solution, in view of the price of coastal land.

In all the examples described above, methods already exist (cages or pens and moorings for aquaculture in the sea, offshore oil platforms, shipping developments, flexible reservoirs/tanks, etc.).

11.4.4 Recycling of waste water and bivalve aquaculture

Filter feeders, including planktonic or benthic animals, are typical of aquatic environments. Their large size and the fact that they are often in a sessile/fixed position offer numerous opportunities for development. We shall restrict ourselves here to the "purifying" aspect.

The utilisation of waste water for feeding bivalves was tested in a pilot scheme in the USA during the 1970s (Mann and Ryther, 1977). In France, the Ecotron Programme produced some promising results on a laboratory scale in the late 1970s, but has since been suspended.

In Israel, the Philippine clam (*Ruditapes semidecussatus*) has been reared in the effluent from marine fish culture ponds. In 13 months, their mean weight went from 1.9 g to 18.3 g despite summer temperatures of between 27 and 31°C and despite densities between 2000 and 5000 per m^2. It is true that the effluent was full of phytoplankton. This combined use reduced the abundance of phytoplankton and provided a highly-valued aquaculture product (Shpigel and Fridman, 1990).

In the UK, mussels are used purely to clean up the water in the river Mersey. The method consists simply of establishing the mussels in the estuary; these mussels filter the water and return to the water "a clarity which it has not had for a long time".

In Chesapeake Bay in the USA, oyster banks are used for the same purpose; artificial reefs have been installed to allow the fixation of oysters which filter the water.

Filter feeders are also used in other countries, mainly for the purification of waste water from shrimp-rearing units where these have become widespread, such as in Thailand (Lin *et al.*, 1993). These filter feeders concentrate chemical pollution, but also consume bacteria and thus act as true biological and chemical filters. We shall return to these filter feeders in Chapter 15.

The use of molluscs for purifying purposes in polluted marine zones (by chemical or bacterial means) can thus be regarded as a treatment method for these pollutants. However, it obviously is not clear to those who treat large amounts of water and propose land-based treatment methods requiring huge investment, which are much more lucrative but less effective in purifying terms (see the colloquium proceedings published by Guillaud and Romana, 1991). Construction of channels and treatment stations in coastal waters is not complicated and can be based on shellfish culture techniques, either on the seabed, or suspended. All this is not very costly. It involves a change in strategy and the installation of rearing units which are not providing food. Integrated into the environment, these technologies are also likely to create employment.

All this evades the fact that shellfish culture zones situated in clean areas (production of filter feeding molluscs destined for human consumption), from time to time play the role of backup purification plant, to the detriment of the consumer and the shellfish farmer, when treatment stations—overloaded, inefficient or absent—are not purifying the water. Clean areas of water would therefore gain from being surrounded by rearing zones which would provide evidence and a buffering effect, rather than the imposition of purification techniques, which in any case do not treat chemical pollution, the most dangerous form.

11.4.5 Biofiltration of domestic waste water

Biological purification by bivalves could be utilised in urban littoral purification plants, after dilution by half with sea water. A 20 g mussel filters about 20 l of water/day, thus 10 mussels (0.2 kg) are sufficient to "filter" the water used by one inhabitant (i.e. 20 tonnes/100,000 inhabitants).

This is obviously a type of culture where food is not the goal. This is a system that allows the clarification and perfect purification of water from littoral communities on its exit from the treatment plant, before being discharged into the sea. The water quality on the beaches should improve.

Another technique, this time involving the treatment of activated sludge, has been proposed by a German study group (Zeeck *et al.*, 1992). It uses a detritivore, the marine and brackish water worm *Nereis diversicolor* which withstands desalination and pollution, for the direct purification of urban waste water treated by activated sludge. The mud is deposited in a tank lined with sand and stocked with *Nereis* (about 2000/m^2). Waste water is introduced into the tank, and the *Nereis* filter the water and extract suspended matter and the organisms within it. The movement of

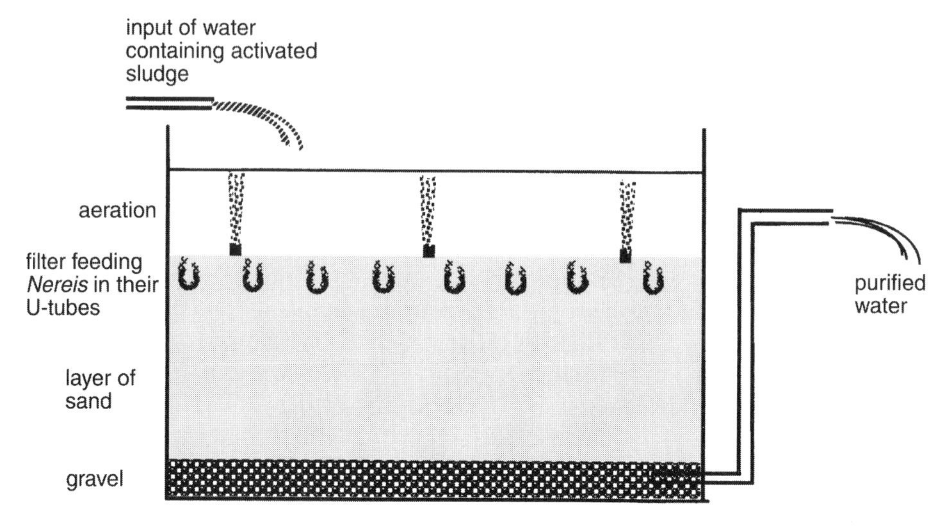

Fig. 11.7. Treatment of activated sludge by the polychaete *Nereis diversicolor*. (Adapted from Zeeck *et al.*, 1992.)

water through the sand bed completes the purification process (Fig. 11.7). The authors claim that this process is a complete purification method.

11.4.6 Waste recycling and fish culture

The largest aquaculture production in the world, that of carp in China (more than 10 million tonnes in 1997, according to FAO statistics published in 1998 [FAO website <http://www.fao.org>], representing two-thirds of the world's freshwater aquaculture production and one-third of total aquaculture production), results from recycling manure from farmed animals, bran, rice, distillery waste, grass, slugs, etc. in fish culture ponds; the process is widespread and is proven. Edwards (1980) pointed out that the interest in rearing fish from waste resides in the fact that 60–72% of the nitrogen and 61–87% of the phosphorus in animal feeds are found in their excreta.

We do not wish to dwell upon these techniques described in aquaculture publications (Marcel, 1990; Jhingram, 1990; Barnabé, 1991), but production levels can reach 10 tonnes/ha without added food. These integrated recycling and aquaculture systems once again demonstrate the intensity of the biological transformation process of matter, within aquatic ecosystems. Lagunage is therefore not the only type of aquatic ecosystem capable of exceptional levels of production where animal rearing is concerned.

In Asia, this type of recycling–rearing system is very widespread, as there is no psychological barrier to the use of waste products as is the case in the West. In Thailand, for example, a farmer can have up to three levels of rearing: chickens at the top, whose droppings are used by pigs situated at an intermediate level, on grids/racks, above a pond full of fish. Edwards (1980) also reported several examples of direct recycling of human excrement in fish culture ponds in China (where the reprocessing of animal and human waste for agriculture is widespread and amounted to 299 million tonnes in 1966) and in Bengal in India. There have been few health problems.

In order to counteract the health obstacles concerning waste water of human origin, trials have been carried out in South Africa of rearing tilapia in tanks, in sea water fertilised with domestic effluent which has been previously treated in a purification plant and disinfected with chlorine (Turner *et al.*, 1986). The natural production induced by this input provides the only food. According to the authors, the introduction of this effluent, also monitored in terms of its physicochemical effects, constitutes a free and adequate source of fertiliser. Fish production has been between 3.9 and 8.1 tonnes/ha for durations of between 210 and 226 days. In Tunisia, at Monastir, waste products from a school canteen have been used for the rearing of grey mullet in sea water ponds.

In Italian valli*, water from two intensive rearing farms for bass and sea bream has been recycled, then in turn used in an extensive farm situated downstream, for production of sea bass, sea bream, grey mullet and eels (Melotti *et al.*, 1992). The results are affected by the fact that there are water quality problems in these shallow ponds (60–80 cm deep), but the extensive fish production (all species combined) was about 72 kg/ha/year.

Although production of marine fish may be lower, these examples show that the potential is the same in both fresh and salt water.

11.4.7 Aquatic vegetation, macroalgae and water purification

In freshwater, the water hyacinth is used to get rid of water pollution (polluted water from papermills, for example). Water hyacinths absorb mercury, cadmium, nickel, lead, gold and silver. Mann Borgese (1986) proposed their collection as soon as a method of metal extraction becomes profitable.

While aquatic plants and marine macroalgae are capable of extracting nutrient salts from coastal waters, it is not yet known how to use them owing to their bulk and the difficulty involved in their elimination or the use of such a biomass. They thus constitute, at present, a nuisance, of which we have several examples (Chapter 6).

Despite this current technological limitation, there is a promising biological process. In Japan, a variety of *Ulva pertusa*, or sea lettuce, has been isolated by the University of Kagoshima. This variety contains more vitamins and minerals than chlorella (whose nutritive value is greater than that of egg white or powdered skimmed milk), and is capable of absorbing 2.7% of its own weight in nitrogen and 0.2% of its weight in phosphorus. It grows five times as fast as normal strains of *Ulva pertusa* and withstands high temperatures as well as less clean water conditions than are required by other species (Anon., 1992). Knowing the difficulties in rearing, collecting and the cost of chlorellae (reared for human consumption in Asia) and the massive proliferation of ulva, there is no doubt that, if it can be achieved, it is a promising source of cultured algae. Ulva contains up to 4.75% nitrogen, or 15–20% protein in terms of dry matter.

In Israel, Vandermeulen and Gordin (1990) have shown that the sea lettuce, *Ulva lactuca*, placed in effluent from intensive fish rearing ponds extracts up to 85% of the ammonia in this effluent, independently of fluctuations in temperature and also in the lack of light.

The sea lettuce, *Ulva lactuca*, when dried and incorporated as 20% of the food for juvenile bass (*Dicentrarchus labrax*), significantly increases its growth (Gabaudan, 1976, unpublished). While the incorporation of soy cake or other added plant proteins has been studied, it is curious that the incorporation of macroalgae, which are freely abundant, has not been studied more in fish diets.

11.4.8 Intensive aquaculture in sea cages and water pollution

The widespread pollution of many coastal areas comes mainly from agriculture and industry, which discharge their waste into the aquatic environment via streams and rivers. River catchment areas are immense and the amount of pollutants ending up in the sea is enormous (see Section 11.3.4), a different order of magnitude to the pollution attributed to intensive aquaculture in sea cages in coastal waters.

Waste products from fish consist mainly of nitrogen, phosphorus and carbon dioxide. This discharge of organic carbon and mineral salts into the aquatic environment by intensive rearing units does not necessarily do damage. They can be harmful or beneficial to the receiving environment and to automatically consider them as harmful pollutants, without considering their characteristics and those of the receiving environment, is not realistic.

It is, however, all a matter of scale. The installation of cage-rearing systems in a small lake (in freshwater, productivity is limited by phosphorus concentration) will modify its functioning. Solbé (1988) thus calculated that the production of 1 tonne of fish, entailing 10 kg of discharged phosphorus, in a lake 500 × 100 × 10 m, could lead to its eutrophication, synonymous with pollution. P concentration would change from 0 μg/l, characteristic of an oligotrophic lake, to 20 μg/l, that of an eutrophic one. The production of 1 tonne of fish would also result in the discharge of 68 kg of nitrogen into the environment. Such situations demonstrate the role which ecological studies must play in the development of small bodies of water. The production and absorption capacities of the ecosystems must therefore be estimated so that intensive aquaculture can be integrated into the management of a given environment, in the same way as addition of fertilisers or more extensive activities.

With regard to the surfaces and volumes involved, unless they are gigantic, intensive aquaculture rearing units in the open sea cause very little modification to the aquatic environment. The reason for this is simple: the quantities of waste products created are several orders of magnitude smaller than those produced by agricultural or industrial activities in the catchment areas. Fish do not discharge phenols, pesticides, mercury or hydrocarbons. In the open sea, the waste products from the fish are recycled in food chains. Has it ever been said that a zone abundant in fish pollutes the sea? This is what the systematic criticism of sea cages leads us to believe.

However, nowadays it is fashionable to be preoccupied with this kind of pollution and scientific studies are flourishing. Cages affect the "underwater prairies" below and this impact has been quantified, but without taking into account any of the other roles the cages may play in the environment (deadening of the effect of waves, input of nutrient salts, etc.) or their socio-economic impact (employment).

11.4.9 Treatment of algal blooms or "coloured waters"

We have examined the problems of massive proliferation of macroalgae and microalgae in Chapter 6, and seen their seriousness.

Some solutions have been put forward: the use of low concentrations of an amino acid (threonine) proposed by Japanese authors does not seem to have gained widespread acceptance; Benlahcen *et al.* (1995) reported "very promising results" from using a few mg/l of copper sulphate together with champagne chalk, but these results were obtained in freshwater and it should be remembered that copper sulphate is toxic above about 1 mg/l and sometimes at 0.1 mg/l.

11.4.10 Oil pollution and new technology

One of the most common and most visible pollution problems both in natural coastal waters and in ports is pollution by hydrocarbons (see Section 11.3.4 above). When the discharge is very significant, such as the running aground of oil tankers on the coast (e.g. Amoco Cadiz, Exxon Valdez) environmental recovery must be worked towards, using appropriate materials such as floating booms, aspiration using a vortex pump, spreading of detergents and absorbent powders, etc., and the physical cleaning of the environment when hydrocarbons reach the shore (brooms, brushes, buckets and spades).

It is also possible to eliminate these substances using biological processes, at least when the amounts involved are not very great. Many oil and biotechnology companies have perfected genetically-modified stocks of bacteria for this purpose: the DNA in these bacteria has been given a gene which codes for the synthesis of enzymes which can break down hydrocarbon molecules. As in all biological processes, the activity of these bacteria is temperature-dependent and also depends on the appropriateness of the enzyme to the hydrocarbon to be broken down, i.e. the specific composition of the hydrocarbon. Thus there are many formulations. The technology for recombinant DNA is not only used for hydrocarbon degradation: dioxins, waste oil and also many other pollutant or toxic molecules can be eliminated in this way. These "recombinant" organisms are also capable of concentrating metals and so might be used for this purpose.

Their use in the marine environment, however, must be considered with caution (precautionary principle) since their dispersal could pose irreversible problems in terms of environmental and public health: these genetically-modified organisms must not be allowed to persist in the environment. For this reason, other avenues of research are being pursued on the isolation and selection of bacterial stocks from the natural environment which are capable of breaking down hydrocarbons; they do not have the same drawbacks as genetically modified organisms.

11.4.11 Unfairness of regulation

Pollution attributed to aquaculture is not limited to real or alleged physicochemical pollution. The presence of cages or pens in the sea also spoils the waterside view (aesthetic pollution), as if the countryside and the use of the public maritime domain is under common ownership.

Such facts, in reality, are overtaken by the "powers that be", as is shown in the following examples.

- In the 1970s, the catches due to underwater fishing in coastal waters were thought to be threatening the fish stocks; in France, for example, the result was strict and often absurd regulations. They did not affect the overfishing, which had nothing to do with the underwater fishermen (Chapter 7), and underwater fishing thus acted as the scapegoat. Today it is the fish farms which play this role, forming a "smoke screen" that masks the chemical pollution caused by industry and agriculture which is much more dangerous for coastal waters.
- Five per cent of the mercury discharged into the Mediterranean finds its way into fish; this has already rendered some pelagic species unfit for human consumption (Cossa, 1995), but how many polluters discharging mercury into rivers have been convicted?
- The transport of aquatic or terrestrial species from one place to another is regulated, but thousands of aquatic species are transported in ships' ballast water, without any controls, between all the ports in the world. These introductions may be responsible for the phenomena of toxic coloured waters (algal blooms) (Chapter 6). An *ad hoc* commission is carrying out research on this subject, another way of saying that all form of regulation is being postponed indefinitely.
- Scientists visiting our laboratory at Sète and taken to see the fishing port, have been shocked at the catching of juvenile bottom-dwelling species by trawlers: little scorpion fish, mullet and other fry, 3–4 cm in size being sold in boxfuls; we eventually had to stop doing this visit. This bottom trawling has almost disappeared, as there are no more fish; now semi-pelagic trawlers catch sardine and anchovy. Again, juveniles and fry are caught intentionally and sold. The same fisherman go further and further out to sea, complain about the decreasing fish populations, but none of them has proposed an increase in mesh size of their trawl nets.
- Fishing for fry is still allowed in France on the Côte d'Azur, where adult fish populations of fish have already disappeared as a result of overfishing. At the same time, research organisations are financed to study these populations, especially the juvenile stages, in relation to recruitment (see Chapter 7). Fishing for fry is claimed by the fishermen to be "traditional". The

situation is the same for glass eels/elvers. In the 1970s, 1200 tonnes/year were caught in France; now catches have fallen to below 500 tonnes and continue to fall. "It's the pollution," say the fishermen, who declare themselves innocent, of course.

11.5 CONCLUSIONS

The self-purifying processes exploited by man in lagunages or fish culture ponds are an integral part of aquatic food chains in the natural environment. These processes function equally well in fresh, brackish or salt water, but with different dominant species. The starting point of the food web is primary production. In the natural environment, the concentration of nutrient salts is limited, which restricts this production. Waste water imports organic matter which, once broken down by bacteria, provides the nutrients used for primary production by macroalgae.

The difference between natural ecological processes is that of intensities. While aquatic food chains function slowly in natural conditions at very low levels of nutrients salts, they are capable of working with a yield several orders of magnitude greater when the environmental conditions (richness in nutrients, temperature, light, etc.) are optimal.

Both the recycling intensity (i.e. the purification rate) and that of new production can be accelerated, since they are components of the same natural process. This potential is exploited in freshwater, for the production of microalgae, purification by lagunage and fish culture in ponds. Examples in coastal waters are still rare, but illustrative. The prospects which we propose are both immense and completely accessible in ecotechnological terms.

Ecotechnological development is thus in an extraordinary and unique situation. It can allow both the generation of production of aquatic organisms which can be used by man and also ensure the large-scale recycling of "biodegradable" pollutants dispersed in the aquatic environment. Waste water provides new interfaces between man and nature and constitutes a major opportunity of ecotechnology. The idea is to use technology less and the complexity of nature more, to transform waste within an ecosystem which is useful to man (Odum, 1989).

It is in the protection of the aquatic environment, severely polluted by human activity, that the development of coastal waters should play a major role. Techniques for purification and recycling of waste water use the same biological processes as those used in fish-farming activities. We shall return to these prospects which continually extend the field of this discipline.

The majority of the human population, gathered on a long, finite coastal strip, is now looking to coastal waters for its future: ecotechnological development is the incontrovertible solution.

In contrast, chemical pollution does not lend itself to management and thus must be controlled at source. It constitutes an insidious threat to the survival of the human species, because of its many effects and the absence of ultimate treatment.

11.6 BIBLIOGRAPHY

Ali S.A. and Khan M.A. Culture prospects of *Macrobrachium rosenbergii* in domestic waste water ponds. In: Proceedings of the National Seminar in Aquaculture Development in India, Kerala University Publication, Thiruvananthapuram (India); 27–32.

Anon. Algue proliférante, une aubaine. *Flash Japon. Soc. Jap. Oceanogr.*, 1992; 44: 3.

Aqua Revue Editorial. Dossier "Eaux usées et lagunage". *Aqua Revue* 54, 1994; 15–32.

Aubert M. *Cultiver l'ocean. La science vivante.* PUF Ed, 1965.

Aubert M. Les risques en matière de pollution microbiologique de la Méditerranée. *Rev. Int. Oceanogr. Med.* LXXXXVII–LXXXXVIII 1990; 6–20.

Barnabé G. Système de collecte du zooplancton à l'aide de dispositifs autonomes et stationnaires. In: Billard G (Ed) La pisciculture en étang. INRA, Paris, 1981; 215–220.

Barnabé G. Harvesting microalgae. In: Barnabé G. (Ed) *Aquaculture*, Vol 1. Ellis Horwood, Chichester, 1990a; 207–211.

Barnabé G. Harvesting zooplankton. In: Barnabé G. (Ed) *Aquaculture*, Vol 1. Ellis Horwood, Chichester, 1990b; 264–271.

Barnabé G. La pisciculture en étang. In: Barnabé G. (Ed) *Aquaculture: Biology and ecology of cultured species.* Ellis Horwood, Chichester, 1994.

Barnabé G. Sea bass. In: Neimann-Sorensen A. and Tribe D. (Eds). *Fishes.* World Animal Science Series. Elsevier, Amsterdam, 1995; 269–287.

Barnabé G. and Perigault C. Collecte et utilisation du phytoplancton produit dans les étangs de lagunage de Mèze: Données préliminaires. *Recherches biologiques en aquaculture*, 1983; 1: 161–175.

Becker E.W. Production and use of microalgae. *Advances in Limnology* 20, 1985.

Benlahcen K.T., Salinières J.B., Faugère J. and Guarrigues P. Proliférations algales dans les eaux continentales et marines: Facteurs impliqués et moyens de traitement. *Equinoxe* 1995; 56: 23–29.

Bérard A. Pesticides. Quels sont les risques? *Aqua Revue* 1994; 53: 12–15.

Cossa D. Poissons au mercure: Une spécialité redoutable. *La Recherche* 1995; 277: 688–689.

Costa-Pierce B. Stirring ponds as a possible means of increasing aquaculture production. *Aquabyte* 1989; 2(3): 5–7.

Curds C.R. and Hawkes H.A. (Eds) *Ecological aspect of used-water treatment.* Academic Press, London, Vol 1 (1975), Vols 2 and 3 (1983).

Curds C.R. Protozoa. In: Curds C.R. and Hawkes H.A. (Eds) *Ecological aspect of used-water treatment.* Vol 1. Academic Press, London, 1975; 203–268.

De Pauw N., Moralès J. and Persoone G. Mass culture of microalgae in aquaculture systems: Progress and constraints. *Hydrobiologia* 1984; 116/117: 121–134.

Doumenge F., Durand-Chastel H. and Toulemont A. (Ed) *Spiruline, algue de vie. Bull. Inst. Oceanogr.*, special edition no. 12, 1993.

Durand-Chastel H. Production and use of spirulina in Mexico. In: Shelef G. and Soeder C. (Eds) *Algae biomass, production and use*. Elsevier, Amsterdam, 1980; 51–63.

Edeline F. *L'épuration biologique des eaux résiduaires*. CEBEDOC Ed., Liège, 1979.

Flynn M.C. and Martin D.F. Inhibition of growth of a red tide organism *Ptychodiscus brevis* by a green algae *Nannochloris oculata*. *Microbios ett*. 1988; 39(153): 13–18.

Fostier A., Barnabé G. and Billard R. État actuel des connaissances dans le domaine de la pisciculture en étang et perspectives. In: Billard R. (Ed) *La pisciculture en étang*. INRA, Paris, 1980; 429–434.

Fox R.D. Construction of village-scale system integrating spirulina production with sanitation and development. In: Doumenge F., Durand-Chastel H. and Toulemont A. (Ed) *Spiruline, algue de vie*. Bull. Inst. Oceanogr., special edition No. 12, 1993; 195–201.

Grès P. Production intensive de Brochetons (*Esox lucius*, L.) nourris de proies vivantes issues de bassins de lagunage naturel. *Doctoral Thesis*, University Blaise Pascal, 1994.

Grobbelaar J.U. Observations on the mass culture of algae as a potential source of food. *South Africa Journal of Science* 1979; 75: 133–136.

Grobbelaar J.U., Soeder C. and Toerien D.F. *Wastewater for aquaculture*. University of the OFS Pub. Series C, No. 3, Bloefomtein, 1981.

Guillaud J.F. and Romana L.A. *La mer et les rejets urbains*. Actes de Colloques No. 11. Infrémer Éd., Brest, 1991.

Iwasaki J. The mechanism of mass occurrence of *Dinophysis fortii* along the coast of Ibaraki Prefecture. *Bull. Tohoku Reg. Fish. Res. Lab.*, 1986; 48: 125–136.

Jhingram V.G. Aquaculture in India. In: Barnabé G. (Ed) *Aquaculture*, Vol. 2. Ellis Horwood Ltd, Chichester, 1990; 988–1016.

Kawaguchi K. Microalgae production system in Asia. In: Shelef G. and Soeder C. (Eds) *Algae biomass, production and use*. Elsevier, Amsterdam, 1980; 25–33.

Keondijan V., Kudin A. and Borisov A. Practical ecology of sea regions. *Geo. Journal* 1992; 27(2): 159–168.

Levesque C. *Environnement et diversité du vivant*. Coll. Explora Pocket Edition, 1994.

Lin C.K., Ruamthaveesub P. and Wanuchsoontorn P. Integrated culture of the green mussel (*Perna viridis*) in wastewater from an intensive shrimp pond: Concept and practice. *World Aquaculture* 1993; 24(2): 68–73.

Mann Borgese E. *The future of the oceans. A report to the Club of Rome*. Harvest House, Montreal, 1986.

Mann R. and Ryther J. Growth of six species of bivalve molluscs in a waste recycling-aquaculture system. *Aquaculture*, 1977; 11: 231–245.

Marcel J. Fish culture in ponds. In: Barnabé G. (Ed) *Aquaculture*, Vol. 2. Ellis Horwood Ltd, Chichester, 1990a; 593–626.

Marcel J. Aquaculture in China. In: Barnabé G. (Ed) *Aquaculture*, Vol. 2. Ellis Horwood Ltd, Chichester, 1990b; 946–961.

Melotti P., Colombo L. and Roncarati A. Use of waste water from intensive fish farming to increase the productivity in North Adriatic lagoons. *Aquaculture Europe* 1992; 17(1): 33–38.

Menesguen N. Présentation du phénomène d'eutrophisation littorale. In: Guillaud J.F. and Romana L.A. *La mer et les rejets urbains*. Actes de Colloques No. 11. Infrémer Éd., Brest, 1991.

Moon R.A. and Martin D.F Potential management of red tide blooms. *J. Environ. Sci. Health* 1979; A14: 195–199.

Odum E. The strategy of ecosystem development. *Science* 1969; 164: 262–270.

Odum H.T. Experimental study of self-organization in estuarine ponds. In: Mitsch W.J. and Jorsensen S.E. (Eds) *Ecological engineering: An introduction to ecotechnology*. John Wiley & Sons, New York, 1989; 291–338.

Ogawa K. The role of bacterial floc as food for zooplankton in the sea. *Bull. Jap. Soc. Scient. Fish.*, 1977; 43(3): 395–407.

Sevrin-Reyssac J., La Noue J. and Proulx D. *Le recyclage du lisier de porc par lagunage*. Tec & Doc Lavoisier, Paris, 1995.

Sevrin-Reyssac J. Utilisation de la chaîne alimentaire aquatique (algues, microcrustacés) pour le recyclage du lisier de porc. *Cahier Agriculture* 1995, 4: 101–108.

Shelef G. and Soeder C. *Algae biomass, production and use*. Elsevier North Holland Biomedical Press, 1980.

Shpigel M. and Fridman R. Propagation of the Manila clam (*Tapes semidecussatus*) in the effluent of fish aquaculture ponds in Eilat, Israel. *Aquaculture* 1990, 90: 113–122.

Shumway S.A. A review of the effects of algal blooms on shellfish and aquaculture. *J. World Aquac. Soc.* 1990; 21(2): 65–104.

Solbé J. Water quality. In: Laird L. and Needham T. (Eds) *Salmon and trout farming*. Ellis Horwood Ltd, Chichester 1988.

Turner J.W., Sibbald R.R. and Hemens J. Chlorinated secondary domestic sewage effluent as a fertilizer for marine aquaculture. *I Tilapia* culture. *Aquaculture* 1986; 53: 133–143.

Vandermuelen H. and Gordin H. Ammonium uptake using *Ulva chlorophyta* in intensive fishpond systems: Mass culture and treatment of effluent. *J. Appl. Phycol.* 1990; 2: 363–374.

Weber P. Sauver les océans. In: *L'état de la planète 1194*. Worldwatch Institute, La Découverte, Paris 1994; 65–92.

Williams W.D. and Aladin N.V. The Aral Sea: Recent limnological changes and their conservation significance. *Aquatic Conservation* 1991; 1(1): 3–23.

Zeech E., Stief A., Ide V. and Hardege J. *Sewage sludge treatment by Nereis diversicolor*. Chemical Ecology Research Group, Carl Ossietzky University, Oldenburg, 1992.

12

Creation of water movement

12.1 NATURAL WATER CIRCULATION SOMETIMES FAILS

We have seen in the first part of this book (Chapter 2), how convection in waters is started up by winds, tides or large marine currents and how, in contrast, the waters can be immobilised by the stratification of surface waters as a result of solar warming. This alternation is considered as fundamental in oceanography, as it determines the process of biological production in the marine environment (Chapter 4). What is important is the simultaneous variation, in time and space, of hydrological and biological factors, resulting in maximal efficiency of ecological processes. In this sense, water movements constitute a form of energy which is essential to the ecosystem and, as such, has been described as the physical processes which feed the fish (Hartline, 1980).

The characteristics of this circulation must be such that they bring together nutrient salts found in deep waters and light in the euphotic zone. This happens around upwelling zones which ensures a third of the fisheries production from 1% of the oceans and in coastal waters which are also very rich.

There are four factors influencing the vertical structure of the water column (Mann and Lazier, 1991):

1. The diurnal pattern of warming and cooling caused by solar radiation during the day and cooling down at night.
2. The input of freshwater, creating a low-density layer at the surface, intensifying vertical stratification.
3. Tidal currents which create turbulent mixing of waters and impede the formation of thermoclines and haloclines.
4. Wind-generated currents which can create coastal upwelling or impede stratification.

These factors can interact in different ways as they are independent of each other. In space, their interaction determines the characteristics of a region or even a given marine site. In time, seasonal variations alter the potential of sites and we have described elsewhere the significance of this for aquaculture (Barnabé, 1991). Excluding tides, all these factors are led, to differing degrees, by climatic variations. Thus when tides are absent or weak (e.g. in enclosed seas such as the Mediterranean, some lagoons, etc.), climatic factors are the only controlling factors of marine production. However, they are still unpredictable (see Chapter 7, Sections 7.2–7.3).

It is known that large-scale climatic events, such as the southern oscillation and "El Niño", interrupt the fertilising upwelling, affecting anchovy fishery production. Normal, but prolonged, climatic events can also have deleterious effects on coastal production. In the Étang de Thau, a 7500 ha lagoon in the North Mediterranean where oysters are cultured, a study by Tournier and Deslous-Paoli (1993) determined a numerical relationship between north-west wind speed and oxygen deficit (clearly adverse for reared molluscs) close to the bottom of the étang: there is no oxygen left when wind speed drops below 2 m/s in summer for more than 36 hours consecutively; an oxygen deficit is the "chronic state" in summer.

This is a direct and numerical demonstration of the impact of a meteorological phenomenon on a biological phenomenon. When wind is absent for too long from this lagoon, in hot, sunny summers, marked by long anticyclonic periods, anoxia spreads throughout the layer of water: mortalities affect firstly vegetation, then fixed fauna and the nekton, or fauna which cannot escape. Bivalves can withstand anoxia by closing their shells, but succumb after a few days. These widespread mortalities are known as "malaïgue" and affect all Mediterranean lagoons; they are disastrous for shellfish farmers and regularly destroy reared mussels and oysters (in 1987 there was mortality of 10,000 tonnes in the Étang de Thau).

The same phenomenon is found in freshwater lakes or marine bays, where the surface waters warm up in the summer, isolating the deep waters which become anoxic below a marked thermocline. In the marine Omura Bay in Japan (about 30 km by 10 km), stratification starts in May and oxygen is deficient every summer; the rising of anoxic water by upwelling due to autumnal winds accelerates the phenomena of coloured waters (Akagi and Hirayama, 1991).

In the marine coastal zone exposed to the open sea, it is not anoxia that poses a problem when the water becomes stratified, as the oxygen demand is lower (because of the smaller biomasses than in culture areas), but nutrients are used up by phytoplankton and production slows down or stops (Chapter 4). Waves and tides regularly destratify these waters, but the process has its limits.

Elsewhere, the turnover of waters can pose problems, even in tidal seas. This occurs especially in Brittany where, owing to the jagged coastline, it has been shown that the same waters come and go with the tide. These waters are polluted with agricultural runoff, partially purified waste water and waste from numerous piggeries produced by the catchment area, and their nitrate concentration is excessive. On bays' seabeds especially those which are very flat, the result is an invasion by ulva, due to the lack of dilution of these waters which are over-rich in nutrients. Thousands of tonnes of these obstructing algae have to be harvested and disposed of each year.

Coastal waters are thus often "stagnant" since the guiding of aquatic production by the climate is subjected to unpredictable variations of the meteorological phenomena which provide energy for water circulation. When this energy is lacking, ecosystem function can be blocked, as has been shown above, and these "natural breakdowns" can lead to the mortality of all flora and fauna in the system (e.g. the "malaïgue" of lagoons in Languedoc).

Management can consist of inputting the missing energy to create water movement, or intervening in such a way as to restore this movement. Frontier and Pichod-Viale (1991) termed imported energy as "artificial auxiliary energy", since it is similar to auxiliary energy. In practice, there are various methods of intervention.

12.2 PASSIVE SYSTEMS FOR THE MANAGEMENT OF THE MOVEMENTS OF COASTAL WATERS

12.2.1 Management of water flow in lagoons used for fishing and aquaculture

Coastal lagoons are bodies of water with communication with the sea limited to one or more channels. The forcing of currents by winds is more important here than forcing by the tide. These lagoons are often close to estuaries (Miller *et al.*, 1990). Under the aegis of the FAO, these authors devoted an 88-page publication to the hydraulic management of these lagoons, emphasising that their productivity could be increased by such management.

The ecological functioning of the lagoons depends on physical phenomena (hydrology), which are themselves dependent on meteorological phenomena. According to Miller *et al.* (1990), water circulation in the lagoons depends on their geometry and that of the channel, on the influence of wind, amplitude of the tides and, to a lesser extent, the amount of freshwater input. The width and size of the channels determine the significance of external influences. The influence of wind is fundamental, since the currents are proportional to the winds (Chapter 2). Relationships between hydrological, climatic and topographical parameters are provided by the same authors.

Thus the presence (or absence) of dominant winds, sunlight and rain are the true masters of biological production. The actions of man on the catchment areas (discharge of waste water, leaching of fertilisers and pesticides used in agriculture and waste products from aquaculture) have obvious repercussions for the receiving environment, but their action is modified by the weather.

Fisheries production in lagoons varies between 2 and 800 kg/ha and their primary production represents 10–20 times that of lakes. In France, in deeper lagoons where rearing of molluscs in suspension is carried out (Étangs de Thau, de Leucate and the Étang de Diane in Corsica), production reaches 100 tonnes/ha/year for developed areas and about 6.5 t/ha for the entire Étang de Thau. Exploited salt marshes constitute another example of large-scale management for specific ends, linked to management of water flow.

The phenomena limiting production in these lagoons are linked:

- to their isolation from the sea by an obstruction, or
- the narrowness of the channels, which limits the entry of water and also juveniles coming from the sea, or

- their low oxygen concentration (often these two factors are associated), or
- to the high temperatures reached during the warm season as a result of this isolation and the lagoons' shallowness.

According to Miller *et al.*, (1990), hydraulic management can include:

- the addition of freshwater
- water exchange with the ocean
- internal circulation.

Instead of detailing the possible effect of these actions, we shall summarise some working examples.

- In Israel, on the Mediterranean coast which has very small tides, the channels to the large Bardawill lagoon (650 km^2) became clogged, leading to high salinity in the lagoon and the consequent reduction in the ichthyofauna. This lagoon was developed: communications with the sea were dug again and maintained, with 15% of the fishery value being charged to finance this maintenance; these taxes were collected by the soldiers who control the only access road to the lagoon. This revival was accompanied by a study on fish population dynamics, which led to a regulation of fishing net mesh size, set at 70 mm. This collection of measures allowed fishery production to be increased to 18 times its former level, and the number of fishermen rose from 607 to 1135 between 1972 and 1978 (Pisanty, 1981).
- We have seen in Chapter 10 that many Italian lagoons on the Adriatic are managed. This is not by chance: because of its shape, there are small tides, about 1 m in size, at the bottom of this sea. It is this water movement and enrichment by the Pô which form the basis of the biological production in these managed ecosystems.
- In Japan, the digging of rearing channels in the intertidal zone allows better water circulation in the intertidal zone and improved production of algae such as *Undaria* and *Laminaria*. A development carried out in the Iwate prefecture of northern Japan accounted for 92 channels over 5 km of coastline (costing about €4 million). A study was carried out on 1200 m of these channels, and it was found that they allowed the production of 6.5 tonnes of algae, 550 kg of sea urchins and 250 kg of abalone, with a value of €28,000 (the pay-back period was nine years). Next to these channels was an integrated development, with 25 spawning reefs for fish, using nearly 10,000 blocks, 25 foot bridges and the construction of 870 m of breakwater using material extracted from the channels, completing the system (Mottet, 1985). This gives an idea of the scale of both the variety of developments and the size of investment which the Japanese are capable of, while at the experimental stage!
- In the tropics, the explosion of penaeid shrimp production in Indonesia, China and Thailand, in Ecuador, the Philippines, Taiwan and elsewhere, has been carried out in lagoons or estuarine areas, or to the detriment of the mangrove swamps. All intermediate types of hydraulic management are found, from water flow through channels caused by the tides, to large-scale

management of ponds, tens of hectares in size, fed by pumping, and where the water is recirculated and artificially aerated. Production thus varies between 220 and 2800 kg/ha/year, but in 1994 they occupied a surface area of 1,147,300 ha, for a global production of 733,000 tonnes (Anon., 1995). This production comes from 51,000 farms and 4400 hatcheries, most of which did not exist 15 years ago.

- In general terms, throughout the world channels, canals and various means of communication have been created in order to control water movements between sea and lagoon. Sometimes, the arrival of continental freshwater, either from an estuary or, in a more limited way, from the streams in a catchment area, results in their mixing with sea water and this enriches but also pollutes them (see Chapter 11). This confers particular characteristics on reared products (oysters reared in oyster beds in the Seudre estuary on the Atlantic, for example) or results in exceptional productivity—the density of clams can reach 3000–4000 per m^2 in the Sacca di Goro on the Adriatic, under the influence of the river Pô.

These types of developments do not require permanent inputs of energy since they utilise the normal energy of currents or tides which, to a greater or lesser extent, renew the waters in the rearing units freely. Initially, civil engineering work must be carried out in order to reduce maintenance. The force behind these activities is the additional energy of the tides, currents or the arrival through gravity of freshwater. This type of traditional development forms the basis of the majority of current aquaculture production units (Chapter 10).

In Europe, in contrast to all other regions of the world, the great majority of these water management developments are reasonably old, especially in the south. Coastal developments set up since 1946 have been almost exclusively for tourism or harbours: new yacht marinas and swimming pools can be counted in their hundreds, but this is not the case for fish ponds.

12.2.2 Deviation of large marine currents

Whipple (1990), reported that, at the start of the 20th century, an engineer named Riker, who was involved in the development of the plans for the Panama Canal, proposed to divert the Labrador current which would converge with, and cool down, the Gulf Stream; Wipple described their mixing as "if, each second, two million tonnes of ice were tipped onto the Grand Banks of Newfoundland and into the warm waters of the Gulf Stream ... it would be a massive blow, from which it would not recover". In order to protect the Gulf Stream from the icy waters of Labrador, Riker envisaged the construction of a 360 km long dyke, starting at Newfoundland and leading towards the Grand Banks. For this purpose, he proposed that obstacles should be placed on the seabed; the Labrador current carries a lot of sand and gravel and these sediments would be trapped by the obstacles, forming a barrier which would be reinforced, bit by bit, until it reached the surface and channelled the waters.

Large-scale developments are in the process of planning and execution in Japan in the inland Seto Sea for "urban developments in marine areas, to ensure food supply and develop a rich aquatic environment to be used for leisure activities and to rehabilitate the damaged aquatic environment" (Ueshima *et al.*, 1991). To this end, the installation of technology to control water flow is essential, since there is a large-scale problem with coloured and anoxic waters as a result of the waters being stagnant. A hydraulic circulation model has been developed since 1973 and has allowed the study of current dynamics and the prediction of diffusion phenomena in the bays and straits. This hydraulic model (230 m long, 100 m wide and 7000 m^2) is to the scale 1 : 2000 on the horizontal plane and 1 : 159 on the vertical plane (to respect Froude's Law). The study on Osaka Bay (60 km by 40 km) was carried out on the closure of the straits and the removal of islands in order to modify natural tidal currents. The implementation is anticipated to occur in three stages.

Also in Japan, the study of a sea loch, Beppu Bay (15 km by 15 km), and the following of floats in the model have shown that the construction of a dyke allows the complete alteration of the lack of circulation of the waters in the bay: various locations of a dyke, tested in the model, improve circulation on the bed of the bay (Fig. 12.1). While the floats remain in the same place in the absence of a dyke, they all move in an almost identical fashion, for the three dyke positions (1,2,3) tested in the model.

The authors emphasise that such developments allow the continual and large-scale control of major movements of water without the use of artificially-applied energy, and conclude that topographical modifications and the implantation of structures constitute adequate tools in a global management perspective.

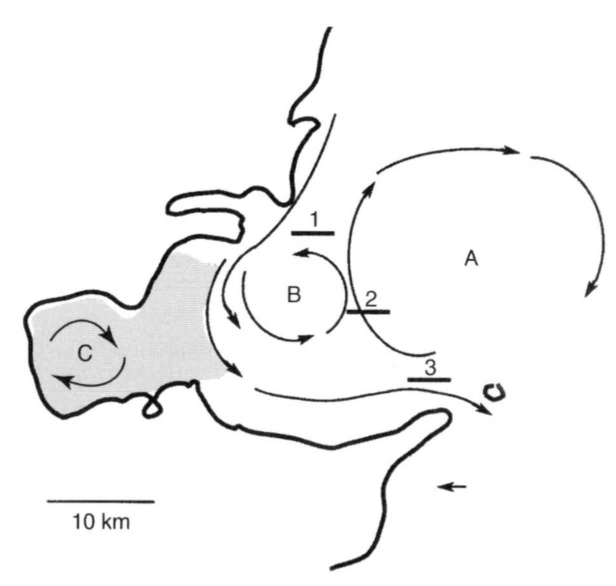

Fig. 12.1. Features of water currents in Beppu Bay. A, B, C: main vortices. Shaded area: virtually stagnant water. 1, 2, 3: position of dykes tested in model.

12.2.3 Creation of upwelling using obstacles

Since the stratification of waters isolates deep, rich waters from surface waters where nutrients are used up, the creation of vertical currents has been proposed and instigated in many cases.

Yet again in Japan, in the Seto Sea, a "deflector" 10 m high and 20 m wide was installed 50 m below the surface in the Bungo channel, to induce an upwelling of water in the eastern part of the channel (Yanagi and Nakajima, 1991). Numerous monitoring studies have been carried out: the horizontal current was subjected to an acceleration, its vertical component increased and created an upwelling of deep water, rich in nutrients, into the euphotic zone and an increase in chlorophyll, i.e. phytoplankton biomass. The situation was similar for the zooplankton, whose density increased from 498 to 871 individuals/m^3 between 1987 and 1989. The benthic biomass increased, but its diversity diminished.

As part of the Japanese "Marineforum 21" project, another deflector obstacle was installed off Ehine, between 1988 and 1992, in a zone where there was a current of a knot at depths between 30 and 50 m. It was the same height as that in the Bungo channel, but was 200 m long. The bloom of the plankton attracts small fish, which in turn attract larger fish.

12.2.4 Upwellings created by the sea's energy

In the coastal waters of the Black Sea, Psenichny and Vershinskij (1985) installed a 14 cm diameter vertical pipe, which carried water from a depth of 22 m, below the thermocline which was 15 m below the surface. Its temperature was 7.8°C less than that of the surface water, but its nutrient content was several times higher, leading to a doubling in phytoplankton production around the surface of the upwelling. During another experiment, the same authors brought water up from 200 m deep, using a 275 mm diameter tube (temperature 15°C below that at the surface). The nutrient concentration was double that of the upwellings in Peru, but the presence of hydrogen sulphide caused odours. The system can produce up to 1.2 g of C/m^3, which is an order of magnitude greater than the production normally observed. In dry weight, production can reach 5 g of phytoplankton/m^3. When the system is activated by 20–40 cm waves, 150–200 l/min are carried to the surface in the first system, and 300–350 l/min in the second. These authors estimate the capacities of large systems based on this system of using wave energy for pumping (not described in their article). They propose its use for rearing filter feeders.

This system has also been used experimentally in Japan. Wave energy is used to ensure the upwelling of deep waters through a 700 mm wide pipe from a depth of 200 m; flow is of the order of 2–4 m^3/min. A 1200 mm pipe with a flow of 60 m^3 is proposed, which would allow the upwelling of 60 million m^3/year. With a resultant concentration of phytoplankton of 5 g/m^3, this upwelling would allow the production of 150 tonnes (dry weight) of phytoplankton/year.

Again in Japan, in Toyama Bay, a state agency has experimented with the pumping of deep, nutrient-rich water. It is brought from a depth of 300 m to the surface, using the temperature difference as an energy source; the experiment aims to increase plankton production using the imported nutrients salts.

12.3 FORCED WATER CIRCULATION AND MANAGEMENT

Passive water management uses the natural movements of the waters caused by tides, marine currents or winds, to influence the ecology of the waters towards results useful for man. Complementary energy added by machines is termed "auxiliary, artificial" energy, since it is similar to auxiliary energy. When this auxiliary ecological energy fails and when it is not possible to use natural energy, inputs must be made; pumping, forced water circulation and water displacement using various machines are quite widespread.

12.3.1 A "traditional" example of pumping and water circulation

In a non-tidal sea, such as the Mediterranean, pumps have long been used to introduce millions of m^3 of sea water into basins, large shallow ponds that supply the already very saline waters of the salt marshes. This particular type of pump in the vertical axis, or pulley-wheel, raises several millions of m^3/h of water several tens of centimetres, in order for the water to reach the basins by gravitational flow and, from there, the salt marshes.

12.3.2 Destratification of lakes and energetic costs of water movement

More recently, the forced circulation of waters—not only in tanks or ponds, but in open water using minimal construction—has been studied for the management of aquatic ecosystems. Displacement speeds are much lower than for traditional pumping or mixing of waste water, for example. "Low energy" pumping is also a term used, since the power used is very low.

Some American workers, struggling against the summer stratification of lakes, evaluated the energetic requirements of forced water circulation necessary for destratification, describing economically viable prototypes. Thus Garton (1978) and then Steichen *et al.*, (1979), described two prototype "circulators", with powers of 1 kW and 0.375 kW. The latter, the more efficient, displaced between 2397 and 3528 m^3 per hour of water vertically and carried oxygen-saturated water from the surface to the deep water. Figure 12.2 gives a schematic representation of this apparatus.

The diffusion tube is 4.88 m high, with a top diameter of 1.07 m and a bottom diameter of 2.13 m. It is suspended from a raft, 4.88 m long by 2.44 m wide, the floats of which are made from 200 litre barrels.

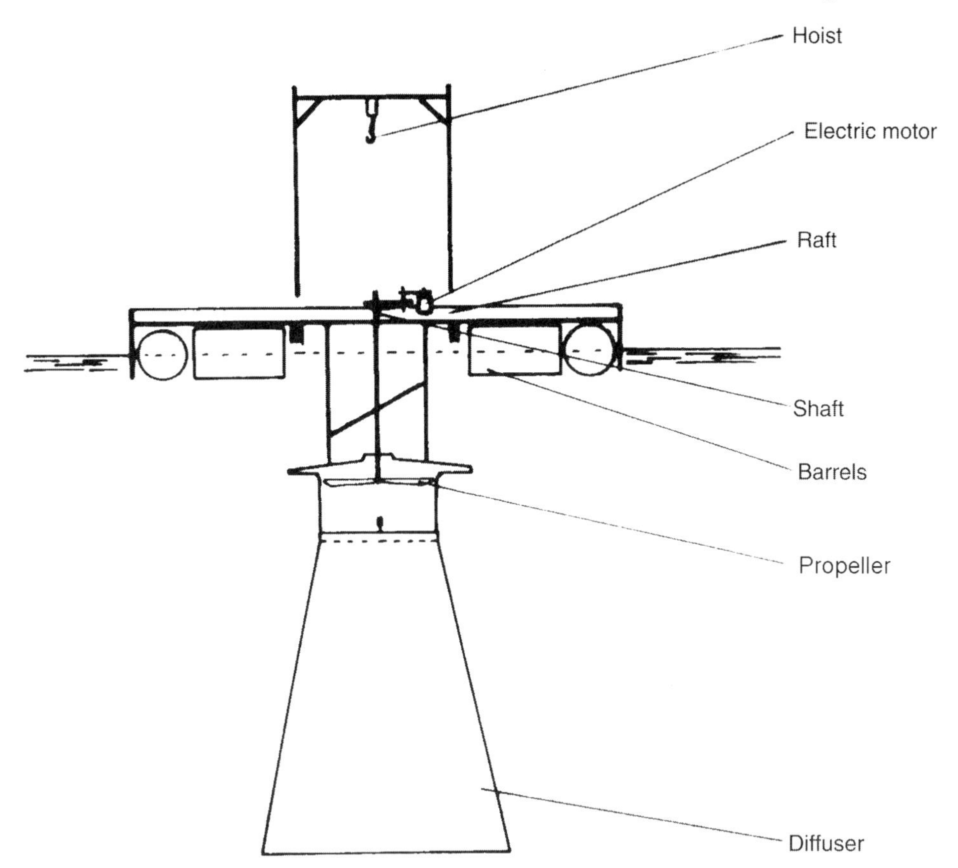

Fig. 12.2. Diagram of circulator used to destratify lake water (adapted from Steichen *et al.*, 1979).
Diameter of screw-propeller: 1.06 m (propeller from an industrial fan or hay-making machine)
Number of blades: 7
Angle of blades: 44° at the hub, 20° at the tip
Diameter of hub: 0.29 m
Propeller speed: 37–44 rpm
Power used: 375 watts

The flow calculated for 1 kW varies between 6392 and 9408 m^3/h, but the transfer of oxygen is exceptional: the circulator takes oxygen-saturated water from the surface and transfers it to below the thermocline, in an area low in oxygen. The transfer rate achieved can be 12.2 kg O_2/kW/h, while conventional aerators rarely exceed 2–3 kg O_2/kW/h.

There are a great many publications about destratification of lakes and the equipment employed, since it is a widespread technique and a "guide" has even been published (Lorenzen and Fast, 1977). Emphasis is placed on the aeration created by the diffusion of compressed air used to raise water from the bottom (air lift), or by contact with the surface; little energy is used, but the horizontal currents created are the same. A comparison of conditions in the same lake, stratified in summer and then artificially destratified by pumping, was carried out for modelling purposes (Krembeck 1988). The growth rate of phytoplankton was increased by 30% and that of zooplankton by 110% after the water mixing; destratification resulted in higher phytoplankton production, which was used directly by the zooplankton.

Horizontal water circulation and the creation of stronger currents have been the subject of industrial research and several "stirrers" are now commercially available. These are submerged electrical motors, moving a vertically positioned, large diameter (up to 2.5 m) propeller, via a reducer, at very low speeds (from 30–50 revolutions/min). The energy efficiency is lower than before (3280 m^3/kWh in the best case), but the horizontal currents created are still detectable ($v = 0.1$ m/s) in the axis of the apparatus, 110 m away. In ponds which require permanent agitation, the manufacturer estimates that a horizontal speed of 0.3 m/s can be obtained with a mixing rate of less than 0.5 W/m^3.

The figures given above are significant: the artificial auxiliary energy which must be invested in order to displace or to aerate water masses is not very large. One kilowatt allows 2000–9000 m^3 of water to be mixed and 2–12 kg of oxygen to be introduced to the same mass of water. These figures also demonstrate the gap, or delay, between research results and industrial applications.

12.3.3 Forced circulation of coastal waters

Industrial equipment for water circulation is mainly used in France on the Côte d'Azur, for renewing stagnant water in harbours which are isolated by dykes and polluted by ships.

In Italy, the construction of breakwaters in order to stabilise the beaches in popular bathing areas, such as Rimini and Cattolica, has led to a reduction in water circulation, leading to the development of algae and unpleasant fermentation. Mobile stirrers situated on rafts inside the breakwaters have resolved this problem (Anon., 1988).

12.3.4 Water movements, aquaculture and fisheries

A variety of equipment is available in aquaculture that combine water mixing and oxygenation, since forced water movement is not an end in itself; this movement must input oxygen or nutrients and take away waste and pollutants. Two main groups can be distinguished in terms of their prime function, circulators and oxygenators or aerators, but these two functions go together, since water circulation also redistributes the introduced oxygen.

An effective management strategy must therefore include both. This problem has been discussed for the management of aquaculture ponds by Rogers (1989). It is known that oxygen concentration of the water is a vital parameter in rearing units, but Justic (1991) showed that it can also serve as an indicator to describe the trophic state of the water.

12.3.4.1 Aerators

A number of types of equipment may be used to ensure the mixing and oxygenation of waters. It is not possible to describe them all, but several main principles are given below.

- Paddle wheels and vertical turbines (installed for aeration at water treatment plants) throw the water up to aerate it by contact with the atmosphere; water displacement is thus limited and aeration is the main aim. This equipment is widespread throughout the world because of its simplicity, its efficiency and its reliability. In China, not normally extravagant with energy and equipment, shallow fish production ponds are aerated and mixed mechanically, since the increased production justifies it. This technique has also allowed the improved survival, growth and production of penaeid shrimp reared in ponds in Hawaii (Rogers and Fast, 1988).
- Horizontal or oblique displacement of water by a submerged propeller is often associated with air being taken in at the surface. The propeller draws in the air and diffuses it into the plume of mixed water; the prolonged horizontal contact between air bubbles and water ensures oxygenation. This equipment is widely used both in fresh and salt water and is manufactured throughout most of the world; there are numerous variants, depending on whether the main aim is to circulate or aerate the water. In some equipment, the amount of air taken up by the Venturi effect can even be regulated, thus favouring either circulation or aeration (see Fig. 10.3).

This type of equipment still requires little power (1–4 kW), since it is adapted for water circulation and aeration in culture zones with different surface areas, but most often these are rather shallow; such waters are susceptible to stratification, even when the depth is between 0.5 and 1 m. In these conditions, it is preferable to increase the number of machines rather than to increase their power. Despite this limited output, a Japanese manufacturer claimed that, with a power of 2.2 kW, a current of 5 cm/s was measured, 93.5 m away from its pump in a 1.6 m deep pond (shrimp rearing pond). For tilapia rearing, Costa-Pierce and Pullin (1989) reported production figures of 5.1 t/ha/year in two ponds in which the water was mixed, in contrast to 2.1 t/ha/year in unmixed ponds.

This low-power equipment is in widespread use. For the correct rate of mixing and oxygenation, it is estimated that a 2 kW/ha pump would be required for fish pond culture.

12.3.4.2 Water pumps and large-scale developments

Submerged pumps are used routinely to circulate stagnant and anoxic water below salmon rearing cages situated in the very enclosed areas of Norwegian fjords. This type of submerged mixing is also widely used in waste water purification. A system of artificial upwelling has also been used in Hawaii.

In the Orbetello lagoon (west coast of Italy, south of Tuscany), which covers 3200 ha and represents 30 million m^3 of water, the fishery was threatened by the

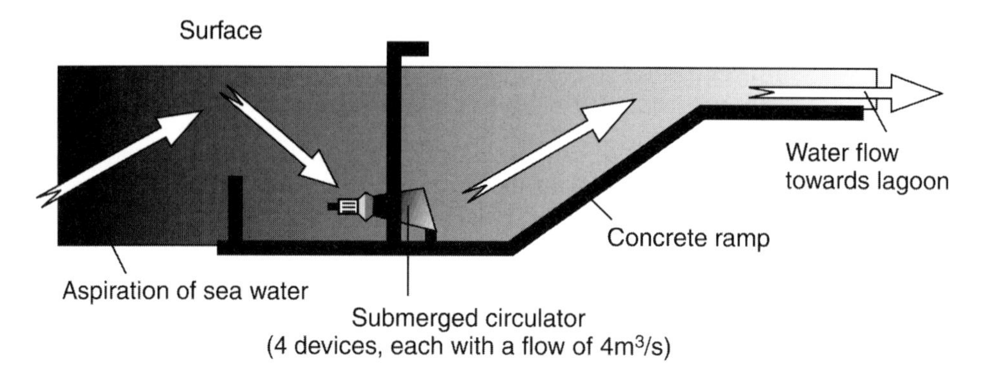

Fig. 12.3. Diagram of circulatory system installed at the entrance to the Orbetello lagoon, Italy.

decrease in oxygen levels in the water and excessive temperatures in summer. The decision to introduce sea water was made unanimously and the public powers financed the installation of four submerged 13.5 kW pumps, as a "first aid" measure (Fig. 12.3), which carried a total of 57,600 m^3/h of water into this lagoon "like a life-giving wave", allowing the saving of the fishery for a short period.

This formula must have been quite effective, since Italy subsequently obtained finance to the tune of a billion lire for hydraulic developments in the Adriatic (northern Adriatic plan). This programme uses similar processes in the Sacca de Goro region (2700 ha) which receives polluted water from the Pô; in 1987 95% of the shellfish culture production was lost a result of this. The digging of a channel to the sea was not effective as the Adriatic tides which proved too weak. Studies led to the requirement to pump sea water (at least 45,000 m^3/h), but in both directions, in order to get rid of the temporarily polluted waters from this area into the sea, and allow sea water to enter. Ultimately, 36 "mixers" or horizontal stirrers had to be installed, most of them assembled in groups of eight in channels dug in the spit of sand which separated the lagoon from the sea. These function without any problems (Minett, 1991).

In Japan, in Ofunato Bay, the construction of a breakwater at the harbour's entrance caused a certain amount of water stagnation; to counteract this phenomenon, a 10 m high, 80 cm diameter cylinder was installed underwater. The injection of compressed air on the inside of the cylinder created a vertical current, which set the waters around the system moving up to a distance of 120 m away. Used from May to November, this water circulation improved oyster and scallop production: in three groups placed 20, 80 and 260 m from the system, the first gained 43 g, the second, 42 g and the third, 28 g in six months. The import of plankton is thought to explain this positive effect of mixing (Anon., 1994).

12.3.5 Intensive rearing in ponds

This is practised both for freshwater and saltwater species and is based on the turnover of pond water, forced water circulation, water aeration and input of nutrients. An Israeli experiment with five ponds, with a surface area of 900 m^2 and a volume of 1000 m^3, provided the following results.

Each pond was equipped with four 1.5 kW paddle-wheel aerators and received new water at a rate of 70 m^3/h (i.e. a turnover of 1.7 times the volume of the pond/day). Each pond was stocked with tilapia hybrids (body weight of 70 g) at a rate of 35, 40 and 50 fish/m^3. Food was distributed using automatic feeders. Production, extrapolated to the hectare, varied from 84 to 130 tonnes for a rearing period of 150 days. The conversion rate varied from 1.6 to 3.2 and daily growth rate of the fish from 2.1 to 2.5 g. The author recognised that these very high stocking rates would create water quality problems in the rearing ponds. This situation is thus an extreme warm water example.

12.3.6 Creation of artificial waterways

It is well-known that at the Olympic games in Barcelona, the canoe and kayak events were carried out in artificial torrents created by pumping, but it is less well known that this technique is used for water management. The example is not typically marine, but concerns a species living in both freshwater and in the sea—the Ayu, *Plectoglossus altivelis*, a delicious little fish (maximum 30 cm long), found in Japan, China and Korea. Japan produces 30,000 tonnes of ayu, equivalent to the scale of trout production in France.

A large Japanese lake, Lake Biwa (674 km^2) has been developed for fry production by constructing two artificial waterways, 650 and 230 m long, 7.7 and 4.5 m wide, respectively. Both are fed at a rate of 2800 m^3/h by two water channels from the lake; one takes water from 5 m above the bottom, the other at 20 m and these waters are mixed to obtain the ideal temperature for reproduction, i.e. 18°C. This arrival of water induces the broodstock in the lake to migrate up the artificial water courses. These fish are thus captured, their maturity checked in a tank and they are introduced to artificial spawning channels (gravel of 5–25 mm in diameter covered by 20 cm of water circulating at 50 cm/s), separated by screens in order to space the spawners out effectively. Egg production reaches 1 million/m^2, 10 times the natural production in the river, with a hatching rate of 90% in contrast to 60% in the wild. Three spawnings can be planned per year. The annual total is about 7 billion fry (the equivalent of 300 tonnes), which represents 70% of the Japanese demand for re-stocking (Anon., 1992).

12.3.7 Management of outer harbours

We have shown above that pumps are used to renew water in pleasure-boating harbours, but there are other possibilities for this equipment.

These outer harbours constitute sheltered areas and can also be developed for aquaculture; we have made such proposals, summarised in Fig. 12.4 (Barnabé, 1979).

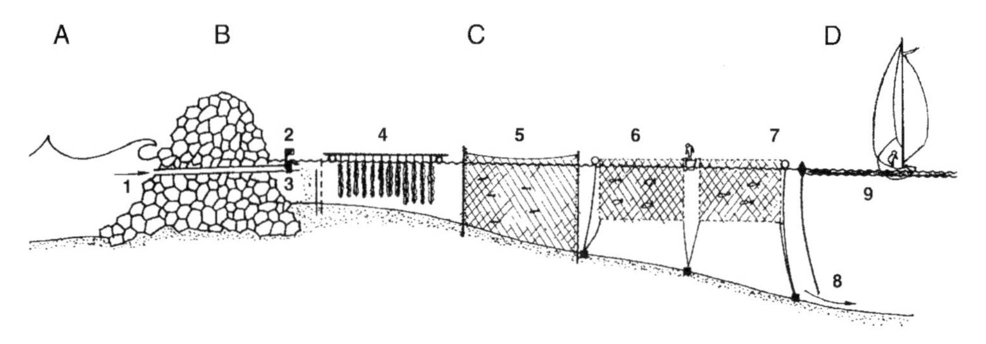

Fig. 12.4. Enterprises possible in establishments in outer harbours.
A: open sea; B: dyke or natural protection; C: rearing zone; D: area reserved for sailing.
1: water intake through or round the dyke
2: induced water circulation (e.g. outboard motor)
3: settling (sand)
4: capture and rearing of molluscs
5: enclosure
6: cages/pens
7: anti-pollution barrier
8: waste water discharge
9: polluted harbour water

12.3.8 Other examples of management

The tidal power station at Rance is the only one of industrial scale in use. It involved the closure of an estuary, which resulted in the disruption of the environment. Twenty years after its installation, Frau (1989) carried out an environmental impact assessment and noted that this development had been a success.

On a larger scale, proposals for modifying climate by artificial mixing of deep marine and surface waters were the subject of a special issue of *Israel Journal of Earth Science* in 1985. In it, Assaf indicated that this mixing would reduce the surface water temperature and increase the accumulation of heat in the surface layers in summer. This accumulation would increase the temperature and, with it, winter precipitation. Each kilowatt invested in this convection in the Mediterranean could generate 20 m^3 of rainfall on adjacent land.

12.4 WATER CIRCULATION AND PLANKTON HARVEST

Many types of fishing equipment are a means of filtering sea water on a large scale in order to retain the largest elements: fish, molluscs and crustaceans are all captured this way. Seine and trawl nets are mesh nets which filter selectively. We have seen (Chapters 3 and 4) how aquatic food chains are organised in relation to size. It can thus be imagined that, with increasingly fine filters, one could access the first levels of aquatic production: the plankton.

The harvest of phytoplankton by filtration is hardly possible in practical terms, on a large scale and economically; the different methods have been reviewed (Barnabé, 1990b).

The harvest of zooplankton is technically possible and has been the subject of many studies and projects (review in Barnabé, 1990). For some families (copepods, mysids, euphausids, scyphomedusae), exploitation is already under way. In 1980, 25 planktonic species were being exploited (Omori 1980). Catches of euphausids (krill) were about 250,000 tonnes at that time. These are not consumed directly by humans; in Norway, trawl nets, either towed or anchored in the current, collected copepods to feed farmed salmon (Wiborg, 1976).

In terms of developments, this harvest is worth while for the provision of plankton required by aquaculture hatcheries to feed crustacean or fish larvae. The mean density of copepod nauplii larvae in oceanic waters is about 17–40/l and between 1 and 7 copepodites/l according to Hunter (1981). From various data, he calculated that the plankton would have to be 1.3 times more concentrated than the mean natural density for an "average" fish larva to have a chance of survival.

It might therefore be assumed that plankton harvest, particularly in the form of concentration, could constitute a worthwhile means of providing the zooplankton required by hatcheries; we have been interested in this problem since 1976, but unfortunately, in the Mediterranean the density of copepods is only 0.5–10/l (with a mean of 1/l), according to planktonology data. Certain lagoons show seasonal richness, but without doubt the richest environments are the lagunage systems for water purification (Barnabé, 1979b).

We have described (Barnabé, 1981b) a low-power (20 watts), autonomous and stationary set-up working on battery power, for the collection of plankton from lagoons, ponds or salt marshes. A propeller situated in a horizontal or vertical tube pumps a hundred m^3 of water per hour into a filter bag which retains the plankton. The propeller is driven by a car windscreen-wiper motor; this equipment has been reproduced in numerous models and is widely used in marine or pond aquaculture (Fig. 12.5).

Plankton collection has thus demonstrated its viability for the provision of planktonic prey required by the larval stages of fresh and saltwater fish (Barnabé, 1984) and several applications have been implemented from the techniques or equipment which we have described (Barnabé, 1981a,b; 1989). Saltwater copepods are sold to the aquarists' market in frozen form (Cie des Salins du Midi in Montpellier) and rotifers and daphnia harvested from lagunage systems (Mèze lagunage) are used as food for ornamental tropical fish. Many hatcheries use harvested plankton either as complementary or the main foodstuff in diverse water systems, for feeding both fresh and saltwater fish.

Basic, inexpensive harvesters that are capable of running continuously, provide a simple solution to the complex problem of obtaining zooplankton. This technology is currently progressing towards the collection of small species (size $< 12\,\mu m$).

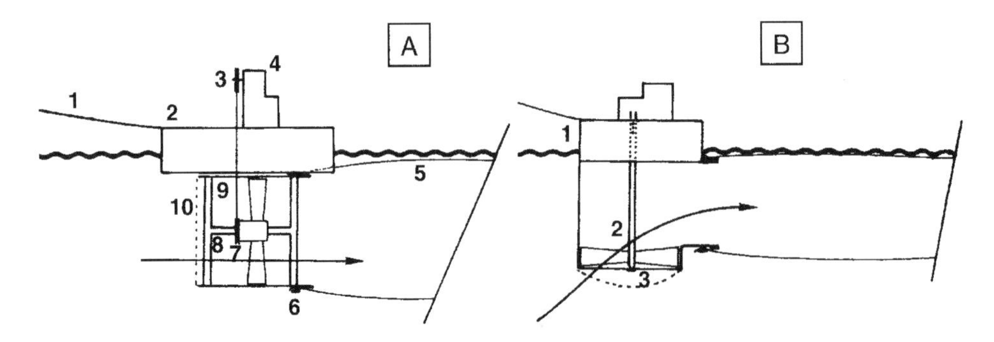

Fig. 12.5.
A. Automatic stationary planktonic harvester with horizontal axle.
1: electricity supply from battery situated on bank and mooring
2: flotation (PVC pipes or expanded polystyrene)
3: motor axle and driving pulley
4: motor (car windscreen-wiper motor)
5: filter bag of bolting material (mesh size 70–200 μm, area 5–20 m^2)
6: tightening flange of filter bag on apparatus pipe
7: fan propeller
8: propeller axle and support
9: drive belt
10: large mesh grid (1–3 cm)

B. Automatic stationary plankton harvester with vertical axle.
1: flotation
2: vertical axle
3: axle end and protection grid

12.5 SOME POSSIBLE NEW APPLICATIONS

12.5.1 Exploitation of shallow Mediterranean lagoons

In France, these lagoons are neglected. Development and then maintenance and management can be implemented to conserve the productivity of these "liquid fields". Once again, the example of the Italian lagoons in the Adriatic can be recalled, since it involves similar aquatic environments. In the Cadiz region of Spain, rearing of sea bream in ancient salt marshes has been redeveloped, from zero production in 1985 to more than 800 tonnes in 1991 (CUPIMAR group) and doubling every year. The forced circulation of water from the sea, linked with various other developments (water fertilisation, removal of macrophytes and re-stocking), was proposed (Barnabé, 1992).

12.5.2 Management of a large lagoon for mollusc culture

The Étang de Thau produces 10,000–15,000 tonnes of cupped oysters every year and is regularly struck by "crises" due to bacterial pollution or lack of food caused by the summer oxygen deficit. The absence of water circulation is often at the origin of these crises; on the basis of data given above for Italian lakes and lagoons, we have proposed (Barnabé, 1990c) a double management of this lagoon. Using the forced exchange of water with the sea, by pumping through channels and by the installation of vertical pumps in the shellfish culture area, would oxygenate the deeper waters and also increase production. Numerous discussions took place, but were not completed.

12.5.3 Automatic harvesting of plankton and krill

The collection of plankton using autonomous, stationary equipment could be applied to krill-type plankton by changing scale. Stationary or drifting platforms equipped with automatic harvesters, guided by computers which would also operate the detection apparatus, could be installed at latitudes where plankton is abundant. Beacons, GPS, etc. could be used to locate the harvesters; the collection of the catch could be carried out by a ship managing a flotilla of harvesters.

12.5.4 Creation of artificial upwelling zones

We have cited several achievements in experimental terms, but attempts have also been made in aquaculture, mainly at Sainte-Croix in Antilles, for the production of algae to be used for feeding clams in earth ponds. This trial failed in economic terms, but there are many available sites in the Caribbean (mainly in Martinique where Georges Claude hoped to install his power station), where the temperature differences between deep and surface waters could be used (the thermocline in tropical waters is situated between 125 and 300 m above the sea bed). It is very likely that a fish farm would be better situated at sea, on a ship or a barge or even in cages. The potential is very real. Programmes involving marine electricity generating or power stations to exploit the temperature differences of the water are equally numerous.

12.6 CONCLUSIONS

The features of natural sites hardly change (apart from pollution), but the development of techniques may allow tomorrow what was not possible yesterday: the development of areas which until now were unexploited and the rehabilitation of neglected sites can become possible through forced water circulation.

In exploited sites, the control of water circulation, whether passive or active, allows a choice: either to conserve the natural ecological equilibria by limiting intervention to times of crisis, or to force these equilibria towards more productive management, similar to extensive aquaculture in ponds.

The forced circulation of coastal waters is no longer something for the future, but is practical now and the possible achievements are many and varied.

12.7 BIBLIOGRAPHY

Akagi S. and Hirayama F. Formation of oxygen-deficient water mass in Omura Bay. *Mar. Poll. Bull.* 1991; 23: 661–663.

Amanieu M. Écologie et exploitation des étangs et lagunes saumâtres du littoral français. *Ann. Soc. Roy. Zool. Belgique* 1973; 102(1): 79–94.

Anon. Des agitateurs pour sauver la saison balnéaire. *Impeller* 1988; 24: 2–3.

Anon. Ayu sweetfish culture. *Fishery Journal* 1992; 38

Anon. *Flash Japon*. No 53; May 1994: 3–4.

Anon. World production of farm-raised shrimp jumps 20% in 1994. *Aquaculture Europe* 1995; 19(4): 37.

Assaf G. Artificial sea mixing. *Isr. J. Earth Sci.*, 1985; 3(2–3): 110–112.

Barnabé G. Utilisation des chaînes alimentaires naturelles et du recyclage des eaux usées dans la production à grande échelle de juvéniles pour l'aquaculture. *Actes de Colloque No 7*. CNEXO, Paris, 1979a; 221–238.

Barnabé G. Système d'aquaculture modulaire intégré en zone portuaire. *Aquaculture* 1979b; 16: 369–374.

Barnabé G. Collecte et utilisation de plancton. In: *Compte rendu des activités scientifiques du Centre de Recherches du Lagunage de Mèze*. Rapport 1980–1981, 1981a: 112–132.

Barnabé G. *Système de collecte du zooplancton à l'aide de dispositifs autonomes et stationnaires*. In Billard R. (Ed) *La pisciculture en étang*. INRA, Paris 1981b; 215–220.

Barnabé G. Utilisation de plancton collecté pour l'élevage de masse de poissons marins. In: Barnabé G. and Billard R. (Eds) *L'aquaculture du loup et des sparidés*. Actes du Colloque, INRA, Paris, 1984; 185–207.

Barnabé G. Harvesting microalgae. In: Barnabé G. (Ed) *Aquaculture*, Vol 1. Ellis Horwood Ltd, Chichester, 1990a; 207–211.

Barnabé G. Harvesting zooplankton. In: Barnabé G. (Ed) *Aquaculture*, Vol 2. Ellis Horwood Ltd, Chichester, 1990b; 264–271.

Barnabé G. Proposition d'aménagement de l'étang de Thau. *Aqua Revue* 1990c; 29: 27–32.

Barnabé G. (Ed). *Aquaculture: Biology and ecology of cultivated species*. Ellis Horwood, Chichester, 1994.

Barnabé G. and Périgault C. Collecte et utilisation du phytoplancton produit dans les étangs de lagunage de Mèze: Données préliminaires. In: *Recherches biologiques en aquaculture*. C.R. Travaux GIS-ARM, CNEXO Éd., 1983; 161–175.

CEPRALMAR. *La diversification des productions lagunaires en Languedoc-Roussillon*. CEPRALMAR, Montpellier, 1989.

Chauvet C. L'aménagement des milieux lagunaires méditerranéens. In: Barnabé G. (Ed) *Aquaculture*. Tec & Doc Lavoisier, Paris, 1989; 857–888.

Costa-Pierce B.A. and Pullin R.S.V. Stirring ponds as a possible means of increasing aquaculture production. ICLARM Contribution No.551. *Aquabyte* 1989; 2(3): 5–7.

Frau J.P. *Usine marémotrice de la Rance. Quels impacts sur l'environnement après 23 ans de fonctionnement*. Congrès de l'Union des Océanographes de France, Paris, 20–23 November 1989 (oral communication).

Frontier S. and Pichod-Viale D. *Ecosystèmes: Structure, fonctionnement, évolution*. Masson Éd., Paris, 1991.

Garton J. Improve water quality through lake destratification. *Water Wastes Engineering*, 1978; 1979: 42–44.

Hartline B.K. Coastal upwellings: Physical factors feed fish. *Science* 1980; 208: 38–40.

Hunter J. Feeding ecology and predation of marine fish larvae. In: Lasker R. (Ed) *Marine fish larvae*. University of Washington Press, Seattle, 1981; 33–77.

Justic D. A simple oxygen index for trophic state description. *Mar. Poll. Bull.* 1991; 22(4): 201–204.

Keondijan V., Kudin A. and Borisov A. Practical ecology of sea regions. *Geo. Journal* 1992; 27(2): 159–168.

Krembeck H.J. Partial destratification of eutrophic lakes: A tool for "ecosystem modelling". *Verh. Internat. Verein. Limnol.* 1988; 23: 801–806.

Lorenzen M. and Fast A. *A guide to aeration: Circulation techniques for lake management*. US Dept of Commerce, National Technical Information Service, Springfield, 1977.

Mann K.H. and Lazier J.R.N. *Dynamics of marine ecosystems*. Blackwell Scientific, Boston, 1991.

Miller J.M., Pietrafesa L. and Smith N.P. *Principles of hydraulic management of coastal lagoons for aquaculture and fisheries*. FAO Fish. Tech. Paper 314, Rome, 1990.

Minett S. Pumping solution to shellfish cove pollution problem. *Fish Farm. Int.*, 1991; 18(5): 11.

Mottet M.G. Enhancement of the marine environment for fisheries and aquaculture in Japan. In: D'Itri F.M. (Ed.) *Artificial reefs: Marine and freshwater applications*. Lewis Publications Inc., Chelsea, MI, 1985.

Omori M. *Étude du zooplancton appliqué aux pêches dans le monde*. Enseign. Oceanogr. Biol. Univ. Paris VI, 1980.

Pisanty P. Pêche et aménagement dans la lagune hypersaline de Bardawil. CGPM, Aménagement des ressources vivantes dans la zone littorale de la Méditerranée. *Etud. Rev. Cons. Gen. Pêches Méditer.*, 1981; 58: 39–79.

Pshenichny B.P., and Vershinskij N.V. Possibilities of increasing marine biological productivity by artificial upwelling. *Aquaculture* 1985; 46: 77–80.

Ringuelet R. Un potentiel jusqu'ici en sommeil: Les étangs littoraux languedociens. *Eaux* 1980; 2: 7–12.

Rogers G.L. and Fast A.W. Potential benefits of low energy waste circulation in Hawaiian prawn ponds. *Aquatic Engineering* 1988; 7: 155–165.

Sacchi C. Les milieux saumâtres méditerranéens: Dangers et problèmes de productivité et d'aménagement. *Arch. Oceanogr. Limnol.*, 1973; 18 (supplement): 25–58.

Steichen J.M., Garton J.E. and Rice E.C. The effect of lake stratification on water quality. *Journal AWWA* 1979; 71(4): 219–224.

Tournier J. and Deslous-Paoli J.M. Variation spatio-temporelle estivale de l'oxygène dans les secteurs conchylicoles de l'Étang de Thau. *J. Rech. Oceanogr.* 1993; 18 (3 and 4): 71–73.

Ueshima H., Tanabe H., Trakada M., Yuasa I., Hashimoto E. and Yamsaki M. Flow control technology for enhancement and diverse use of the marine environment. *Mar. Poll. Bull.* 1991; 23: 743–746.

Wiborg K.F. Fishery and commercial exploitation of *Calanus finmarchicus* in Norway. *Cons. Int. Expl. Mer.* 1976; 36(3): 251–258.

Whipple A.B.C. *Les courants marins.* Ed Time-Life, La planète terre, 1990.

Yangi M. and Nakajima M. Change of oceanic conditions by the man-made structure of upwelling. *Mar. Poll. Bull.* 1991; 23: 131–135.

13

The hatchery-nursery and the management of recruitment

13.1 REPRODUCTION AND CRITICAL STAGES FOR MARINE SPECIES

13.1.1 Obtaining juveniles in traditional aquaculture

The traditional activities of biological exploitation of coastal waters, such as fisheries and shellfish culture, form part of the natural functioning of aquatic ecosystems. There is no control of the renewal of exploited populations and both occupations leave the replacement of individuals to nature. Traditional aquaculture, however, obtains juveniles to be reared (spat for molluscs and fry for fish), but does not control the reproductive process; this has limited this activity to the growing-on of species whose juveniles can be obtained easily from the wild.

When modern aquaculturists wanted to produce certain pre-determined species, the first problem was to obtain individuals to be reared. The fishery of juveniles from the wild could not or could no longer provide them, because of their scarcity, their variability in abundance, or quite simply because catching them was prohibited or limited in order to protect the adult fishery. This has happened in Japan—catches of juvenile yellowtails for rearing have been subjected to a quota of 30 million individuals. Fish farmers are thus aiming towards the complete control of reproduction and the life cycle.

For the aquaculturist, these reared individuals are the equivalent of the recruits which become part of a fishery (Chapter 7). It is thus clear that aquaculture must manage reproduction, rearing of eggs and larvae up to the juvenile stage ready for growing-on, using a particular kind of infrastructure, namely the hatchery.

13.1.2 Fecundity of marine species and pelagic strategy

We have seen (Chapters 3 and 7) that the fecundity of marine species is enormous. This fecundity is in parallel with the very low survival rate of pelagic larval stages.

This phase can be likened to a dispersal or passive migration stage within the plankton, subjected to water movements. Fecundity is perceived as an ecological capacity for adaptation, which counterbalances what some people have called the "weak ecological link" and others, the "critical stages": the abundance of the issue would thus compensate for the low chances of survival. This is the "pelagic strategy" discussed elsewhere (Bakun, 1989 and Chapter 7 of this volume). In ecological terms, this characteristic is also found in parasites which have a very low chance of survival.

Algal spores and animal eggs and larvae represent forms of pelagic dispersal. We have seen (Chapter 4) the problems which these larvae will have to confront to survive. The relationships between biology and water dynamics affect this planktonic life (Chapters 1, 2, 4 and 7): this all affects recruitment.

Most fish eggs, for example, are very small (0.7–1.5 mm diameter, with a weight of 0.4 to several mg). Mussel, oyster and sea urchin eggs have a diameter of 100 µm. These eggs are released and fertilised in open water, are completely passive and rarely receive any parental care, but their varying buoyancy allows them to be dispersed through the water by currents (Chapters 1, 2, 4 and 7).

Fish eggs generally hatch after between 1 and 30 days (depending on temperature), but most often after 1–5 days. The larva is less than one milligram in weight; a tiny creature a few millimetres long which is very different from the adult, to such an extent that it is not possible to differentiate larvae from the same family of fishes. This complicates even further the study of these stages at sea.

After hatching, the larvae are not very mobile but can migrate vertically in the plankton, although they remain dependent on the hydrodynamics of the waters (Chapter 4). Detailed data and references to numerous studies on the larval life of reared species can be found in the literature (Barnabé, 1990, 1991; Nash and Novotny, 1995).

This planktonic larval life lasts from a few weeks to three months in fish, several weeks in shrimps or molluscs, but nearly a year for lobsters. It is possible to predict the route that might be followed by the larvae within coastal or oceanic water masses, carried by currents throughout this long planktonic phase.

These planktonic stages (eggs and larvae) possess energy and food reserves inherited from the mother, in the form of yolk. When the egg of a marine fish hatches into a larva, it is not adapted for independent life; it has no functioning eyes or mouth and its organs are only rudimentary. The yolk-sac (alevin) stage of the larva is brief—often less than three days for fish larvae and sometimes one day for crustacean and mollusc larvae. It is during this yolk-sac stage that the organs become functional. The mouth opens, the eyes become pigmented, the larva orients itself and reacts to light and the locomotory organs (cilia of sea urchin or mollusc larvae, fins of fish) develop. This physiological and morphological development, which requires 1–5 days on average, renders the larvae adapted for feeding by capture of the planktonic prey present in their environment.

The larva is now ready to feed itself and has only a finite slot of time to find food since its energy and nutrient reserves, and thus its autonomy, are limited. If it does not succeed in capturing prey in its pelagic environment in this time period, it will not survive.

The food items must be living and compatible in size with the mouth size of the larva. They must be present in sufficient quantities that the larva does not expend more energy in capturing prey than it attains on its ingestion. Some larvae are phytoplanktonophages and are adapted to catching prey of less than 20 μm (mollusc larvae). After a short phytoplanktophagous phase, shrimp larvae become zooplanktonophagous. Fish larvae are mostly zooplanktonophages, but their prey must be mobile in order to stimulate the larva, inducing attack and ingestion; the size of preferred prey grows with the increasing mouth size of the larvae. These specific and changing requirements complicate the problem: planktonic prey must be varied. Figure 13.1 (Person le Ruyet *et al.*, 1986) shows the succession of prey items used by turbot larvae in a hatchery.

In order for larval survival to be significant, it is thus necessary to synchronise the larval development stage with the presence of adequate plankton. As this is not always the case, and because larvae are subjected to predation by other species, survival is low: this survival has been assessed in wild marine fishes and is less than 5% for sole eggs (Riley, 1974). For the plaice, Bannister *et al.* (1974) estimated this mortality at 80% per month during the pelagic phase (egg and larva), then at 40% per month for juveniles less than a year old, falling to 10% per year for adults. According to Nash (personal communication), 5000 eggs produce one adult plaice. Average survival rates of larvae of freshwater fish species have been estimated at 5.3% and those of marine species at 0.12% (Houde, 1994). Hunter (1984) showed the significance of predation on the mortality of cod larvae. Dufour and Galzin (1993) estimated the survival of coral reef fishes during their journey from the oceanic environment to benthic life on the coral reefs in Moorea (Polynesia): 1 million eggs result in the recruitment of 100 larvae and one single adult. An extensive review of data and the problems posed by the larval pelagic life of fishes can be found in Hunter (1981).

Although the spawning periods appear to be synchronised with the existence of adequate planktonic prey populations, all these data come together to demonstrate that larval survival is low in the natural environment. It is also variable, since the combination of hydrodynamic and biological parameters can favour survival (abundant food, few predators) or in contrast reduce it (unfavourable physicochemical parameters, absence of food, high levels of predation). Several examples of this have been given throughout this book.

13.2 THE HATCHERY-NURSERY REVOLUTION

13.2.1 Management of reproduction in the hatchery

In contrast to the rearing of terrestrial animals—mainly birds and mammals, whose fecundity is often limited—marine cultures and rearing units involve large numbers of organisms: algae, molluscs, crustaceans and fish. All these species have very high fecundities (Chapter 3) and sexual reproduction is the norm. Spermatogenesis and

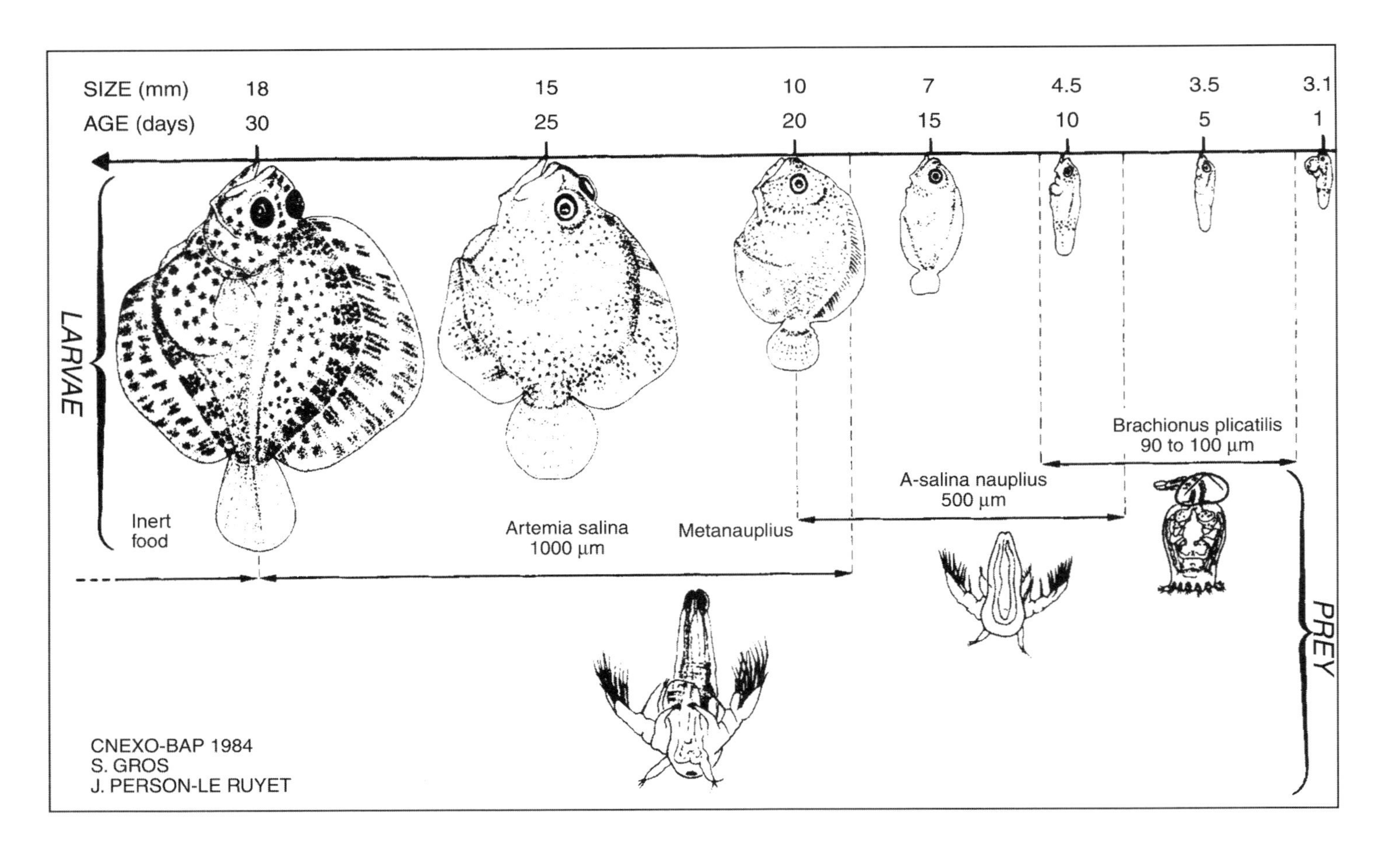

Fig. 13.1. The succession of prey items used by turbot larvae in a hatchery.

oogenesis are dependent on external factors and show a periodicity which is usually annual, at least in temperate zones.

When reproduction is controlled in captivity, very large amounts of fertilised eggs can be obtained. The reproductive potential in terms of fertilised eggs produced by a pair of spawners is enormous in these species (about 200,000–400,000 per kilo for fish, 100,000 for a pair of shrimps and several million per pair of adult oysters or mussels).

In marine species, broodstock are most often maintained and their descendants reared in the same system, forming the basis of aquaculture: the hatchery-nursery. There are numerous biological and technical reasons for this. The biological processes are dependent on physicochemical factors (temperature, salinity, light, etc.), or dependent on other biological entities for their food or reproduction. Therefore the hatchery is, above all, an environment that is controlled and manipulated to optimise all the production phases of the reared species. It is generally situated on land, under cover and the tanks can be thermoregulated in relation to ambient conditions and the requirements of the reared species. Tanks of breeding crustaceans or fish are placed in a room where photoperiod can be controlled. Depending on the breeding species, the rearing site and its climate and availability of water, the possible configurations for a hatchery are very variable, as adaptation to the environment makes each one a different case. The dimensions and complexity also differ, from the family-run or artisanal hatchery, which is often "self-built" and which produces juveniles for its own requirements, to the industrial-scale hatchery capable of providing tens of millions of young molluscs, fish or crustaceans, plus everything in between.

Reproduction in captivity, but under conditions similar to those in the wild, has been routine for a long time in pond fishes and salmonids and has been carried out with molluscs, penaeid shrimps and some marine fishes since the 1970s. In contrast, many species still remain difficult to manipulate biologically, which limits their rearing potential (tunas, yellowtails, eels, crayfish, etc.). Captive spawning is achieved by placing broodstock into tanks of variable volume, relative to their nutritive requirements and forming an environment intended to reproduce the conditions for maturation as found in the natural environment. These conditions have been described elsewhere, for the majority of reared species (Barnabé, 1990, 1991; Nash and Novotny, 1995). In several cases, it has been possible to control spawning by hormonal induction in marine species, as well as ensuring the last stages of sexual maturation.

The ultimate stage of this type of investigation is obviously to obtain on-demand reproduction in captive broodstock throughout the year. This has been achieved in the goldfish (*Carassius auratus*), but the unlocking of spawning of some marine species is also practised on a production scale based on the results of many experimental studies; understanding the physiological mechanisms which regulate reproduction constitutes the scientific basis for these biomanipulations.

13.2.2 Rearing larval stages in the hatchery-nursery

In addition to managing the maturity of the broodstock, the aim of the hatchery-nursery is to maintain eggs and larvae. In the hatchery, numerous offspring may be reared (because of their tiny size) in small, but carefully monitored, volumes of water. Environmental conditions can be optimised (physicochemical characteristics, absence of predators), but it is especially the feeding of larval stages that poses numerous problems since these tiny animals require the provision of very large numbers (several hundred to several thousand per individual per day) of living planktonic prey which the hatchery must rear or harvest; prey size must also increase with the mouth size of the larvae.

Microalgae must be provided for mollusc larvae and zooplankton for crustaceans and fish. Rearing these zooplankton also requires microalgae. Thus the hatchery-nursery must not only maintain the broodstock but also control a planktonic food chain (Chapter 3) for the rearing of larval stages.

As we have seen, phytoplankton (made up of microalgae) represents the starting point of a food chain in which it constitutes the first link (primary production). These microalgae provide food for molluscs (larvae and adults). Specific culture rooms are devoted to algal production in the hatchery. In mollusc hatcheries this algal culture can take up more than three-quarters of the surface area.

Only two species of zooplankton, a rotifer, *Brachionus plicatilis* and artemia, *Artemia salina*, are reared in hatcheries using these microalgae; in their turn, these are prey for either fish or crustacean larvae (see Fig. 13.1). Again, a dedicated culture room is used for each of the prey species, which require specific temperature conditions. Despite many attempts, the most numerous animal species on the planet, the copepods which populate the oceans, have never been reared in massive numbers but they are often harvested for feeding crustacean or fish larvae. Recently it has become possible to rear rotifers without algae, using specific foods made from brewers' yeast, fish oils and various protein, vitamin and mineral supplements. Several manufacturers distribute such foods in a dried form which simplifies storage and allows different uses. Lyophilised microalgae are also available commercially. These foods are expensive but competitive in relation to the costs (infrastructure, production and man hours) of an algal culture suite; the trend is towards the suppression or limitation of algae in the rearing of larval fish, which also reduces the time for which the algal production suite is needed (two weeks). This simplification has allowed many hatcheries to significantly increase their capacity for juvenile fish production.

Crustacean and fish larvae can also be reared in tanks in special rooms where light and temperature can be modified in relation to requirements.

These feeding problems are not found in the salmonids where the eggs are much bigger and result in larger and more robust fry endowed with significant vitelline reserves: when these are nearly absorbed, the fry is capable of feeding directly on a dried, synthetic food. The control of the availability of juveniles has been carried out for a long time for these species, which explains why they currently provide such large fish production. Salmonid hatcheries are therefore different from those for marine species.

The marine hatchery-nursery is modelled on a planktonic food web, each link being cultivated separately. Only when the algal culture has reached its optimum is it given to the molluscs or rotifers, but under the best environmental conditions unique to these species. In turn, these prey are then distributed in the rearing tanks where conditions (temperature, salinity, oxygen concentration, pH, water movements, clarity, etc.) are optimised for the larvae.

In this framework of optimal environment and feeding conditions, and in the absence of predators, the survival rates bear no comparison to those found in the wild, as reported above: they are between 10% and 40% for marine fishes and more than 80% for freshwater fish and these results improve continually as a result of technical progress.

The hatchery is at the centre of current problems and prospects in modern aquaculture and it is estimated that several thousand hatcheries were constructed in the 1980s. A description of the unique biotechnology methods used for both the control of reproduction and larval rearing can be found in aquaculture publications (Barnabé, 1990, 1991; Nash and Novotny, 1995).

13.2.3 Billions of juveniles

The controlled spawning of marine species and the rearing of planktonic stages which characterise them have been achieved in the past two decades. Until now the production costs, for uses other than the production of high-value species, have been excessive, but hatchery techniques have been diversified and production and selling costs have decreased (in Europe, the selling price of bass fry halved over three years, that of turbot fell to a fifth of its original price and oyster larvae ready for attachment (spat) are valued at €0.2 per thousand). Thus, productions which were faltering or non-existent 20 years ago now amount to millions or billions of individuals of reared molluscs, crustaceans and fish.

Our experiences, putting into practice the techniques for reproduction and larval rearing of bass and sea bream (review in Barnabé, 1990, 1991), and many other studies carried out elsewhere on other species both in the wild and under controlled conditions, have shown that the ways that food chains work, and also the behaviour of a fish or a mollusc, are broadly comparable without being exactly the same. What is possible today for a few economically significant species may be achieved in the future for the majority of other species.

This has resulted in a demonstrable increase in aquaculture production (including algae). From six million tonnes in 1976, it increased to 18 million tonnes in 1992 (FAO, 1994) and to 25.46 million tonnes in 1994 (with a value close to €44 billion) again according to the FAO (Anon., 1996). For 1995 (the statistics are published three years later), this organisation reported a production of 27.8 million tonnes, with a value of €50 billion. This progressive increase in production is a long-term trend in aquaculture. Between 1984 and 1992, aquaculture production increased annually by 9% in volume and 14% in value which compensated for the decline in fish catches (FAO, 1995).

About 150 aquatic species are reared, of which 89 are fish, 23 crustacean, 35 mollusc, 4 algae, and various other species such as frogs, turtles and sponges. For 34 species, production exceeds 20,000 tonnes. In Japan, 18 species were reared in 1993, and production varies from 2000 tonnes of penaeid shrimp to 245,000 tonnes of cupped oyster. The number of individuals released in Japan in 1990 were: 16

million Japanese sea bream; 500 million prawns; 3 billion scallops and 2.5 billion salmonids.

It is estimated that in Japan one-third of total production of the coastal zone (fish, shellfish, etc.) is used for restocking.

In 1998, the FAO (Fisheries Division) estimated, on the basis of data provided by member states for 1996, that the production of juvenile fish in hatcheries has reached 58 billion, or nearly 180 million young fish per day (FAO website <http:// www.fao.org.).

13.3 THE PROSPECT FOR THE EXPLOITATION OF MARINE RESOURCES

13.3.1 The synergy between hatcheries and restocking

The science and techniques established in the hatchery-nursery are wide-ranging and potentially useful for all marine species. Recently, man has thus found himself in a new situation in terms of exploitation of marine resources: fisheries production reached its maximum in 1989 and has since decreased (see Fig. 7.2). Even if there is a slight recovery, the evidence suggests that this activity has reached a plateau: fish catches often concern species which are high in the food chain, thus reducing the yield of the already weak natural food chains; most stocks are overexploited and the turnover of caught individuals, i.e. recruitment, is neither understood nor controlled (see Chapter 7).

In contrast, we have seen in preceding chapters that the production of marine ecosystems is far from reaching its full potential. If these perspectives are combined with those of the hatchery, capable of providing juveniles that can become recruits, one can imagine what productive developments of lagoon and coastal waters are at our disposal. Intensive or extensive rearing, restocking of limited areas and reinforcement of natural populations all become possible.

13.3.2 Critical theories contradicted by facts

All these potentialities have been contested. A paradigm of fisheries management declared that populations which are regulated by density-independent factors (such as climate) cannot benefit in any effective or predictable way from restocking programmes. The argument is that, if a population is controlled by density-independent factors, releases from hatcheries will have a negligible effect, both during good years (since individuals produced naturally will be abundant), and in bad years (since the individuals used for restocking will be subjected to the same high mortality rates as the wild individuals).

Another criticism concerning hatchery programmes is that they assume that the environment is capable of meeting the requirements of the released individuals, implying a capacity for sustaining higher population densities; these environmental capacities are finite, dynamic and difficult to evaluate with any certainty. At the end of the 1960s it was stated that "all these and other considerations have led to a crisis of conscience in the fisheries such that they cannot simply assume that hatchery programmes have a reasonable chance of survival in nature".

We have seen some edifying examples in preceding chapters, but the examples below demonstrate how far all these proposals are from reality.

Hatchery species are also likely to be more fragile since they are the progeny of too few parents. The possibilities of transfers between countries, or even between continents, are totally ignored, as is the always possible return to using fish from the wild as broodstock in the hatchery; in contrast with natural stocks, the hatchery allows all genetic combinations.

13.3.3 Revision of the idea of critical state

We have seen that the experimental study of recruitment is in progress (Chapter 7), but new light has been shed on the pelagic strategy and critical stages by research carried out using mesocosms on planktonic food webs utilised by species reared by man.

The survival of fish larvae is often very variable, but has reached 90% in some larval-rearing experiments with sea bream (Barnabé *et al.*, 1995; Gianacourou, 1995), which shows clearly that it is not the larval stage which is critical but the conditions in which the larvae are placed. Thus mortality rates are not programmed, but constitute a species' adaptation to uncertain survival conditions (absence of food, predation, etc.).

When the larvae's requirements are met, survival rates in the hatchery are high and it could be supposed that this corresponds to high recruitment levels in the wild. While it is pointless in economic terms to search for survival rates close to 100%, these have been obtained experimentally (Gianacourou, 1995).

13.4 EXAMPLES OF RESTOCKING AND TRANSPLANTATION

13.4.1 Preliminary attempts

Transplantations of mollusc spat have been carried out for more than a century. They are based on the fact that growth of certain molluscs is significant in sites where they do not reproduce; thus the requirements for growth and reproduction in aquatic species can differ. A fish such as the European sea bass, for example, spawns at 12–14°C, while its optimum temperature for growth has been recorded at 22–24°C.

Fish restocking is also nothing new. At the start of the century, Garstang, a British ichthyologist, remarked that large numbers of plaice could not survive in the small area of the coast of Holland where they were found in high densities; he caught thousands of them, marked them and transported them to the Dogger Bank. These marked fry grew three times more quickly than those which remained in the overpopulated zone. An analogous experiment was carried out in Denmark with similar success (Whipple, 1990).

Between 1910 and 1950, marine fish hatcheries were installed on the east coast of the USA and also in Norway. Their aim was to repopulate the coastal areas with millions of fish larvae produced from captive broodstock placed in a large tank where they spawned naturally. The newly-hatched larvae (cod, haddock, plaice, mackerel, etc.) were released into the sea without having the slightest effect on the natural populations. These individuals had to get through the critical stage of first feeding, at the start of exotrophy,* and remained subject to predation, the main cause of mortality during larval life (cf. Hunter, 1981). These hatcheries closed in the 1950s (Shelbourne, 1964; Stottrup, 1996). This example agrees with reports that recruitment in the natural environment is independent of the size of the spawning stock (Chapter 7).

These failures can be understood in the light of recent results. In Hawaii, during experiments on restocking with grey mullet, *Mugil cephalus*, 141,618 individuals of which were marked in 1990–1991, Leber (personal communication) showed that survival is very dependent on the size at release and that 60 mm was considered the minimum.

13.4.2 Some restocking operations around the coasts of Japan

13.4.2.1 The scallop

The traditional coastal fishery for scallops (*Patinopecten yesoensis*) in northern Japan (Sarufutsu) produced about 10,000 tonnes at the end of the Second World War, but this production fell progressively and in 1970 was practically nil. Aided by scientists, the Fishermen's Cooperative of Sarufutsu, the main harbour involved, started a study of the seabed and in 1970 decided to engage in restocking operations using spat produced in other areas. In 1971, 14 million juveniles were released over 38,000 hectares and current restocking levels are between 0.5 and 1 million juveniles/year. Catches on these seabeds have risen rapidly and have stabilised at between 25,000 and 30,000 tonnes since 1980 (Fig. 13.2). In 1990 (Anon.), 29,000 tonnes were produced, representing an income of €35 million, the profit divided between the 185 cooperative members being about €38 million (profit margin $\approx$51.7%). A similar restocking operation of the entire littoral zone of the Okhotsk Sea has been carried out, where fishing by dredging (after three years growth of the shellfish) produced between 180,000 and 200,000 tonnes (Doumenge, 1995). Additional biological data can be found in Saito (1984).

13.4.2.2 The abalone

The situation is identical for the abalone, *Haliotis discus*. Catch records (1000 tonnes in the 1960s) started to decline in the 1970s, threatening the viability of the fishery. The fisheries cooperatives drew up a management plan using the restocking of fishing zones and their management (preservation of the fields of brown algae which

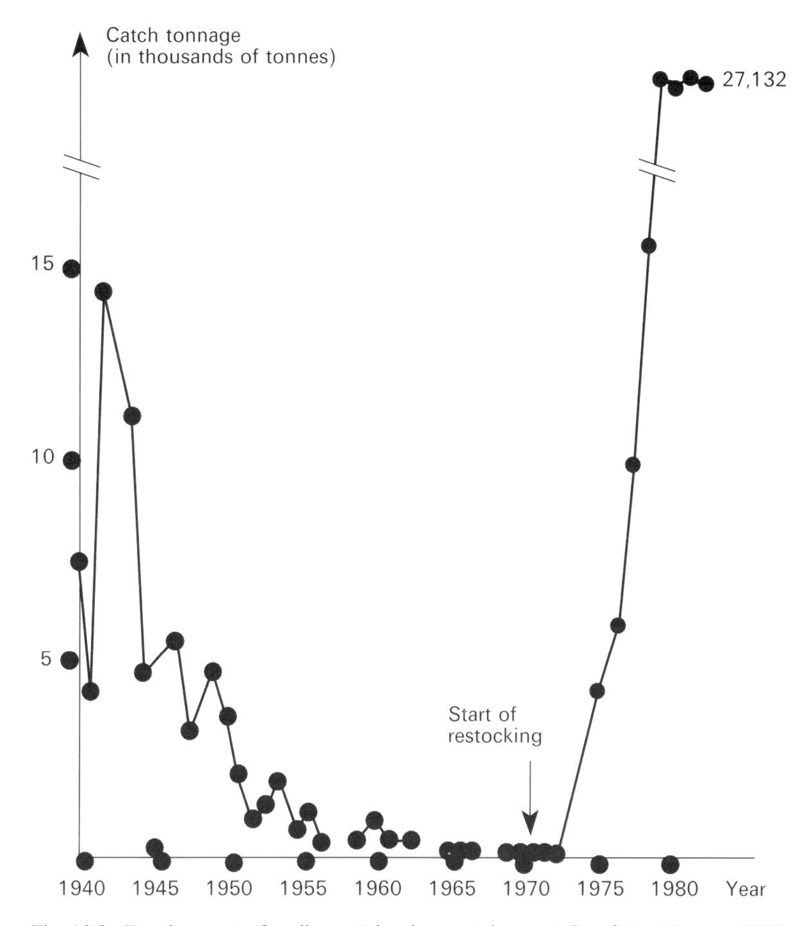

Fig. 13.2. Development of scallop catches in coastal seas at Sarufutsu (Anon., 1990).

constitute the food of the abalone and control of catches). A hatchery constructed using public funds and producing 6 million 15 mm abalone/year completed the set-up. The juveniles were grown on by the cooperative until their release at 30 mm. Capture was allowed two years later at a size of 90 mm. Recapture rates were about 38%. The investment for such a programme (€0.66 million in 1992) produced a revenue of 170% of the market value of the catch (€1100 million) (Bourguignon, 1995). This author, like the preceding one, stressed the importance of the driving force provided by the fisheries cooperatives in the initiation and conduct of these operations.

Other species have been subjected to the same process: algae, penaeid shrimps, bass and the freshwater "ayu", as well as several pelagic species, released in fishing zones equipped with fishing reefs (see Chapter 14).

This system has its faults, but the Japanese coastal fishery has been stable since 1925 and produces between 2.5 and 3 million tonnes.

13.4.3 Hatcheries for restocking Pacific salmon

The examples given are again Japanese, but the same activity is also carried out in Canada, the USA and the CIS; however the Exxon-Valdez oil tanker disaster in Alaska has been a blow to activity in this area.

In Japan, artificial spawning has resulted in increases in salmon catches in the coastal zone: 50% since 1978 and 70% since 1985. Since 1985, the number of returning salmon has exceeded 40 million (Hiroi, 1989). The management of the resource is carried out via the device of coastal fishery management. Releases of chum salmon, *Onchorhynchus keta*, were 2 billion/year from 1980 to 1988, catches in fixed traps increasing from 63,000 tonnes on average between 1979 and 1984 to 85,000 tonnes between 1985 and 1989, reaching 115,000 tonnes between 1989 and 1991 (Doumenge, 1995). Expenditure devoted to chum salmon is of the order of €0.1 billion/year, the catch value being about €0.9 billion.

Production due to restocking exceeds that of the fishery and the gap is especially amplified when the potentialities are different (Fig. 13.3).

With regard to the Pacific salmon in Alaska, this restocking contributed to 80% of the catch in 1991 (Doumenge, 1995). Releases have reached 1.4 billion since 1988. The same author reported the release of 2 billion salmon by the Russians to sustain the coastal fisheries of Kamtchatka and the Kourillesin during the same period.

The care taken in the spawning and release of young salmon bears no relation to the releases of hatched larvae carried out at the beginning of the century, since the rearing period prior to release now lasts four months. This is also the case for most

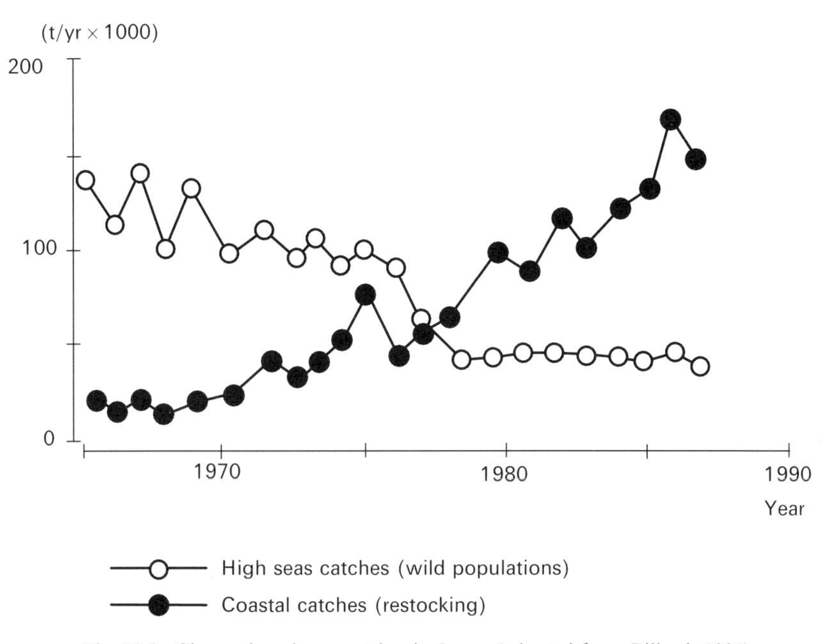

Fig. 13.3. Change in salmon catches in Japan (adapted from Billard, 1995).

marine species. Nowadays, marking and tracking are used to ensure that the young produced in the hatchery are sufficiently robust to meet the requirements for intensive growing-on and also restocking. Survival of these individuals is thus now comparable to those of juveniles hatched in the wild, which has been disputed. It is no longer the single problem of recruitment which does not correspond to the facts in fisheries theory, but many of its assertions.

Hiroi (1989) provided the following information. The conditions in their river of origin are vital for salmon maturation, but seasonal factors must also be taken into account in determining the degree of maturation. Temperature during the maturation process is maintained at between 6°C and 12°C by adding spring water to river water and current speed in the fish tanks must be less than 10 cm/s, with a regular turnover of water.

After spawning, the eggs are activated by being put in contact with water and the spermatozoids are activated by the female's coelomic fluid. Incubation is carried out at 8°C. After hatching, alevins remain immobile on the bottom and water currents must be less than 3 cm/s. The water is warmed by 0.2–3°C using infrared radiation provided by radiant electric heaters situated 40 cm above the fry-rearing troughs, or in the bottom of them. The current speed in the troughs is a determining factor for the first-feeding fry, so that they gradually learn how to swim. Released fry are fed for several days in the stream to allow them to adapt to catching natural food. Chum fry must be released at a weight of 3 g for their downstream migration*.

Despite this progress, it would be wrong to believe that all restocking or reinforcement operations are necessarily destined to be successful. The holding capacities of natural ecosystems are significant but not unlimited and there have been some failures in sea-ranching, as well as elsewhere in aquaculture. Sea-ranching, however, constitutes a demonstration of the power of hatchery techniques, since it is capable of modifying the population of very large ecosystems.

13.4.4 Transplantation of clams in the lagoons of the northern Adriatic

In the Adriatic, production of the Manilla clam started at the beginning of the 1980s with the restocking of some concessions using spat provided by a French hatchery. It remained insignificant until 1985, but now exceeds 20,000 tonnes: this species reproduces very rapidly and can reach densities of 3–4 thousand individuals per m^2. This production resulted in falling prices elsewhere and finished the rearing of this species in France.

The Italians also had problems in rearing this species. Production sites were threatened by pollution, but initiative was certainly not lacking; they applied for funds from the IMP* (Integrated Mediterranean Programme) of Europe for the "hydraulic invigoration of valli and lagoons" and obtained 190 billion lire or €145 million, for various developments (Manoli, 1991). Approximately 1200 fishermen and fish farmers making a living from this activity in the Goro region. This transplantation was accompanied by hydraulic developments aimed at restarting water circulation (see Chapter 12); thus a complete system of management was achieved.

13.5 LARGE-SCALE PERSPECTIVES

13.5.1 Reinforcement of stocks of overexploited species

We have seen (Chapter 7) how cod stocks have been overexploited, especially in Canada where fishing has been prohibited. The CCFI (Canadian Center for Fisheries Innovation) developed hatchery rearing for cod, scallops, Arctic salmon and yellow-tailed plaice. Together with the government, this centre studied the feasibility of adapting cod culture techniques to increasing the stocks in the northern part of the country. If it is successful, such research would help in the reconstitution of overexploited fishery zones (Anon., 1995).

13.5.2 Marine ranching as a substitute for intensive aquaculture

In Norway, the production capacity for intensively cage-reared salmon was close to 300,000 tonnes in the 1990s. This type of rearing has its limitations, however, and in ecological terms it has been demonstrated (Folke and Kautsky, 1989) that the production of 1 tonne of salmon is indirectly dependent on the primary production of 1 km^2 of sea (using food mainly made from fishmeal), which includes intensive aquaculture in cages as one of the activities which is dependent on fisheries. It would therefore be interesting to put a limit on the industrial fishery of herring and capelin by releasing hatchery salmon into the sea where they could exploit these resources and be caught again by fishing.

Norway encourages sea-ranching and devoted nearly €56 million to it between 1990 and 1998 (through the PUSH programme for Atlantic salmon, cod and lobster). Released marine fish do not go far: less than 10% of marked cod were recaptured more than 5 km away (Anon., 1993).

Another solution to the problem posed by the use of fishmeal is the incorporation of plant products into feeds. The Japanese have succeeded in incorporating up to 70% plant products in a feed for yellowtails, which is as efficient as fishmeal-based feed. Development in this area is predicted to continue; already, pelagic fish which could only be fed on fodder fish (fresh or frozen fish) have been adapted to feeding on dried granules (yellowtails and dolphin fish).

13.5.3 Private fisheries based on aquaculture

Needham (1990), an international specialist in salmon farming, proposed in an almost analogous report the transformation of this activity by integrating the fishery with management of the ocean in a fisheries system based on aquaculture. His proposal was based on the fact that the fishery, in terms of its use as a common resource, was outmoded: there was overexploitation, since the fishermen caught the fish but did not produce any, and the other, predictive part of fisheries theory was never able to suggest future abundance. His proposed solution was to privatise access to salmon fishing grounds in the entire Pacific and to link this privatisation with the restocking of salmon in these areas using the hatchery-nursery system as a stakeholder, as is done in sea-ranching, the private fishery financing the hatcheries. In 1989, salmon hatchery profits in Alaska exceeded €38 million (this programme was temporarily halted by the Exxon-Valdez disaster). Such an approach may be more profitable than rearing in cages since there are no cage or feed costs etc. for the growing-on phase. When the fishery is privatised and limited to the littoral zone, its profitability is better (reduced exploitation costs). Restocking regularises recruitment and so one has a production and management system for a renewable resource which has never been attainable through fisheries management.

New Zealand has privatised its fisheries and, in Japan, coastal zones managed using restocking are exploited exclusively by members of fishing cooperatives which finance and participate in these operations (see Section 13.4.2). The installation of artificial reefs and fish concentration devices (see Chapter 14) in coastal waters in the Exclusive Economic Zone (EEZ), constitutes a form of resource allocation to the artisanal fishery, to the detriment of other users (trawlers). Fishing quotas (see Section 7.2.4) instituted by the nations concerned constitute another method of regulating access to resources.

13.5.4 Management of natural resources and juveniles

13.5.4.1 Sedentary molluscs

We have seen that the chances of survival for eggs and larvae in the wild are very low. It is sometimes possible for man to intervene in a very simple way to improve this survival; this is what has been done empirically for a long time with molluscs, where the provision of substrates has allowed the capture of larvae, i.e. their attachment to a solid substrate when their planktonic life is complete and they must attach and start their benthic life.

This method has far more widespread applications than in shellfish culture. It would be possible to envisage the real management of the coastal zone and the installation of new food webs. In view of their significance, we devote a whole chapter to this type of development (Chapter 15).

13.5.4.2 Growing on of juveniles caught in the wild

Many marine fish farms are based on the abundance of juveniles in the natural environment and their ease of capture. This is the case when a particular behaviour of these stages allows them to be caught *en masse* for rearing purposes—for the Japanese yellowtail, juveniles are caught under drifting algae (sargassum). It is one of the highest producers in marine fish culture, with production of more than 150,000 tonnes/year (cf. Doumenge, 1990).

Many other fish species are caught as juveniles in seine nets along the beaches, before being grown on (milk-fish in south-east Asia, grey mullet, sea bream and bass all around the Mediterranean). Italian valliculture*, which continues to flourish, is based on the capture of these species which are grown on in private lagoons or valli (cf. Ravagnan, 1978).

We have seen (Section 13.1.2) that the chances of survival for these stages are very low in the wild. The capture of these juveniles, the majority of which would die, and putting them into rearing systems for growing on or restocking, constitutes an alternative to the hatchery-nursery system, on condition that all precautions are taken to avoid vastly reducing these populations.

Taking into account the chances of survival for the larvae of coral reef fishes (Dufour *et al.*, 1995), these authors proposed the collection of live larvae during their migration towards the reef (using capture nets placed on top of the reef), and then placing them in areas devoid of predators in order to increase certain stocks. They noted that "The well-thought out exploitation of coral reef fishes for food purposes requires controlling the mortality rate of larvae at metamorphosis and their first hours of life on the reef ... Since the natural density of fishes is often below the holding capacity of the reef, it should be possible to artificially increase the recruitment rate in suitable sites." They also proposed extensive aquaculture in lagoons, using fish larvae collected during colonisation.

13.5.4.3 Capture and rearing of crustaceans

For many species with a long or complex larval cycle (e.g. crayfish and eels), the technical and economic viability of rearing larvae in a hatchery-nursery system can pose problems. Results obtained for rearing larvae of several species of spiny lobster (Kittaka, 1994) and his current work remain experimental.

When larval behaviour allows, management can improve larval survival. This point was made with reference to the spiny lobster (Phillips *et al.*, 1994).

At the end of a larval phase lasting more than 300 days, the young stages of spiny lobster become benthic and acquire adult morphology and lifestyle. It might be assumed that the absence of available habitats would limit recruitment; in fact, it is noted that on soft littoral shores, concentrations of juveniles are found on boats' mooring ropes or fish cages, rather than in habitats situated in the open water. The absence of adults caught in these areas shows that these juveniles are in practice condemned by the absence of suitable habitat. Young spiny lobsters have even been observed on the mooring lines of fish-concentrating equipment anchored in depths of over 1000 m in the Caribbean.

It is possible to capture these juveniles for growing-on and restocking and developments have been trialled and/or achieved (Phillips *et al.*, 1994). After capture, attaining marketable size takes around 18 months in the Antilles (Coton, 1987). A study of the capture and growing-on of red spiny lobster is underway in Martinique. Many other coastal crustaceans which have pelagic larval stages could have their survival improved in this way.

13.5.5 Adaptive management of large oceanic systems

In Spain, the fishermen have tried to get rid of hake, dogfish and rabbit fish in order to attempt to create a shrimp fishery (but without restocking).

In the Yellow Sea, having overexploited the stock of demersal fish, the Chinese tried introducing shrimp to the seabed; success was limited (Tang, 1989). According to Sherman and Alexander (1986) and Sherman (1990), other large ecosystems have been the subjects of such "adaptive development": fisheries of the Benguela Current in South Africa, Antarctic fisheries and the ecosystems of the Great Barrier Reef. The starting point for these actions is the fact that what has happened to the biomasses of these ecosystems is the result of human predation, which widens the choices for their future ecological management.

Examples of management, restocking and reinforcement of populations are now numerous and varied: Doumenge (1995) reported numerous other examples of this type of activity; he also cited some failures, notably that of turbot restocking in Japan, which contrasts with a great success for intensive aquaculture in Europe (although the species are different).

As emphasised by Stottrup (1996), restocking with hatchery-reared juveniles must be well planned and managed, by taking into account the factors controlling population dynamics and also by managing the protected habitats. Fisheries regulation, habitat manipulation and restocking must be carried out in a concerted fashion; this is true integrated management.

The common denominators between fisheries and aquaculture are just as numerous in technical terms. Loverich (1991) noted that equipment on boats (pulleys and winches) and also the practical experience of the fishermen (lifting and handling the nets) and experience of the open sea, render them capable for adaptation to off-shore aquaculture and predicted collaboration between fishermen and fish farmers.

13.6 CONCLUSIONS

The methods used in the hatchery-nursery have completely changed aquaculture's prospects and allowed the production of many marine species at significant economic levels (millions of individuals and tens of thousands of tonnes) during the past two decades. This is a new stage in the history of humanity, to be able to reproduce and rear these species and it is also the realisation of enterprising work carried out, often since the start of the twentieth century.

Fisheries studies are not without their uses since the regulation of catches requires knowledge of the population dynamics of exploited species with ever-increasing accuracy; in contrast, fisheries theory no longer corresponds to the facts and has become an outmoded paradigm. A new framework of ideas must be described to take into account the realities found in fisheries, those of aquaculture and restocking.

Hatchery-nursery methods have revolutionised fisheries' perspectives:

- Fisheries as a method of gathering alone would not disappear but would need to be regulated to survive. It has been seen (Chapter 7) that this regulation cannot be achieved in the framework of the common good, since "liberty applied to common property causes ruin to all". In addition to this is the crisis in fisheries' theory which poses a scientific problem; the collapse of some fisheries (especially those of cod), considered as examples of management, and the revival of fisheries whose decline was predicted, such as that of the lobster in Canada (Griffin, 1996), removes all credibility from official predictions and accentuates the split between scientists and practitioners (Chapter 7). It has been seen that the variability in recruitment that determines the abundance of cohorts of recruits between each fishing year, is actually unpredictable since it is linked to the hydrodynamics of the waters, which in turn depend on the climate. The last two categories of phenomena are ruled by the law of deterministic chaos: they are not predictable, thus rendering intensity of recruitment unpredictable.

- Despite their variability, biological processes determining recruitment are always present in seas and oceans, through hydrodynamic and climatic uncertainties, when fishing pressure has not exterminated the populations. A method of fisheries management which would succeed in protecting broodstock, spawning sites and nursery grounds, as well as manage the better exploitation of adult populations, could constitute a prerequisite for changing to the proposed paradigm. We have seen (Section 7.2.5 and Section 9.11) that the criteria for such management are known and their straightforward application would increase fisheries production by about 10 million tonnes in a few years. We note also that calling on research could aid fisheries; for example, mesocosms allow modelling of food webs within ecosystems.
- The current state of the fish stocks shows, however, that the straightforward management led by fisheries scientists, which has been carried out for several decades, is not a complete solution; to decrease variability in recruitment, the use of hatchery methods becomes indispensable. This alternative is already used in France in oyster production (300,000 tonnes/year): when natural recruitment by "capture" of larvae in the sea is sufficient, there is no need for hatcheries. In contrast, when recruitment is low, hatcheries provide the millions of juvenile oysters required for production. The examples given above for the salmon demonstrate what is involved in the large-scale restocking of fish and work to control reproduction and larval rearing in many other species (cod, haddock and spiny lobster) are in progress, with the same aim.

Currently, man uses tools that could allow the control of recruitment in many exploited populations in very large ecosystems, which would be new and fundamental. This opens up all sorts of possibilities within the management framework for coastal and oceanic waters and in that of fisheries science and management, which are currently difficult to imagine.

Despite these prospects, there are many problems, for example, interferences between natural and introduced populations and the impacts of fisheries and aquaculture on population genetics. All biological and ecological aspects of restocking must therefore be studied carefully before being implemented—the precautionary approach is the basis for success.

Within this framework, the integration of aquaculture concepts and methods, particularly those of the hatchery-nursery with fisheries management, becomes the new, incontestable pattern for the exploitation of aquatic living resources.

13.7 BIBLIOGRAPHY

Anon. Working toward equalized income in a fishing village. *Fishery Journal*, 1990; 34: 6–7.

Anon. Note. *Fish Farming International*, October 1993; 26–27.

Anon. Ranching may revive depleted cod stocks. *Fish Farming International*, April 1995; 14.

Anon. World farm production is boosted again by China. *Fish Farming International*, 1996; 23(7): 5.

Bakun A. L'océan et la variabilité des populations marines. In: *L'homme et les ressources halieutiques*. Ifremer Éd, Plouzané, 1989; 155–188.

Bannister R.C.S., Harding D. and Lockwood S.J. Larval mortality and subsequent year-class strength in the plaice (*Pleuronectes platessa*). In: Blaxter J.H.S. (Ed) *The early life of fish*. Springer-Verlag, Berlin, 1974; 21–37.

Barnabé G. (Ed). *Aquaculture: Vols 1 and 2*. Ellis Horwood Ltd, Chichester, 1990a.

Barnabé G. Rearing bass and gilthead bream. In: Barnabé G. (Ed). *Aquaculture: Vol. 2*. Ellis Horwood Ltd, Chichester, 1990b; 647–683.

Barnabé G. (Ed). *Bases biologiques et écologiques de l'aquaculture*. Tec & Doc Lavoisier, Paris, 1991.

Barnabé G., Giannacourou A. and Ben Khemis I. *A comparative study of cost effectiveness with growth performance of sea bream* Sparus aurata *larvae reared under two different climatic conditions in France and Greece, using three types of mescosms*. Confidential Final Report, Contract No. Aq. 429, Commission of the European Communities, Research Program in the Fisheries Sector, 1995.

Bourguignon G. Les coopératives de pêche moteur de l'aménagement et de la gestion des pêches côtières au Japon. Le cas des pêcheries d'ormeau. *Les Carnets de la SFJO*, No. 4. March 1995.

Coton O. *Capture et élevage de post-larves de langouste*. Panulirus argus *en Baie du Robert* (Martinique). Doctoral Thesis, ENSA, Rennes, No. 87/8 H1O, 1987.

Doumenge F. L'interface pêche/aquaculture: Coopération, coexistence ou conflit. *Norois* 1995; 42(165): 205–223.

Dufour V. and Galzin R. Colonization pattern of reef fish larvae to the lagoon of Moorea island, French Polynesia. *Mar. Ecol. Prog.*, 1993; 102: 143–152.

Dufour V., Planes S. and Doherty P. Les poissons des récifs coralliens. *La Recherche* 1995; Vol. 26 No. 277; 640–647.

FAO, Marine Resources Service. *Review of the state of world marine fishery resources*. FAO Fisheries Technical Paper No 335, FAO, Rome, 1994.

FAO. *Review of the state of world fishery resources: Aquaculture*. Fisheries circular, No. 886. FAO, Rome, 1995.

Fernandez C. *Voyage d'étude sur la filière aquacole japonaise*. Les Carnets de la SFJO, Soc. Fr. Jap. Oceanogr., Paris, 1995.

Folke C. and Kautsky N. The role of ecosystems for a sustainable development of aquaculture. *Ambio* 1989; 18(4): 234–243.

Gianacourou A. *Réseaux trophiques planctoniques utilisables pour l'élevage larvaire de la daurade* (Sparus aurata L. 1758). Thèse Biologie des Populations et Écologie. Univ. Sc. Tech., Languedoc, Montpellier, 1995.

Griffin N. American lobster catches defy gloomy predictions. *Seafood International*, April 1996.

Hiroi O. Activités des écloseries artificielles de saumons au Japon. In: *Aquaculture: Examen des données d'expériences récentes*. OEDC, Paris, 1989; 37–52.

Houde E.D. Difference between marine and freshwater fishes: Implications for recruitment. *ICES J. Mar. Sc.*, 1994; 51: 91–97.

Hunter J. Feeding ecology and predation of marine fish larvae. In Lasker R. (Ed) *Marine fish larvae*. University of Washington Press, Seattle, 1981; 33–77.

Hunter J.R. Inference regarding predation of the early life stages of cod. In: Dahl E., Danielsen D.S., Moksnes and Solemdal P. (Eds) *The propagation of cod* Gadus morhua *L.* Flodevinger rapportser 1984; 1: 533–562.

Kittaka J. Culture of phyllosomas of spiny lobster and its application to studies of larval recruitment and aquaculture. *Crustaceana* 1994; 66(3): 259–270.

Loverich G. *Offshore farming has a place for the hunter*. Fish Farming International, 1991.

Manoli P. Il pim Adriatico settentrionale. *Laguna* 1991, 1: 6–13.

Nash, C. and Novotny A.J. (Eds). *Production of aquatic animals: Fishes*. Elsevier Science, Amsterdam, 1995.

Needham T. Canadian aquaculture: Let's farm the oceans. *World Aquaculture* 1990; 21(2): 76–80.

Ozaki T., Mitsuashi K. and Tanka M. Scallop culture and its supporting system in Mutsu Bay. *Marine Pollution Bulletin*, 1991; 23: 297–303.

Phillips B.F., Cobb J.S. and Kittaka J. *Spiny lobster management*. Blackwell Scientific, Oxford, 1994.

Ravagnan G. *Elementi di valliculturra moderna*. Edagricole, Bologne, 1978.

Riley J.D. The distribution and mortality of sole eggs *Solea solea* (L.) in in-shore areas. In: Blaxter J.H.S. (Ed) *The early life of fish*. Springer-Verlag, Berlin, 1974; 39–52.

Saito K. Ocean ranching of abalones and scallops in northern Japan. *Aquaculture* 1984; 39: 361–373.

Shelbourne J.E. The artificial propagation of marine fish. In: Russel F.S. (Ed) *Advances in marine biology*, 2nd ed., 1–83.

Sherman K. *Large marine ecosystems' global units for management: An ecological perspective*. ICES, Biol. Oceanogr. Com. L/24; 1990.

Sherman K. and Alexander L.M. (Eds). *Variability and management of large marine ecosystems*. AAS Selected Symposium. Westview Press, Boulder, CO; 1986.

Stottrup J. Stocking: Actual situation and prospects for the marine environment. *Aqua. Europe* 1996; 20(3): 6–11.

Tang Q. Change in the biomass of the Yellow Sea ecosystems. In: Sherman K. and Alexander L.M. (Eds). *Biomass yields and geography of large ecosystems*. Westview Press, Boulder, CO; 1989; 7–35.

Whipple A.B.C. *Les courants marins*. Ed. Time-Life, La Planète Terre, 1990

14

Interfaces, substrates, reefs and fish aggregation devices

14.1 INTERFACES IN THE AQUATIC ENVIRONMENT

We saw in the first section of this book how contact between water masses with different characteristics creates discontinuities such as a pycnocline, thermocline or front (Chapter 2). The water's surface constitutes contact zones between the water, air and another fluid or a solid, such as gas or oil bubbles (gaseous discharge from plants or animals, or bubbles created by waves), the periphery of living organisms and also the surfaces of inert particles, however small they may be. We have seen (Chapter 3) what the levels of microscopic particles can be in the open sea. We shall describe these interfaces which are very numerous in the aquatic environment.

For example, oil globules are found, especially in the first four metres below the surface; as studied by Marumo and Kamada (1973) in the East China Sea they represent 11 mg/m^3 of water. Blue-green algae, pennate diatoms, corals, bryozoans, barnacles and copepods live on these globules.

"Artificial" particles are not yet used to exploit this microbial production in the open sea, but another application shows that this would be possible: biological water filtration systems use substrates through which water is run to purify it. Bacteria attached to the substrate ensure the operation of the purification process. In aquaculture and in a water purification plant, these substrates are used in the form of a fixed bed. Recently, a new process was proposed by a manufacturer in which units of plastic material, several centimetres long and wide, are placed in the water to be purified, which is stirred using large stirrers with propeller blades in the contact tank. Growth of bacterial biomass on these units is limited by the erosion created by their contact, which maintains the process at optimal efficiency. This is an example of the role of interfaces in open water.

An interface can also be stable, such as the sea bed or a cliff face under the water, a ship's hull or its moorings, or a jetty. The presence of these submerged substrates leads to ecological and biological effects, on the scale of both the organism and total biological productivity. The depth of the sea bed, for example, its consistency, type (muddy, sandy or rocky), its topography (slope, whether its surface is smooth or rough) determine the nature of the benthic species and, ultimately, their richness. A pier or boat's moorings can provide a habitat for a multitude of living organisms which, without it, would not survive beyond their planktonic phase.

At the level of contact between living organisms and the aquatic environment, exchanges of matter and energy take place (Chapter 4), but many other organisms live attached to substrates or in direct contact with the sea bed or with suspended particles; they require both substrate and water to live. The example of attached shellfish is well known (e.g. mussels and oysters), but many aquatic bacteria live attached to the sea bed or on suspended particles, and this is also the case for a multitude of species that constitute the benthic fauna (see Chapter 3).

The contact zones found in the sea are thus very varied, since there is such a wide variety of organisms. The thermocline and ocean fronts are rich pelagic areas, in biological terms, as are the surfaces of suspended particles (see Chapter 3). We shall restrict ourselves here to contact between the aquatic environment and solid substrates, since this involves both the richest interface in biological terms, and the one which best lends itself to development. Artificial reefs, works to protect the shore, rearing networks for algae and molluscs and fish aggregation devices (FAD), constitute the main categories of development involved, but all submerged constructions (off-shore platforms, pipelines, rearing cages, ships' hulls and moorings, whether floating or sunk, breakwaters, etc.), have an identical role to play. Artificial reefs and FADs are specifically made for this purpose and we shall examine them in this chapter; Chapter 15 will be devoted to the particular example of benthic filter feeders and to the management aspects involved. This arbitrary division has been imposed because of the breadth of the subject.

14.2 DEFINITION AND ROLE OF ARTIFICIAL REEFS

Artificial habitats or reefs can be defined as natural or man-made objects, submerged in aquatic ecosystems. Artificial reefs and other similar structures all have the same objective: increasing the productivity of the environment (FAO, 1990). Their direct physical effect is to provide a substrate for attached organisms, to increase the structural complexity of habitats by providing space in a vertical direction and to modify wave and current action.

In ecological terms, artificial habitats are used to increase fisheries production, and also to influence the biological cycle of organisms or the ecological functioning of ecosystems, or to protect and/or conserve habitats. The principles involved are those of ecological engineering, or ecotechnology (Chapter 8) and it is widely accepted that the effectiveness of artificial reefs acts in three main stages:

- aggregation of fish by thigmotropism* or attraction to food;
- production of a biomass of invertebrates (especially molluscs) requiring a solid substrate and also fish, particularly juveniles, due to reduced mortality

by predation (as a result of the provision of refuges) and growth in a favourable trophic environment;
- protection against capture; the fields of posidonia and juvenile fish on soft sea beds are thus made inaccessible to coastal trawlers.

These concepts are not new; the use of artificial reefs is very old and found almost throughout the world, with various aims:

- limitation of harmful effects on the environment and restoration of habitats;
- limitation of fishing in navigation channels or creation of protected zones;
- reduction of fishing on well-defined stocks;
- prevention of trawling in defined areas;
- control of beach erosion;
- creation of breakwaters;
- increasing fish catches;
- creation of spawning grounds;
- creation of artisanal fishing grounds;
- creation of areas for sport fishing;
- creation of areas for recreational diving;
- creation of areas for scientific experimentation;
- creation of new habitats within reserves;
- creation of areas for mollusc culture (by ensuring recycling of nutrients).

This list shows two distinct functions, but with two common features: physical protection and a biological role. We shall not elaborate on the former which relates to the interplay of forces—resistance to the forces of the sea (waves and currents), or the traction of a trawler—this is a matter for civil engineering.

In biological terms, the past use of artificial reefs has provided a huge amount of knowledge about practical aspects. Their use constitutes a multidisciplinary activity which is distinct from fisheries, since they have numerous applications. The biogeographical theory of island populations has been used to study the colonisation of these reefs (Bohnsack *et al.*, 1991).

Many studies relating to artificial reefs concern their role in fisheries, through the fish populations themselves. A detailed account of all these aspects can be found in publications edited by D'Itri (1985) and Seaman and Sprague (1991), but the use of these reefs is even wider, as they are often used as the mainstay for development of coastal waters (see, for example, the review by Mottet, 1985; report of the working group on artificial reefs and mariculture of the CGPM, 1989 (published by the FAO in 1990); and also Tokuda, 1994, Bulletin of Marine Science, No. 55, parts 2–3, 1994).

For fisheries, artificial reefs constitute a form of development and management considered as a beneficial solution to the difficulties encountered by the professionals (levelling off of landings by trawlers despite their increased fishing power, decrease in catches by small artisanal fisheries and conflicts over resource use), to which can be added various forms of degradation caused by human activity that affect the coast (shoreline construction, destruction of vegetation, retreat of the shoreline, etc.).

We shall not examine here the structures that are used in the first instance to modify water circulation, such as redirecting currents or creating upwelling, etc. While it is not possible to separate such a role from the artificial reef structure which creates it, specific developments for creating water circulation have already been described (Chapter 12). Benthic filter feeders, which are often associated with reefs but which also colonise other structures, are the subject of the next chapter.

14.3 TYPES AND USE OF ARTIFICIAL REEFS

While there are many descriptions of artificial reefs, a large number are incomplete; they mainly concern reefs constructed specially to attract fish and only the fisheries aspect of their use is considered. Several volumes would be required to provide a complete description of all the types of artificial reefs used throughout the world; many descriptive diagrams can be found in the works cited above. Tokuda (1994) provides the addresses of nine Japanese reef manufacturers and lists 20 different types for algae and fixed invertebrates alone; each kind can be obtained in different sizes, varying from 0.5–80 tonnes in weight.

In addition to the architecture of the reefs, the depth at which they are immersed has a determining role, but we shall attempt to classify them in relation to their function, although this remains arbitrary. We shall include structures installed for other purposes but which play exactly the same role as purpose-built reefs.

14.3.1 Artificial reefs and recruitment

We have seen (Chapter 3) that the majority of marine species start their lives with a planktonic larval stage, constituting a form of passive dispersal. This characteristic concerns both pelagic* and benthic* organisms and the success of these planktonic larvae regulates the population abundance of most of these species.

14.3.1.1 Benthic species

The transition from planktonic life to life on the sea bed, whether fixed or mobile, poses problems for the larvae owing to the hydrodynamic phenomena they must overcome (Chapter 4). Hard or stable sea beds in coastal zones are almost entirely covered in fixed organisms (the benthos) and new arrivals in the form of planktonic larvae or algal spores must either use the hard parts of benthic organisms (shells), colonise virgin areas if these are still available, or die at the end of their larval stage if there are no suitable sites for attachment or the waters are too deep for them to reach the bottom. On the sandy sea beds of the Golfe du Lion, flat oysters attach to the bottom (2–10 m deep) during the summer. These oysters reach a size of 2–4 cm in diameter in September–October, but the first autumn storm throws them up on the beaches, because of the lack of stable substrate to which they can attach.

This concept of stability obviously depends on the intensity of the "forces of the sea" (Chapter 2): on well-protected mussel beds in Holland (waves <1 m high), juvenile mussels attach on sand-gravel beds and, other than in exceptional storms, this spat is not carried away by the sea.

In the open sea, the vast majority of larvae of benthic or demersal* species are produced in their billions and are thus destined to perish owing to the lack of substrate. This absence of available, virgin support constitutes one of the main causes of mortality for larval stages of attached species. Some of these species thus devote a large part of their energy (estimated at 75% of metabolisable energy in the oyster (Héral, 1990)) to producing offspring that are doomed to die. This is a well-known paradox, but some ecologists interpret it as a contribution to the global functioning of the ecosystem. The structuring of food webs by size means that gametes and larvae are introduced into the microbial loop as a form of recycling energy towards the bottom of the chain (Chapter 3).

Many coastal demersal fish have a preferred habitat (fish from rocky or sandy shores, etc.), without which they cannot live. If they do not find these at the end of their planktonic phase, they will die and this problem is linked to recruitment success (Chapter 7). The paucity of fish in the Mediterranean has been attributed to this lack of adequate habitat at the end of the planktonic stage (Fage, 1913): the rarity of continental shelves in this sea means that, at the end of the larval phase, the species requiring coastal waters do not find suitable habitats, the great depths being very close to the shore. The fishery activity in the Golfe du Lion, which has an extended continental shelf, argues for this hypothesis, but the flat fish farms and farms for bottom-dwelling fish have demonstrated and defined the requirement for such a substrate at the end of the larval phase. One of the possible uses for artificial reefs is thus to facilitate the recruitment of attached species or those that require a particular type of habitat.

The "nursery" reefs installed in Japan have been described by Mottet (1985): in the Seto Sea, 130 concrete block reefs (2.3 × 1 × 1 m) were placed on a soft sea bed, around 4150 m^3 of rocks set up in three mounds (Fig. 14.1). The blocks and the rocks provided a substrate for algae and the young of various species frequented this area which was previously a desert. Figure 14.2 shows a block designed for the recruitment of abalone, proposed by a Japanese manufacturer.

Buckley (1991) showed that recruitment of the rock fish *Sebastes sp.* depends on the availability of habitats; studying the colonisation of artificial reefs in the Puget Sound area (Washington), he showed a positive correlation between the size of holes and that of the fish which occupied them, above a size of 30–40 mm. These data have been used to construct a recruitment reef made up of 700 tonnes of 8–10 cm-sized rock, spread over 13000 m^2, 25 cm high, to avoid creating hiding places for predators. These reefs are submerged at depths of between 6 and 15 m (Fig. 14.3).

We have shown (Barnabé and Chauvet, 1992) the role played by piles of rocks submerged in shallow waters in the presence of juveniles in the underwater reserve at Monaco. The role of habitat availability in recruitment is well established especially for the recruitment of coral reef fishes (Shulman, 1984; Sale and Ferell, 1988). The latter authors showed that the highest mortalities of juveniles of several species occurred in the two weeks after their transition to benthic life. Like Buckley, we can only regret that fisheries management and ecological studies of juvenile cohorts

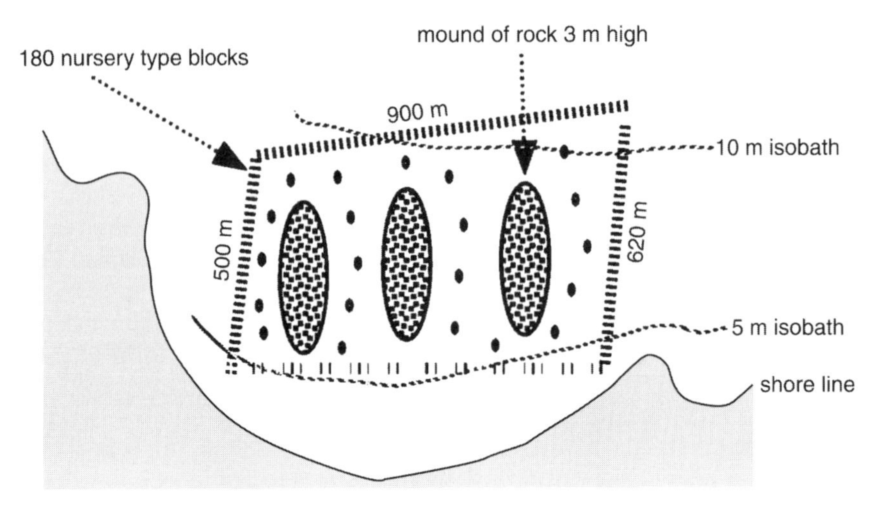

Fig. 14.1. Fish nursery managed using artificial reefs (adapted from Mottet, 1985).

have neglected this recruitment-substrate association. Dufour *et al.* (1995) showed, for the recruitment of coral reef fishes, that the absence of habitats at the end of their pelagic phase contributed to 99% of larval mortality. The influence of depth is emphasised by Moffit *et al.*, (1989) in Hawaii, who demonstrated that the depth of the reef is more important than its structure or type (material); deep reefs gather fish without increasing production.

The presence of aggregations of fish larvae around modules made of cement (Fig. 14.4) submerged on sandy sea beds at depths between 9 and 17 m was shown in Languedoc, close to Sète (Barnabé and Cantou, 1996 and Plate 5, colour section).

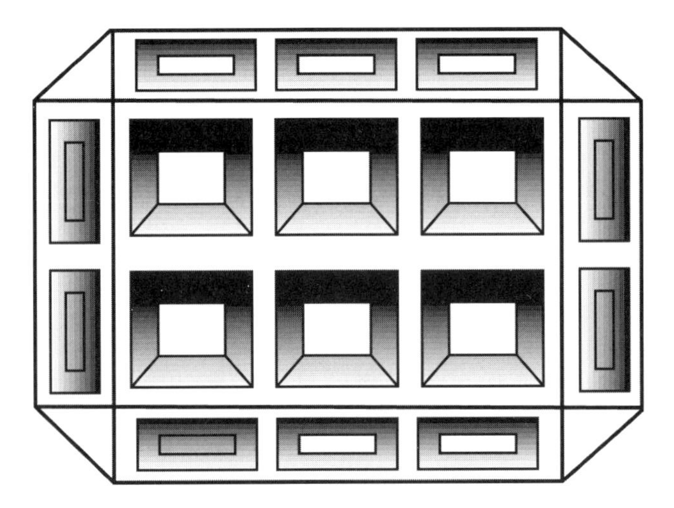

Fig. 14.2. Recruitment and pre-growing on reef for abalone (type A from Terrax Co. Ltd), weight 0.5–4 tonnes. The cavities are used as shelter by the abalone, and algae on which they feed attach to the surfaces (reef surfaces are often treated with iron sulphate to encourage algae attachment).

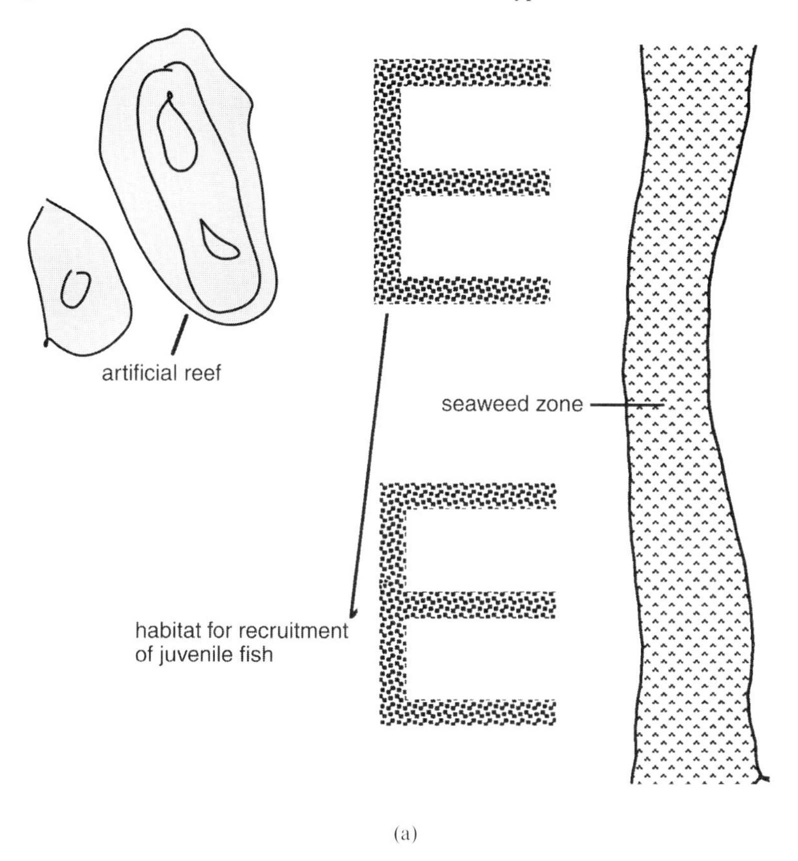

(a)

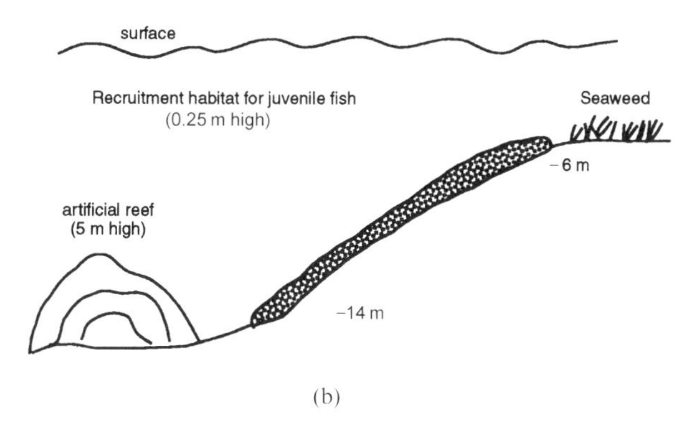

(b)

Fig. 14.3. (a) Plan view of artificial reef used for recruitment of rocky shore fish (adapted from Buckley, 1991). (b) Transverse view of artificial reef used for recruitment of rocky shore fish (adapted from Buckley 1991).

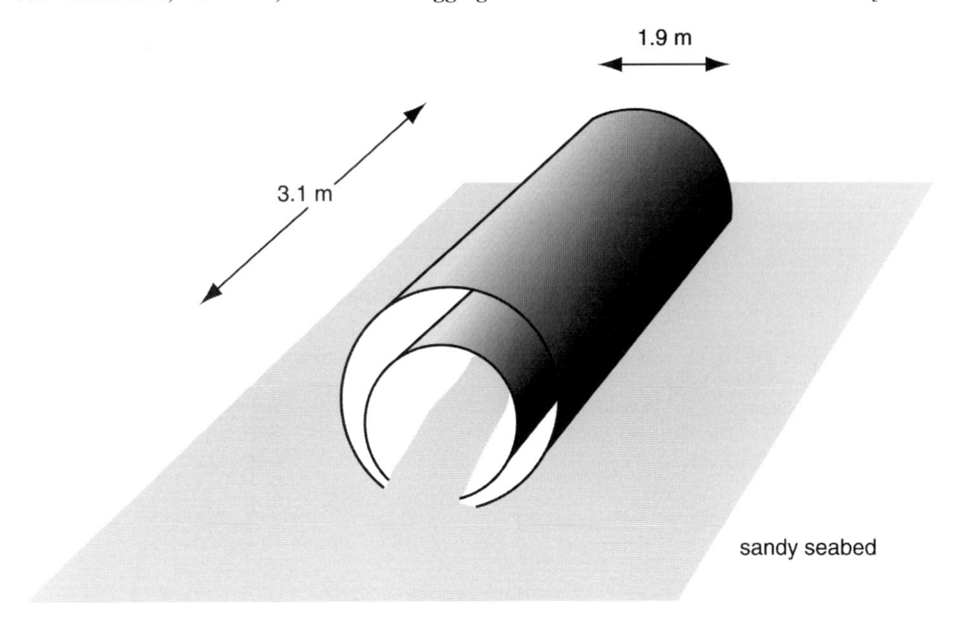

Fig. 14.4. Reef module used in Languedoc to protect littoral fishing zones against trawling.

For crustaceans, work by Riclet (personal communication) has shown that at the end of their larval phase red crawfish (*Panilurus argus*) attach to fishing boat moorings, which are covered in dirt and algae, on the Caribbean coast of Martinique, with 20–40 individuals per 20 m long mooring. Ellis (1991) indicated that a small, submerged collector, made up of various substrates, can catch 300–400 juveniles of this species per year and per collector in Florida and Antigua, where recruitment is continuous. At the end of June 1996, again on the Caribbean coast of Martinique, the maintenance platform for the tourist submarine "Mobilis" was colonised by hundreds of thousands of blue crab (*Callinectes*) larvae.

The protection of littoral sea beds for clam restocking in Japan ranges from the construction of submerged breakwaters made of blocks of rock to barriers made of sand bags to break the current. This is illustrated by Mottet (1985), from whom we borrowed Fig. 14.5. The same author reported the relatively complex and costly development (channels, blocks, etc.) installed for algae and the associated culture of sea urchins.

14.3.1.2 Pelagic species

The production of fish biomass as a result of the recruitment of pelagic larvae on a reef structure, in this case an oil platform, has been studied by Love *et al.*, (1994), following a study based on underwater tracking and marking. For these fish, the structure offered a habitat for recruitment and the transition to benthic life and also for growth before a dispersal phase as small juveniles.

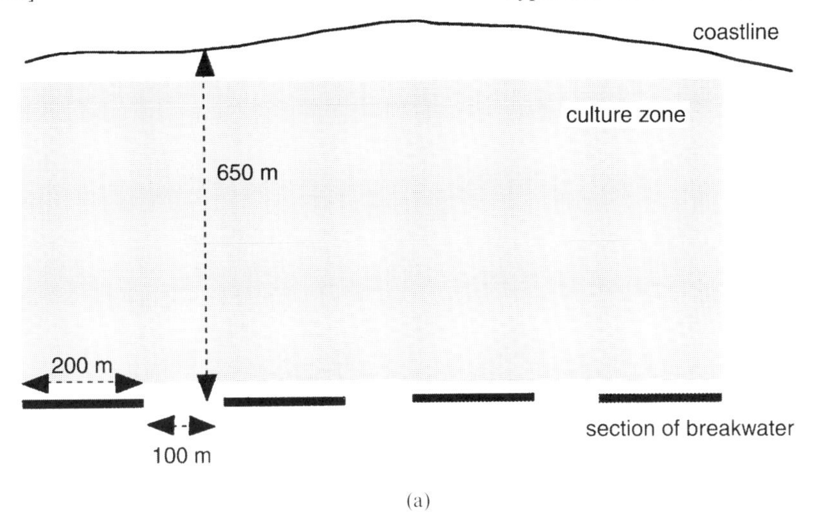

(a)

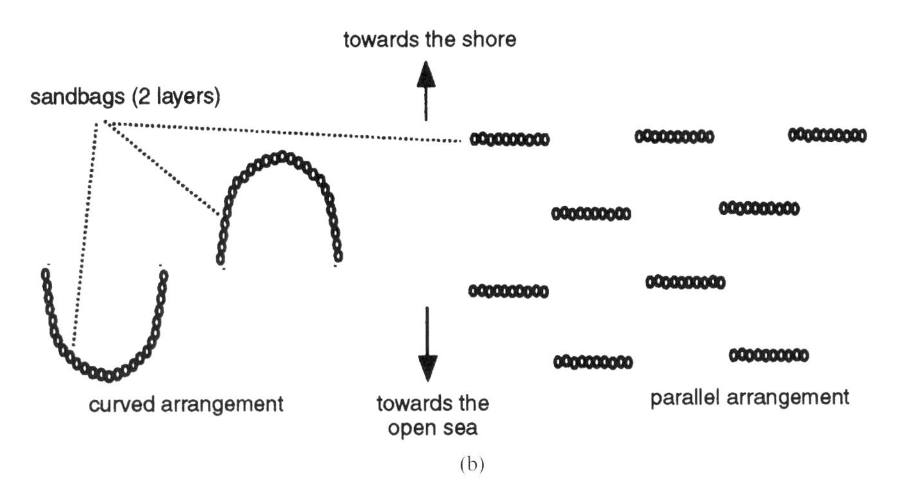

(b)

Fig. 14.5. (a) Culture bed for burrowing bivalves, protected by submerged breakwaters (adapted from Mottet, 1985). (b) Barriers against currents and waves used in the management of natural burrowing bivalve intertidal nurseries in Japan (adapted from Mottet, 1985).

Shoals containing thousands of juvenile bogue (*Boops boops*) and barracuda (*Sphyraena sp.*) have been observed on artificial alga beds situated close to reefs made of tyres, which have been compared to fish aggregation devices (Spanier, 1989).

Oil platforms have also been studied from the point of view of their attractiveness in comparison to that of an artificial reef (Scarborough and Kendall, 1994): some can constitute recruitment zones, while others attract adult populations, because of their location. Beets and Hixon (1994) studied the distribution of grouper populations on different types of artificial reef and concluded that those of a suitable structure were likely to increase the local abundance of these species.

The installation of artificial reefs increases the amount of virgin substrate available and without such a substrate, the biomass developing there would not exist. Attached algae increase the biomass by photosynthesis and filter feeders feed on plankton in mid-water. Thus new production develops which depends on the presence of the solid structure of the artificial reef.

The development of molluscan, crustacean and fish resources by the installation of habitats allowing recruitment of planktonic stages is already established in many sites where the natural richness of planktonic larvae has been proven. This artificial recruitment often results in an increase in the adult stock, which further increases the abundance of larval stages (snowball effect). This phenomenon has been observed in attached molluscs in Languedoc; since the installation of long lines for rearing in the open sea, mussel spat* has become much more abundant and colonises all submerged structures from the surface to a depth of 25–30 m. The juveniles even develop on soft sea beds close to long lines and dredging is now carried out on this type of previously-bare sea bed, since the mussels were not found at depths greater than 12 m.

14.3.2 The role of reefs as a means of protection: breakwaters, jetties and dykes

When walking along the sea shore, one often encounters this type of management system of simple protection. It involves the protection or management of the landward side of the shore; we, however, are interested in the role played by the submerged parts of these structures.

The shallowest artificial reefs are the breakwaters designed to protect harbours, beaches or other segments of the shore. As noted by Stephens *et al.* (1994), after monitoring for 18 years, the oldest and most efficient structure for increasing fish populations was not conceived with this goal. The breakwater which they studied had more abundant and more diverse fish populations than neighbouring solid substrates. These breakwaters, made of rocks and situated in shallow water, favoured recruitment of many fish found on rocky shores; this was also shown by Pondella and Stephens (1994) during five years studying the same site.

A breakwater can provide an exceptional habitat for fish. This was also noted during a study in Monaco, started in 1987 (Barnabé and Chauvet, 1992; Barnabé, 1995).

The impact of the construction of rocky habitats in an intertidal zone (0–2.5 m deep) and the installation of an artificial beach has been studied by Cheney *et al.*, (1994) within the framework of developments designed to compensate for the loss of intertidal zones following the construction of a marina in Puget Sound (Washington); blocks of stone were colonised by algae and provided habitats for many benthic species.

In the Maldives, the rehabilitation of coral flats destroyed by mining explosions (to provide construction material) is under way, to combat erosion and the floods associated with rising sea levels (Chapter 1). These flats are made up of less than 2.5% living coral, with 7 associated fish species and around 20 individuals per species for every 50 m^2 (as opposed to 35 species with 150 individuals/species in adjacent, undamaged areas). One year after installation, artificial reefs installed in damaged areas bring 35 species, each represented by 150–300 individuals, for the most topographically complex reef and 20 species, each with 100 individuals, for the simplest form of reef (Clark and Edwards, 1994). Recruitment of coral on these reefs can be observed after six months.

This type of protective management has long been proposed in response to the rising sea levels associated with the greenhouse effect (Titus, 1986). Although some feel such measures to be alarmist, this author showed that it would be wise to include action against this threat from now on (under the same heading as storms) in construction programmes for housing, drainage and defence against the sea on all shores.

The structures used for the construction of breakwaters are well known, whether natural stone, tetrapods or something else, but it is less well known that they are sometimes used in the open sea to shelter aquaculture areas, in Japan for example, or to form culture areas for brown algae (kelp). Mottet (1985) described the advantages of such reefs, including their arrangement and stability. We have seen (Fig. 14.5) that breakwaters can be used to protect the nursery grounds of various species.

The Japanese are testing a breakwater reef called "OES truss" in Oigawa (Shizuoka) which is capable of absorbing typhoon waves. It is pyramidal in form, placed on the sea bed and is made up of a framework of concrete tubes, the upper parts of which support discs of varying sizes. The whole structure weighs 33 tonnes and occupies $500\,m^3$. The breakwater effect serves to protect aquaculture and fishing zones, and its very jagged architecture could provide a good artificial reef for restocking (Anon., 1993).

14.3.3 Reefs as coastal protection and in the recycling of wastes

The approach by Collins *et al.*, (1994a) combines protection of coasts against storm damage, restoration of habitats and improvement of fisheries; they suggest the use of stabilised waste for this. They noted that defence techniques against the sea have developed from strong concrete walls to softer options, including obstacles submerged in the open sea in order to reduce the energy of the waves reaching the coast (Fig. 14.6); such developments can integrate habitats to allow culture of algae, molluscs, crustaceans and fish. The same team (Collins *et al.*, 1994b) studied the possible spreading of heavy metals from coal ash produced by power stations and used in the construction of artificial reefs ($0.4 \times 0.2 \times 0.2$ m blocks). With the exception of a loss or redistribution of 5% cadmium and an enrichment of the surface of the blocks with manganese and chromium, all the other heavy metals were stable (Cr, Cu, Pb, Mn, Ni, Zn). Using other approaches, other authors (Shieh and Duedall, 1994) provided results indicating that there is no release of heavy metals by such reefs.

A comparison of communities of species colonising the surfaces of artificial reefs made of concrete and stabilised ash, carried out in Florida by Nelson *et al.*, (1994) revealed no difference between the two types. A similar study carried out in the UK

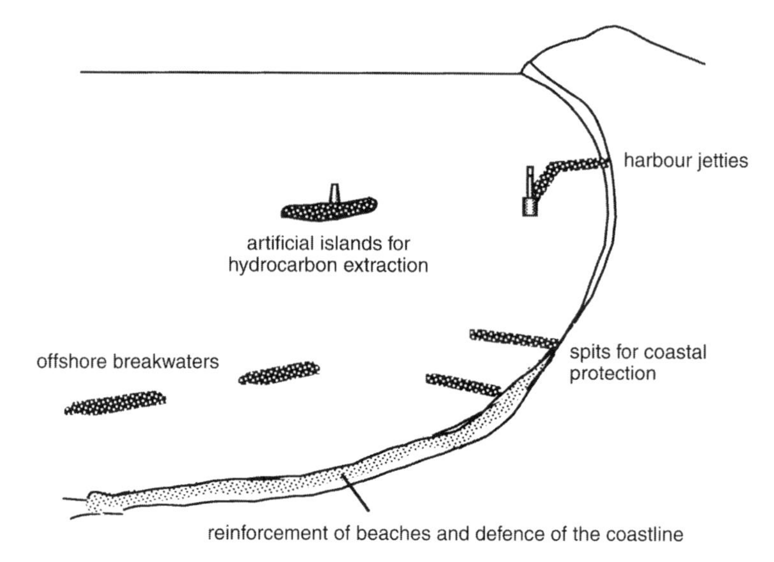

Fig. 14.6. Potential uses for stabilised waste in the marine environment (from Collins *et al.*, 1994).

gave the same results (Jensen *et al.*, 1994), as did another trial in Italy (Sampaolo and Relini, 1994).

Tyres are widely used in the construction of artificial reefs since they constitute an abundant, cheap form of industrial waste which is not toxic and with an unequalled life. A study on their use has been carried out in Australia by Branden *et al.* (1994) and many other examples can be found in Seaman and Sprague (1991). Old concrete electrical posts, assembled into modules, have been suggested by the Société Sotrape Sud (Marseille) in France.

In Thailand, car tyres and cubes of concrete are used on soft sea beds to attract fish which do not usually frequent these areas. In Malaysia, 500,000 tractor tyres have been used for the same purpose: they are assembled in pyramids of 1000 tyres, tied together with polyethylene rope. Some large reefs of 30,000 tyres have also been constructed. We shall see numerous other examples of the use of old tyres in the rest of this chapter and in the following, because it is a cheap substrate (waste) which is durable and is not a hazard to human health (Seaman and Sprague, 1991). The main features of the different tyres used are shown below:

	Int. diam. (cm)	Ext. diam. (cm)	Width (cm)	Weight (kg)
Lorry and bus	50–80	99–104	23–28	35–60
Van	32–40	55–80	13–23	4–30
Car	30–35	53–65	12–18	4–9

14.3.4 Reefs for algae and invertebrates

These can be made of natural stone or be purpose-designed reefs.

14.3.4.1 Reefs made of natural rock

This type of reef is installed where favourable, solid substrate is lacking in areas where the sea bed is made up of soft sediments. The oldest Japanese reefs date from the 1600s; in 1870, about 100,000 40–50 cm diameter stone blocks were put in place for the culture of an alga. This technique of using rocks is still used for the culture of *Laminaria, Undaria* and *Gelidium* to this day, while specialised reefs for this use have been proposed.

Within the framework of "compensation" management, these types of reef have been installed at the time of construction of a nuclear power plant in Japan, following five years of preliminary trials (Yorouchi *et al.*, 1991). Blocks of natural schist with a maximum length of 70 cm were placed on three sites covering 6, 5 and 1 hectares, the reefs forming 3 m high, 50–100 m wide mounds, in 10 m deep water. The populations of algae monitored by the authors were as varied as on natural substrates and became the chosen habitat for many commercially-exploited species.

It should be noted that the reefs designed for algal culture in Japan are also used to attract abalone and sea urchins. For algal culture, a simple layer of rock on a sandy sea bed is sufficient, but abalone and sea urchins will be present if the rocks are piled up to a height of 0.5 m. Such reefs are several tens of metres wide. The collection of algae is relatively mechanised, allowing it to be done on a large scale.

In areas where storms can displace and disperse such blocks, they are arranged in cages of synthetic netting, 4 m long by 1.2 m wide and 0.6 m high, "futon cages" filled with rocks 20–50 cm in diameter (weight 2.9–4.4 tonnes). Partial silting up is a problem in very turbulent areas but, despite this, this type of structure is used in its thousands (10,000, for example, to delimit a 27 ha area for kelp). Mottet (1985) provides numerous details about this technique, including the arrangement of the "futon" relative to the sea bed and so on. The submersion depth of these structures varies from 3–10 m. In certain cases, pipes onto which *Laminaria* spores have already attached are placed on these reefs; 20 kg/m of pipe can be attained at harvest.

The turnover of substrates for algal culture, or their cleaning, improves algal production. A clean block produces 70 kg (393 fronds), while a block submerged for six years produces 4 kg (13 fronds). This example demonstrates the importance of virgin substrates in the sea in the generation of new production.

Various methods are used for cleaning (dynamite, coating in cement, treatment with soda, cleaning under pressure by divers, etc.). The integrated culture of algae, abalone and sea urchins is currently carried out in Japan on purpose-built reefs, of which a great number of forms exist (Tokuda, 1994).

Even on rocky shores, the sea bed can be rendered almost entirely sterile by the development of encrusting calcareous algae; the sea urchins present are not marketable since their gonads do not develop and the abalone are slow growing. These "marine deserts" created by periodic warm currents are only naturally recolonised in some years. To reclaim one of these areas (3 ha), culture of *Laminaria* on nets was carried out. The algal growth broke up the structure on the sea bed and sea urchins and abalone consumed the algae. Thus 3800 m of long lines in 1972 and 4200 m in 1973 allowed the collection of 3.4 tonnes of sea urchins and 3.2 tonnes of abalone. In

1974, the natural growth of *Laminaria* on the sea bed (without long lines) occurred again and 3.1 tonnes of abalone and 6.6 tonnes of sea urchins were harvested. Another estimation evaluated the biomass of algae necessary to obtain 2.3 kg of sea urchin gonad at 100 kg.

When populations of juveniles of commercially-valuable species are insufficient, artificial seeding can be carried out. Here, again, the management possibilities provided by restocking are used (Chapter 13). Areas covered in algae also play the role of nursery for juvenile fish.

14.3.4.2 Reefs purpose designed for algae and invertebrates

There are many different types of these and a publication has been devoted to them in Japan (Tokuda, 1994). Reefs for algae, abalone and spiny lobsters often take the form shown in Fig. 14.7 and are made by many Japanese manufacturers. There can be specific accommodation, notably for lobsters (Fig. 14.8).

For algae, the upper part is often flattened (Fig. 14.7). These reefs are often preferred to natural blocks of rock because of their reduced weight, their structural complexity provides shelter for invertebrates and juvenile fish, their height above the sea bed, and so on. In Europe, these structures correspond to the designation and representation of artificial reefs. The diagrams in this chapter give an idea of their diversity of form depending on function. The publications cited provide diagrams of tens of different reef modules, designed for these uses.

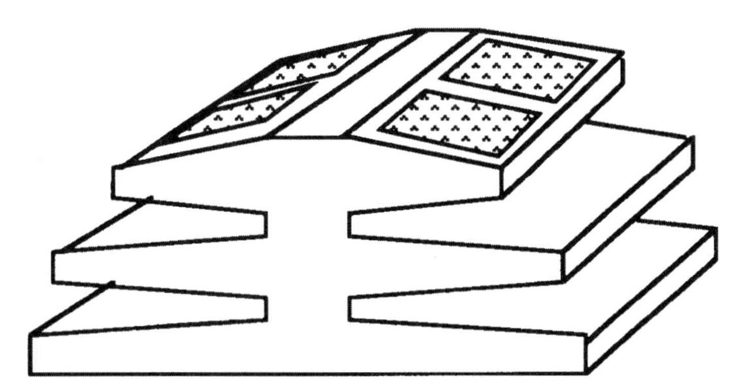

Fig. 14.7. Reef used in Japan as a substrate for culture of algae and shelter for abalone and crustaceans. Plastic honeycombed substrates can be attached to the top for colonisation by small organisms. Seven manufacturers produce this type of reef in Japan (after Tokuda, 1994).

14.3.5 Reefs for the protection of managed sea beds

Alongside these reefs designed for direct colonisation by invertebrates, others are specifically designed for the protection of soft sea beds, managed for the restocking of clams and prawns as in Japan. We have already seen several types designed to favour recruitment (Fig. 14.3) or to protect nursery grounds (Fig. 14.5).

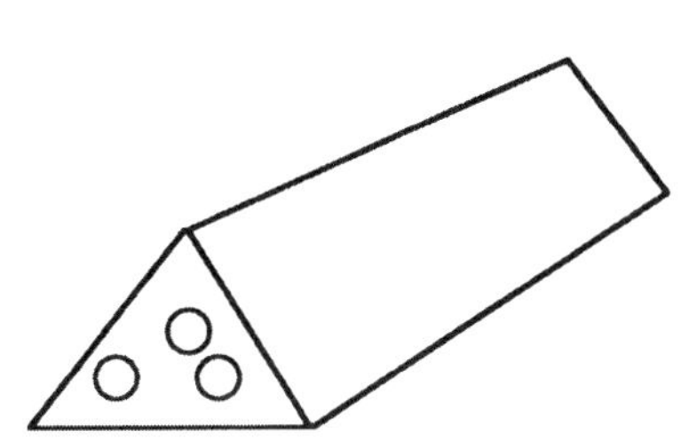

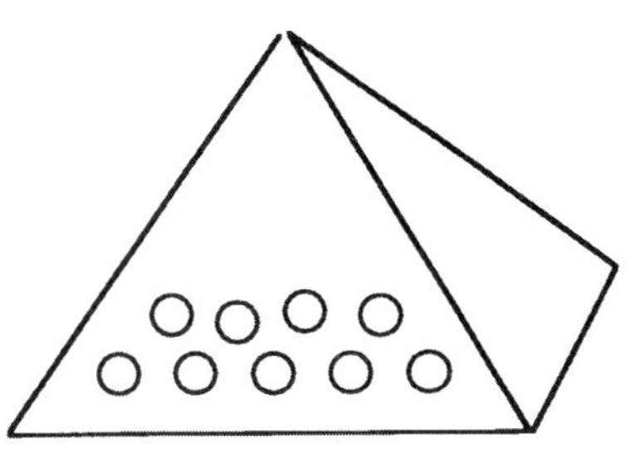

Fig. 14.8. Crayfish and lobster reefs used in Nagasaki (adapted from D'Itri, 1985).

Clams, even adult ones, prefer moderately strong currents which bring them food, but they must avoid waves which dig them out of the sediment and throw them up on the beaches. Once again, breakwaters in the sea can increase survival. The breakwaters are made of blocks, which allow water circulation. Positioned at a depth of at least 1 m, they are covered at high tide. Various other methods of protection are described by Mottet (1985).

Juvenile shrimps produced by a hatchery are released in ponds with sandy bottoms when they reach a length of 10 mm. They can also be released in intertidal zones where they are only temporarily protected. The shrimps move thereafter to the open sea and contribute to fisheries catches; such aquaculture methods for restocking have been described elsewhere (Doumenge, 1989, and Chapter 13). To protect shrimp fishing grounds, significant construction work has sometimes been carried out. One example are the breakwaters in Nosuke Bay, Hokkaido. They are nearly 1 km in length, composed of nine sections each 100 m long, spaced 20 m apart and implanted 1 m in the sea bed. They protect 340 ha of seaweed beds constituting a shrimp habitat, at a cost of €2–3 million (Mottet, 1985).

In France, a reef specially designed for the protection of sea beds against trawling has been installed. It involves a structure of steel reinforced concrete (60 mm long, 20 mm thick elements, with rounded ends) which means that they could be added to the cement mixer and the reef could be constructed by casting the concrete. This type of reef is patented (Inventor, F. Augias, 1987; patent No. 87148-82, dated 27-10-1987). It was designed to "dissuade" trawling and Fig. 14.9 shows its form. This "Sea Rock" reef is made up of a truncated pyramid, 2×2 m base surface area and 0.5×0.5 m upper surface area. This conceals a horizontal concrete plate, situated 10 cm above the sea bed in such a way as to create a suction effect. Although it is light, when knocked over by a trawler, it causes damage because of its rough surface. Several hundred have been installed in the coastal waters off the Camargue (Tocci, 1994). They are reputed to be very effective and dissuasive.

14.3.6 Artificial reefs designed specifically for fish

This use of artificial reefs is the best known and most widespread in the world. It is known that fish gather in areas where a topographical change marks the relief of the sea bed and wrecks constitute the oldest form of reef.

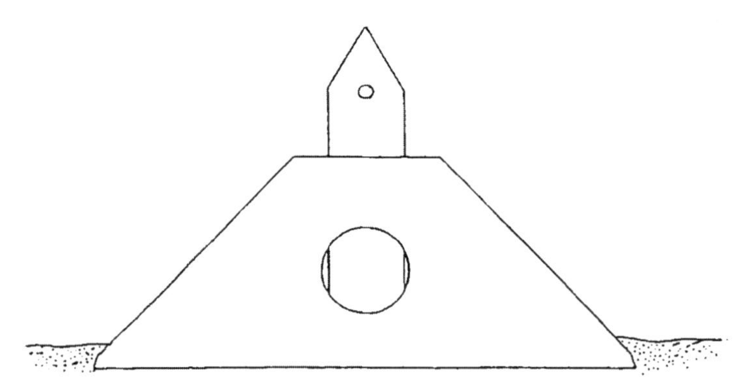

Fig. 14.9. Structure of the anti-trawling "Sea Rock"; side view (base $2\,m^2$, height 1.3 m).

The distribution of fish around reefs has been mapped out, and in practical terms three main categories of fish frequent reefs: pelagic species or surface migratory species, bottom dwelling migratory species and resident, non-migratory species (Fig. 14.10).

The behaviour of pelagic fish around submerged substrates has been compared to the search for a spatial marker in a three-dimensional environment which has none. The fish detect the reefs well beyond their visual limits and return there after being taken several hundreds of metres away. Location occurs from the hydrodynamic perturbation perceived by the fish, or from noise emitted by species present on the reef. This detection is still effective at a distance of 1500 m. Fish catches around reefs are most abundant up to 200 m away and their influence is still perceptible between 200 and 800 m away from it.

14.3.6.1 Architecture and structure

The architecture of reefs designed to attract fish is very varied (see Mottet, 1985; Seaman and Sprague, 1991). A specialised study on the structure of habitats and artificial reef design has been carried out by Bohnsack (1991). Reef architecture most often combines a vertical relief and cavities: concrete cylinders or hollow cubes constitute the basic shapes; the cubes are often between 1 and 1.5 m across, the cylinders between 0.6 and 2 m in diameter; they are often used in a group. Units with more complex structures and of up to $650\,m^3$ in volume are now used. The cavities in the reefs should not be greater than 1.5 m in order to attract fish. The presence of vertical panels makes them more attractive than "skeletal" structures. The abundance of fish is directly proportional to the structural complexity and volume of the reefs. Reef development creates a vertical relief from sea bed habitats, which enhances the attraction of the reef to fish.

In Fig. 14.11 we have reproduced diagrams of some reefs currently used to attract fish, used in France and elsewhere. Figure 14.12 represents an imposing artificial reef which is in widespread use in Japan: the Jumbo reef is made of reinforced concrete, its base measures 8×5 m and it is 7 m high. There are many other reef designs, of

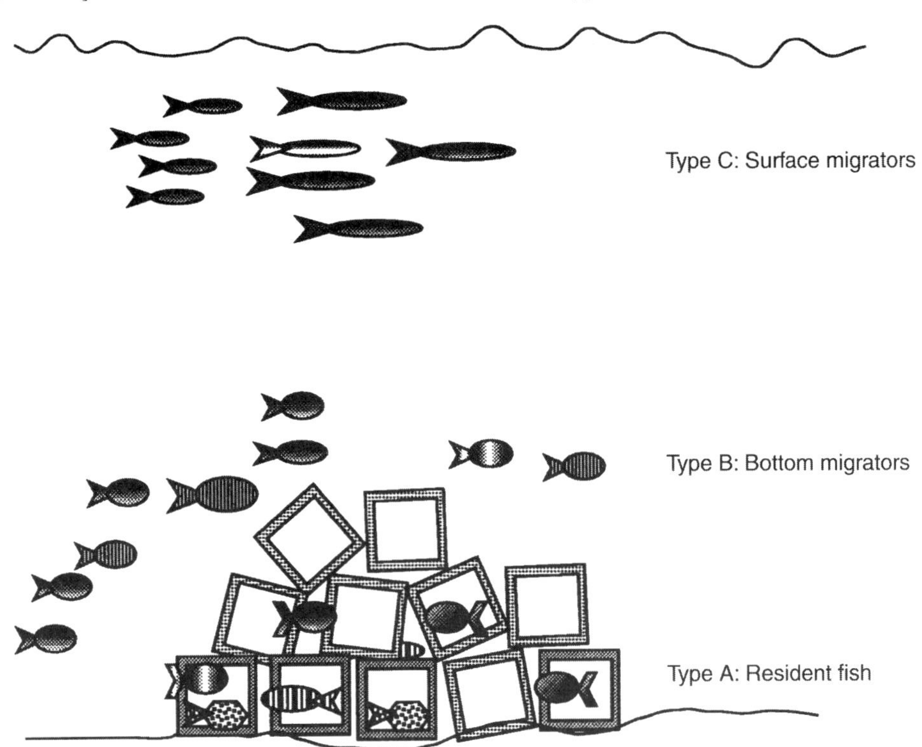

Fig. 14.10. Classification of fish according to their affinity for a reef structure (adapted from Nakamura, 1985).

different architecture, diagrams of which can be found in specialised publications, especially those of D'Itri (1985) and Seaman and Sprague (1991).

It is also known that fish biomasses and densities are higher on artificial reefs than on natural ones; according to Bohnsack (1991), this could be a result of the greater structural complexity of the former. During cage rearing in Hong Kong, faster growth rates were noted when old tyres were placed in the cages, serving as shelter for the cultured grouper. In other species, it is protection against predators which is vital in the reef structure: a study in Israel (Barshaw and Spanier, 1994) concentrated on predation on the Mediterranean slipper lobster, *Scyllarides latus*, comparing the interior and exterior of artificial reefs, mortality was 7% in the reef and 77% outside it.

Artificial reefs in current use are strong structures. According to Bell and Hall (1994), cyclone Hugo had little effect on the artificial reefs in South Carolina, submerged at depths of 5–33 m, despite waves that reached 16 m in height and a surge in water level of 6 m. Damage was light, but one small reef was displaced by 1.9 km, and a wrecked ship, 140 m long and submerged under 33 m of water, moved. Benthic communities of invertebrates and fish were little affected, with no observed harmful effects; in contrast, turbidity in the vicinity of the reefs increased greatly for

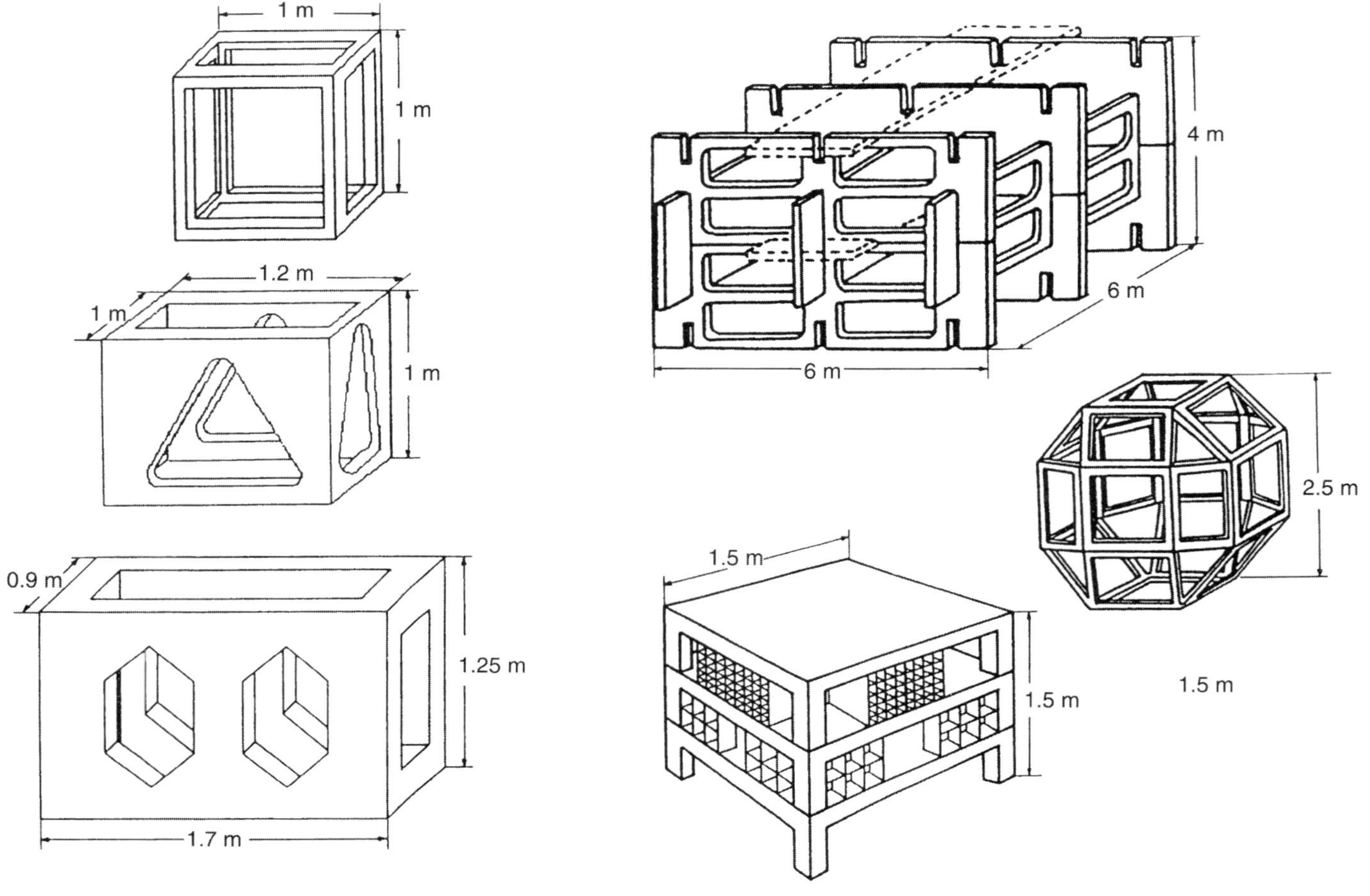

Fig. 14.11. Different models of submerged artificial reefs used in France.

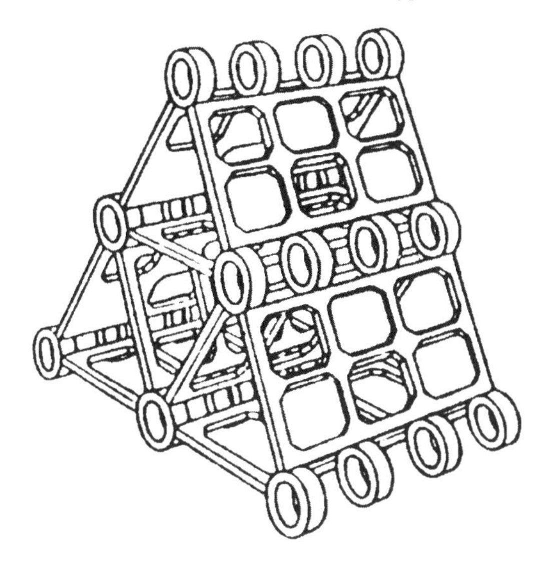

Fig. 14.12. Large "Jumbo" reef, base 5 × 8 m, height 7 m. The reef is complex in structure because the more complicated it is, the more fish it will attract.

a year, as a result of the arrival of estuarine muds in coastal waters. The results are identical in Japan, despite the frequency of typhoons; the life expectancy of the majority of reefs is estimated at 30–40 years, but it varies in relation to the construction material.

Mottet (1985) described the large types of structures used in Japan and their impact on currents and provided many illustrations of the main commercial reefs. Grove (1985) discussed the problem of module size, cavity size, grouping of reefs and their optimum size. In addition to the architectural aspect, he integrated the importance of hydrodynamic phenomena created by artificial reefs in the presence of a current (Fig. 14.13) and reviewed data (sometimes contradictory) on reef architecture recommended by different Japanese authors, which can be summarised as follows:

- The height of the reef must be 10% of the water depth, constituting the optimum for the downstream wave which attracts the fish. The height of the reef is more important for pelagic fish, while the width is more so for benthic fish.
- The minimum advisable size is 1000 m^3, preferably 2500 m^3 for a group of reefs and 400 m^3 for a solitary reef. A reef complex can be between 50,000 and 160,000 m^3, with groups spaced out, 300–500 m apart. They must be positioned across the current and 600 to 1000 m from natural rocky areas; gently sloping sea beds are best. A reef complex can also associate groups which are 3 km apart; in this case, it is not only the reefs which constitute the managed zone: all the space between them becomes a managed area and functions as a new ecosystem.

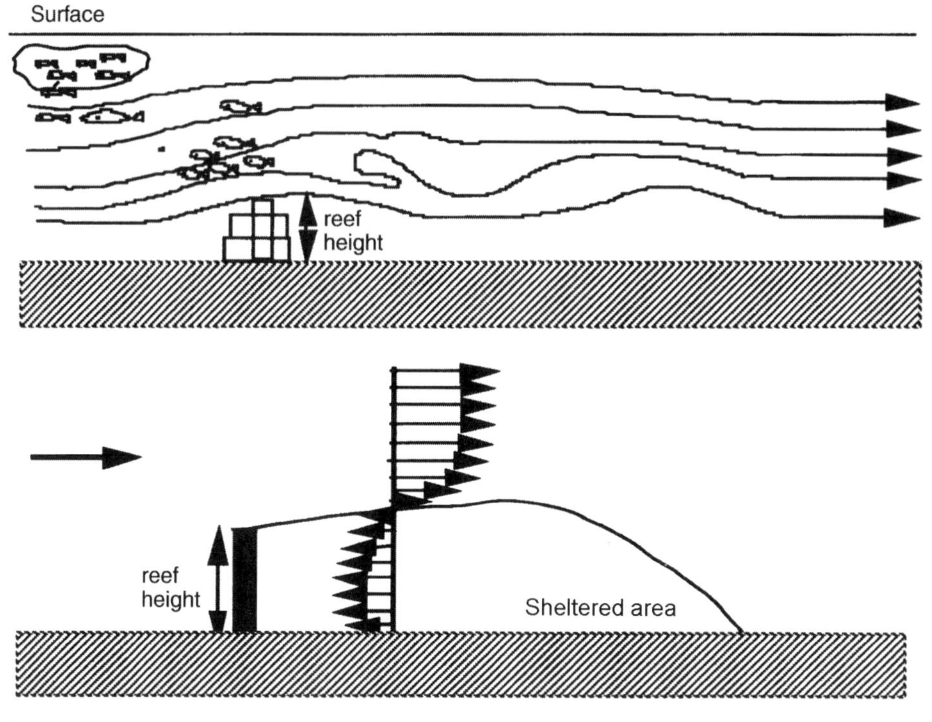

Fig. 14.13. Hydrodynamic changes created by the presence of a reef (adapted from various authors). Top: Diagrammatic representation of the hydrodynamic perturbations created by the presence of an obstacle in a current (adapted from Nakamura, 1985). Bottom: Example of a shaded zone and modification of speed profiles created by the presence of a reef.

- The structure of the reef also affects the attraction and maintenance of fish populations. Potts and Hulbert (1994) indicated that fish abundance is directly proportional to the volume of the structure and also its complexity. This concept is also expressed by the term "roughness" (Harmelin-Vivien *et al.*, 1985).

14.3.6.2 Operation

Artificial reefs provide vital shelter, while food may be acquired elsewhere in the surroundings. This is the case for the black grouper which can move several hundred metres away from its refuge to hunt, or herbivorous white bream which graze on the seaweed beds but return to their rocky shelter at the slightest disturbance. This new biomass of benthic fauna would not be present without a suitable habitat.

From this point of view it can be added that these reefs constitute zones of protection and refuge, both for spawners (increasing concentration of broodfish) and for spawning, as noted by Boudouresque (1995) and Bombace *et al.*, (1994a, b). Shoals of larvae are frequently observed close to the reefs, in the backwash of the current.

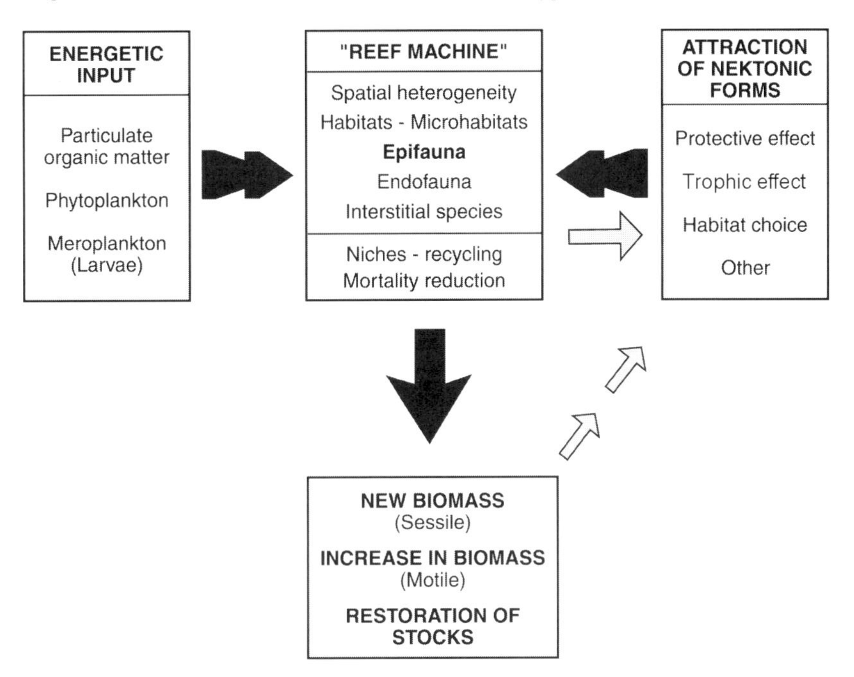

Fig. 14.14. Functioning of the "reef machine" (adapted from Bombace *et al.*, 1993).

The results obtained by Johnson *et al.* (1994) in California were unequivocal: at least 70% of the diet of reef fish is produced by the reef itself and an additional 20–25% from the surroundings. It is not necessary for the reef to provide food, but it must be positioned close to places where food is abundant. For most reef fish, suitable prey are 100 times more abundant in the reef habitat than in neighbouring sandy habitats. Estimated fish production on the reef is about 649 kg/ha/yr, while production of fish trawled from a sandy sea bed is as little as 73 kg/ha/yr, or nine times less. Marking experiments have shown the fish's fidelity to the reef.

It has thus been proven that the construction of rocky or similar habitats allows the establishment of species that live as adults on rocky shores (scorpion fish, grouper and also quite a number of sparids and blennies, etc.) and which are not found on the monotonous sandy sea beds which constitute a large part of the littoral shores. This has also been noted by Bombace *et al.* (1994a) in five different areas of the Adriatic in which artificial reefs were installed: their effectiveness was particularly obvious far from natural rocky zones, with an increase in fish abundance, both in species richness and diversity, and the appearance of species previously absent in the original sandy area. The authors noted a correlation between this increase and the dimensions of the reefs. These authors schematised the role and function of such a "reef machine" in the increase of benthic and nektonic biomasses (Fig. 14.14).

Ody (1987) found that the densities in terms of numbers of individuals or biomass were comparable on artificial reefs and natural rocky zones. In contrast, they were clearly greater than those obtained on posidonia beds, mainly in terms of biomass, which could be 400 times greater on artificial reefs.

Despite these very positive results, some reefs situated in waters which are very low in nutrients have very poor colonisation with algae and benthic invertebrates, since mineral salts and plankton are too low in concentration to be sufficient to maintain populations of fixed organisms which would provide food for the fish. Artificial reefs installed in such areas remain bare and unpopulated for a long time. When fish are found around these reefs, it is obvious that there is insufficient living material for them to be feeding on the reef; these fish have hatched elsewhere and stay around the reef simply because they are attracted to it and so there is aggregation by the reef of a biomass produced elsewhere.

This characteristic of oligotrophic* waters forms the basis for the debate about the real role of artificial reefs for fish populations. Their productive efficiency has been much discussed—do they produce the biomass or concentrate a pre-existing biomass of fish? We shall examine this problem on the basis of the data given above.

14.3.6.3 Production or redistribution of biomass?

The debate is very wide, since the theories explaining the gathering of fish on reefs are numerous:

- The shelter theory states that the fish search for sheltered areas or hiding places provided by the reef.
- The feeding theory explains that the fish feed on living organisms which grow on the reef.
- The thigmotactic theory is based on the attraction of the fish to stationary objects in their environment.
- The turbulent current theory states that the fish are attracted towards the eddies which form around the artificial reefs (see Fig. 14.13).

Several of these factors are likely to combine, depending on the specific conditions at each site, the species being considered, and so on.

Discussions have been held on the effect of reefs on the impact due to aggregation and on that owing to new production but, as stated by Polovina (1991), it is more important, in biological and fisheries terms, to distinguish the following three types of impact:

- artificial reefs can redistribute biomasses of exploitable fish without increasing them;
- they can gather a previously unexploitable biomass and increase its exploitation without increasing the biomass or the size of the stock;
- reefs can increase the total biomass of fish.

The same author went on to discuss these various impacts.

- The redistribution of exploitable biomass is characteristic of flatfish that have been marked and tracked (by radio tracking) in Shimamaki Bay (Hokkaido).

They migrate towards the reefs; there is no increase in the total biomass. In addition, overfishing on the reefs can reduce this biomass. A similar situation has been reported for artificial reefs in the Gulf of Thailand, but the reefs redistribute the resource—artisanal fishermen fish on the reefs, while trawlers and seine net fishers cannot. Reefs can reduce mortality since catches by artisanal fisheries are very likely to be lower than those of trawl or seine net fishing. Thus there is transfer and allocation of the resource to the artisanal fishery.

- Increase in the exploitable biomass, but not of the stock, is illustrated by the case of a reef that can concentrate individuals which were previously not fished because they were too spread out; the reef thus increases catches without an increase in fishing effort. Fish aggregation devices (FAD) illustrate this phenomenon; catches of skipjack (*Katsuwonis pelamis*) increased from 10,000 tonnes in 1970 to 266,000 tonnes in 1986, with the use of "payaos" in the Philippines, but more than 90% of the fish caught were less than a year old. Predation of large tuna round fish concentration devices is greater than in the pelagic environment with no FADs. Excessive fishing at FADs could lead to overfishing of recruits.
- Increase in stock size is reported when artificial reefs provide substrates capable of generating additional food, providing shelter from predation or habitats for attachment. Data such as those of Johnson *et al.* (1994) and Bombace *et al.* (1994a, b) demonstrate the function of artificial reefs in the production of fish biomass and could be increased. Oil platforms in Louisiana represent 90% of the solid substrates in coastal waters (a platform on a seabed 40–60 m deep represents 1 ha of solid substrate). This ecosystem differs entirely from soft sea beds; it involves solid substrate ecosystems that have a large impact, not only on the abundance and composition of fish populations, but also on the socio-economic aspect of fisheries activities. Previously, this fishery for species on solid substrates did not exist.

Stock increase has also been estimated for octopus populations in Japan (at Shimamaki, Hokkaido Island), after the installation of 50,000 m^3 of reefs. The increase was about 90 tonnes, or 1.9 kg/m^3 of reef/year. It has been proved that this increase was not due to population displacement.

An estimate of biomass increase can be made by comparing the fishing yield from natural habitats and the existing stocks on artificial reefs; on coral reefs, the fishery varies from 1–18 t/km^2, with 5 tonnes being the average (Polovina, 1991). The biomass on artificial reefs is, on average, seven times that on natural reefs in the same area (this could be due to their great spatial complexity). If artificial reefs can support 10 times the exploitable biomass of coral reefs, this provides a mean value for fish catches of 50 t/km^2 or 0.05 kg/m^3 for a 1 m high reef.

Another approach to evaluating new production of an artificial reef is to use the biomasses on the artificial reefs to estimate the potential fisheries yield from this biomass. The margin of biomasses on tropical artificial reefs has been estimated at between 26 and 1266 g/m^2 by various authors; with an average 650 g/m^2 of biomass, fishery catches (mortality coefficient, M) could reach 35% for tropical fish, which grow quickly and have a short lifespan, that is 228 g/m^2 potential production or 0.2 kg/m^3/year. Comparison of these figures with actual catches shows that these are about 5 kg/m^3 in the Philippines and between 5 and 20 kg/m^3 in Japan; catches far exceed production owing to reefs and must therefore be attributed to aggregation.

We have also demonstrated (Section 14.3.1, above) the role of artificial reefs in recruitment and the FAO has no doubt that reefs contribute to the protection of juveniles which modifies the mortality pattern and stock biomass increases. This conclusion was reached during the first session of the working group on artificial reefs and mariculture of the General Council for Mediterranean Fisheries held at Ancone in Italy (27–30 November 1989).

It would therefore seem ill-founded to deny the effectiveness of artificial reefs in terms of biomass augmentation. However, in France, the study by Duclerc and Bertrand (1993) provided Ifremer, the Institute in charge of this area on a national level, with the excuse to abandon all activities concerning reefs. A collective statement repeating the conclusions of this study was judged to be not proven (Barnabé *et al.*, 1996). The position taken by this Institute was all the more curious, since it carries out installations of FADs, another form of artificial reef, overseas. This attitude has changed recently. The low level of interest in reefs in France has resulted, in scientific terms, in a limited number of publications (nine publications in contrast to 743 in the USA).

As with lagoon management (Chapter 10), France is well behind its neighbours and rejects European aid. Spain and Italy have both recently installed 200,000 m^3 of reefs with European aid, while there are only 30,000 m^3 of reefs in France (installed more than 15 years ago). Some regional communities/bodies continue to manage artificial reefs with funding from their own pockets; the town of Agde in Languedoc has developed a fishing area protected from dredging 16 km^2 in area (Fig. 14.15), and submerged artificial reefs have been constructed or planned in almost all of the French Mediterranean.

14.3.6.4 Fish reefs throughout the world

Japan has 90% of the world's submerged reefs—20 million m^3 spread over about 6400 sites with an area of 1800 km^2. The goal is the creation of new fishing grounds where previously there were none. Japanese fisheries prefectures have their own builders for reef construction, with barges and installation experience (Ceccaldi, 1988). Patents have been taken out for 130 models of reefs with different functions. Between 1980 and 1987, 10,000 hectares of reefs have been installed and close to 20,000 ha from 1987 to 1994. According to Polovina (1991), 9.3% of Japan's coastal waters less than 200 m in depth are covered in reefs. Catches vary between 5 and 20 kg/m^3, far exceeding the production due to the reefs and must be attributed to aggregation. The impact of reefs on fisheries is thus significant—in Wakayama prefecture, 25–45% of the hook and line fishery occurs on the reefs.

The United States comes second, with a million cubic metres of reefs in place, but their distribution is very patchy with Florida possessing half the country's artificial

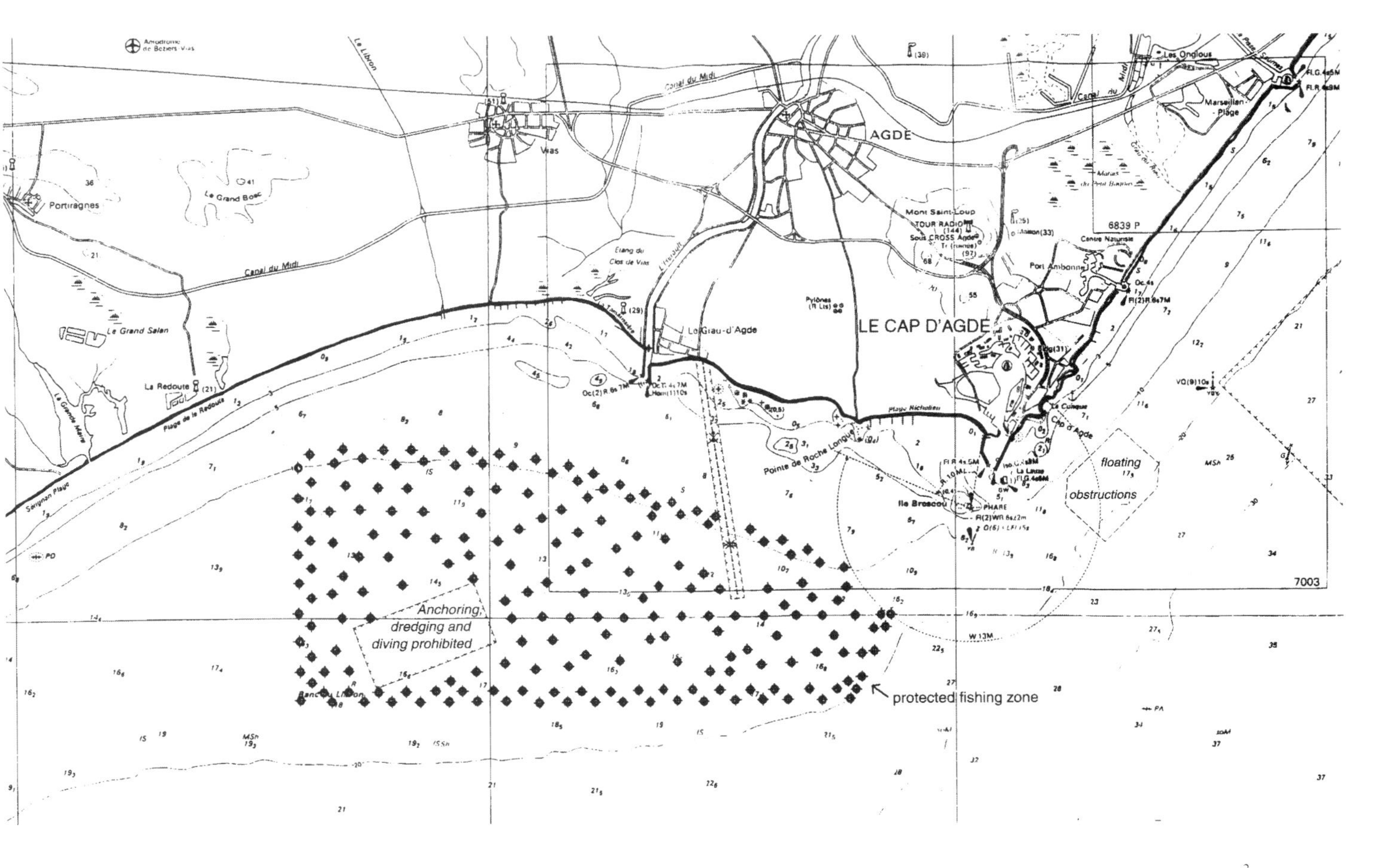

Fig. 14.15. The protected fishing zone at Cap d'Agde. Two hundred modules as shown in Fig. 14.4 are spread over an area of 16 km^2.

reefs (a little more than 300). Many studies on artificial reefs have been carried out in the United States.

According to Polovina (1991), there are 25,000 reefs in the Philippines. Citing the example of 16,000 bamboo frames installed along 40 km of coastline, he stated that the value of catches had covered the costs of these reefs in the first year (cost, ~€3/m^3, and catch = 8 kg/m^3/year).

In Malaysia, we have seen (Section 14.3.3) that tractor tyres are used as artificial reefs. Sri Lanka also uses lorry tyres to construct artificial reefs, but this activity is not well developed.

In Cuba and in Mexico, artificial reefs made of interwoven pieces of wood, raised about 15 cm above the seabed, are used to attract spiny lobsters (*Palinurus argus*) and to aid their capture. These are collapsible reefs—the fishermen shake the reef and catch the spiny lobsters using a net. About 120,000 reefs are used in Cuba, with a catch of about 7000 tonnes of spiny lobsters.

Reefs are not developed solely in rich countries. On the Kerala coast of India, the installation and monitoring of reefs made of concrete and bamboo assembled in modular structures, have shown that they constitute economically viable options for artisanal fishery communities who want to provide fishery habitats (Cruz *et al.*, 1994).

In the Mediterranean and in the Adriatic, Bombace *et al.* (1994a) showed that reefs increase the catches using nets by 2.5 times in comparison to neighbouring seabeds and that the cost of constructing the reefs is paid off three times in seven years (see also Bombace *et al.*, 1994a,b). Other examples can be found in works by D'Itri (1985), Seaman and Sprague (1991) and in the proceedings of a colloquium on the subject, published in a special edition of the *Bulletin of Marine Science* in 1994.

A couple of anecdotal illustrations of reefs: reefs made of tyres may make good nursery grounds, but old Christmas trees are preferred to tyres by fish in Virginia, USA. The best reef in Japan has been found to be an old bus, giving better results than concrete in Kagoshima.

14.3.7 Artificial reefs with a food supply

Reef fish are caught by traditional fishing methods that are often inefficient for species which are close to the reef or living in it. Line fishing is often the only practicable method. Zhuykov and Panyushkin (1990) proposed attracting these species into an open area by feeding them on the sea bed following a sound signal and catching them thereafter by lifting a conical net placed on the bottom. Experimentally, the demonstration is convincing.

In Israel, Spanier (1989) provided low-value fish as food at reefs made up of rows of piled-up tyres, filled with concrete, at a rate of 6 kg/week, for nearly two years. Comparison between this type of reef and a control, unfed site, showed the presence of 42 species of fish in the enriched site, in contrast to eight on the control reef. The abundance of large fish up to 90 cm long such as grouper (*Epinephelus marginatus* and *E. alexandrinus*) increased progressively during the feeding period. It was noted that the recruitment of other, herbivorous, species was not linked to feeding. During two years of monitoring, 617 Mediterranean slipper lobster (*Scyllarides latus*) were caught and up to 34 spiny lobster (0.5–1.3 kg) were observed on the reef during a single observation period.

14.3.8 Reefs for fisheries based on aquaculture and domestication

In Japan, regular provision of food for wild fish populations is practised in several favourable zones, using structures designed for this purpose (Fig. 14.16). Feeding is used to domesticate species such as yellowtail or some species of mackerel until they reach an adequate size for fishing. The common factor in this type of management is that the species involved are those from the open sea, the juveniles of which feed in sheltered coastal waters before returning to the open sea after their first year of life.

Domestication consists of distributing low-value fodder fish such as sardine, anchovy or sprat in a defined marine area, to maintain shoals which move to a fixed location for a given period, during which the fishermen make catches from this shoal by line fishing. The conditioning and fishing site is marked by buoys moored to the seabed between 30 and 50 m apart.

From this initial scheme, management has evolved towards a fishery based on aquaculture using juveniles produced in a hatchery. This is the case for striped jack mackerel which has a typical shoaling behaviour. A shoal of juvenile fish orients in relation to a "base", from which it does not stray. By adding the learned behaviour of regular feeding to this innate behaviour, the shoal can be made resident in a given area.

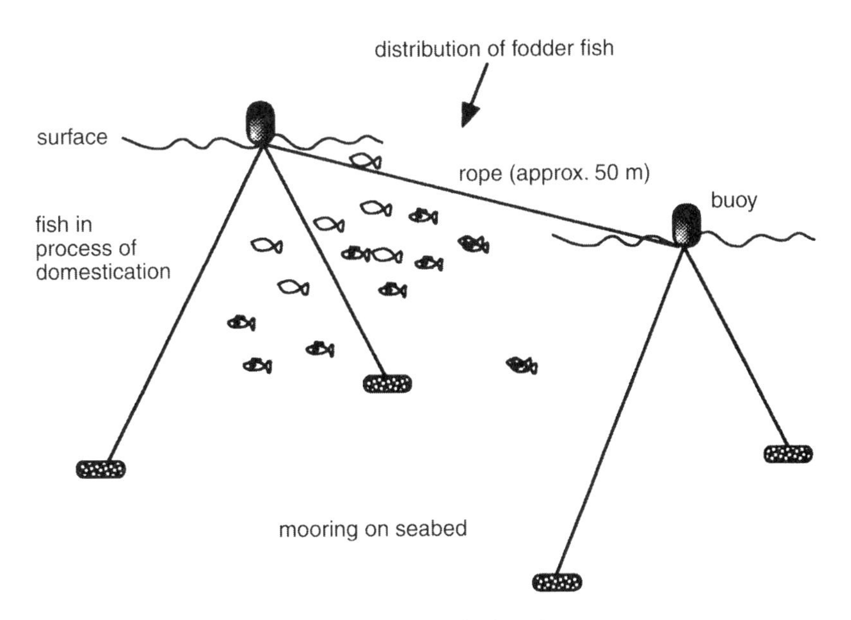

Fig. 14.16. Domestication site.

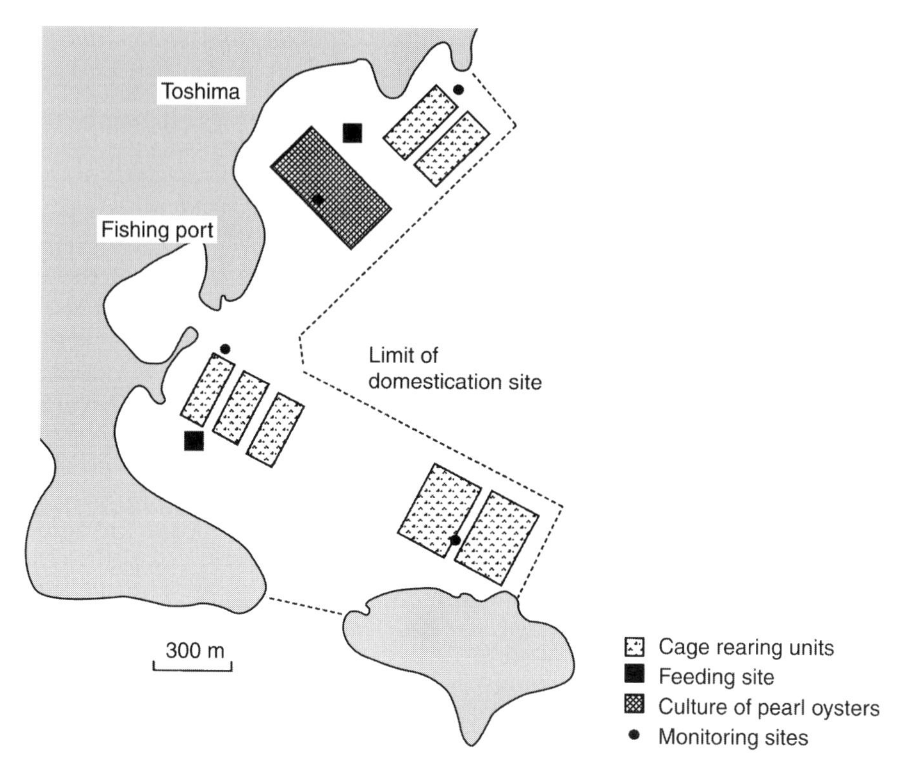

Fig. 14.17. Organisation of domesticated sea beds in the Toshima region.

Juveniles produced in the hatchery are used to stock cages when they are about 3 cm long and reared to a length of about 17 cm (five months).The cages are moved to the feeding site (which has many cages) and rearing continues for about a further month (Fig. 14.17). The cages' netting is lowered towards the seabed so the fish are free, but frequent feeding is carried on at the surface from a 9 m × 9 m raft which also creates shade. After several days, feeding frequency is reduced to a few meals and continued in this way for six months. The fish feed on the distributed food at the feeding site, but consume the food which falls from other cages between meals. To start with, the fish do not go more than 60 m away. Later, they move around the entire culture area, without leaving it. Line fishing is used to catch fish greater than 1 kg, but the recapture rate is low (5.7%). In contrast, the second year after release, fish between 1.5 and 2 kg, migrating southwards, are caught in seine nets (recapture rate of 11.7%). Monitoring has shown the presence of such individuals at the cages, but their capture is difficult. Experiments are continuing (Anon., 1994a).

14.3.9 Reefs for purification of coastal waters

Experiments on this type of reef, which is 300 m long by 30 m wide, have been carried out in the USA (Bartol, 1994). The aim was to restore the oyster fisheries by using a reef which was not permanently submerged, but situated in the intertidal zone (like the old, natural oyster banks) in such a way as to be regularly exposed to the air at low tide, thus limiting predation and sensitivity to disease. The reef was composed of cleaned old shells and was monitored from 1993. However, first results showed that capture of juvenile oysters was higher in zones that were permanently submerged.

In Chesapeake Bay, reefs of tyres were installed to restore oyster populations, not for harvesting, but to filter out particles and nutrients (reported by Polovina, 1991).

A study carried out by Govorin *et al.* (1994) in the Black Sea showed that installing mussels (*Mytilus galloprovincialis*) on the seabed in coastal waters contaminated by urban waste water reduced the level of heterotrophs by 44%, and the level of contamination with bacteria of faecal origin by 44–48%.

In the Gulf of Finland, Antsulevich (1994) proposed a purification project for the waters of a shallow estuary, by installing artificial reefs. Naturally-abundant filter feeding bivalves could attach to these and purify the estuarine water by bioaccumulation of toxic substances.

A similar method, which has been adapted to biological filtration of marine waters, has been used by Shpigel and Blaylock (1991) in Israel. Sea bream rearing resulted in excessive phytoplankton production in the ponds; adding Pacific oysters, *Crassostrea gigas*, reduced the phytoplankton concentration by 50% at the outflow from the ponds and allowed the production of marketable oysters in 14–18 months, at temperatures of 27–30°C. This work is still experimental. Purification of marine water can be carried out by other species; a species of ulva, *Ulva rigida*, has been used in tests in the Canaries to remove ammonia produced by sea bream rearing.

Akai *et al.*, (1991) proposed the purification of waters in closed bays or areas with little turnover, by the construction of dykes made of porous material (stabilised gravel or rubble, etc.). These filtering dykes would purify the waters through tidal movement. These dykes would also serve as substrates for filter feeders, thus increasing their efficiency (Fig. 14.18).

14.3.10 Reefs for tourism or leisure

Artificial reefs are known mainly for their use in fisheries, where fish are the object of direct exploitation within the framework of managed fisheries. Artificial reefs are also used as part of the management of areas reserved for tourism or leisure activities such as diving. Artificial reefs used solely as diving sites are subjected to lower fishing pressure than those used for fisheries purposes, since diving excludes harvesting activity. Such reefs create much more significant "economic returns". An evaluation by Brock (1994) showed that the gross income from commercial fishing only represented 4% of the total tax on the profits of diving clubs operating at the same site.

In the USA, line fishermen have constructed jetties which they exploit for their own use, and in the Mediterranean several diving clubs have invested in artificial reefs, often made of wrecks. Where fish are abundant, tourists will follow: divers gather where iconic fish such as the grouper are numerous and relatively tame. The over-frequentation of these reserve areas by divers has proved a thorny problem in various places (several thousand divers/day in the Mèdes islands in Spain and several

1. Current state: little chance of sea water purification

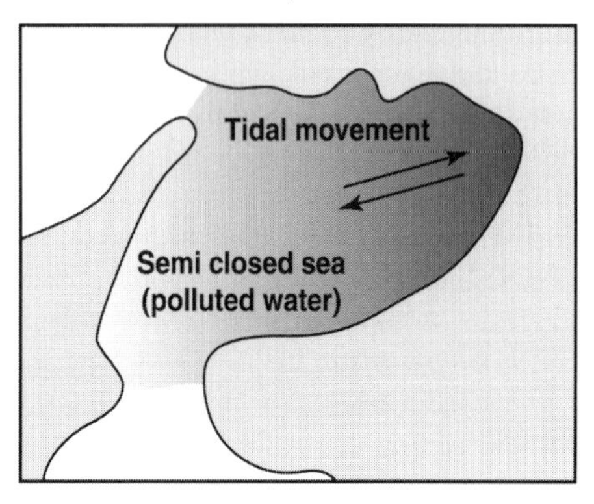

2. Installation of a sink with filtration walls at the coast or in the sea

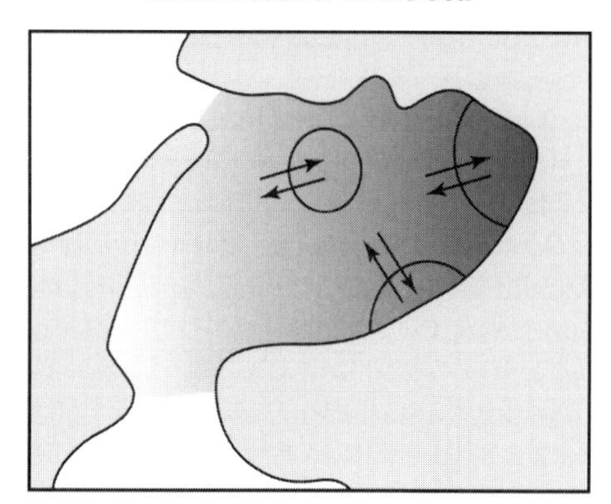

3. Situation at start of high tide

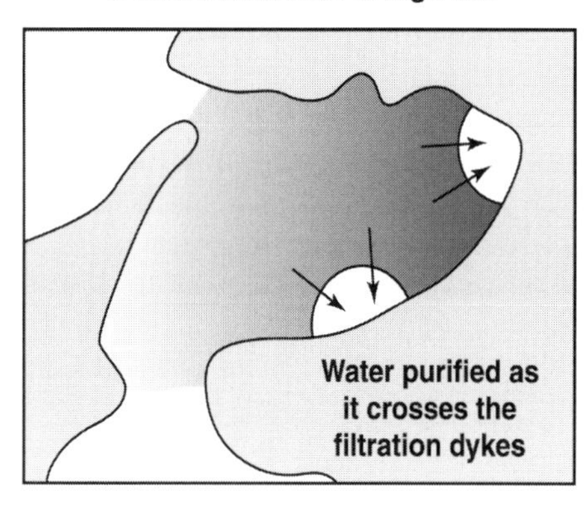

4. Situation at start of low tide

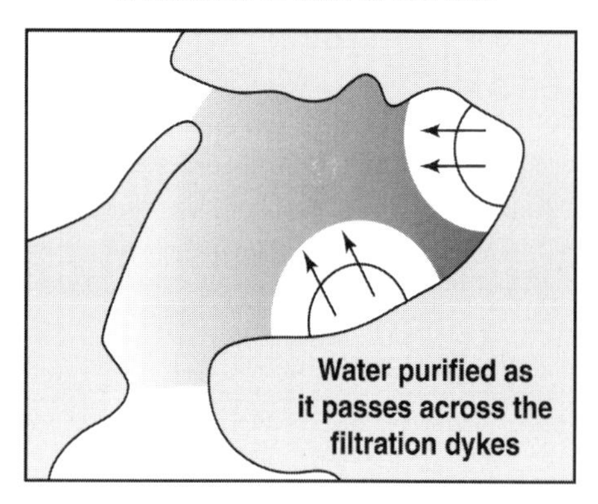

5. Purification after several tidal cycles

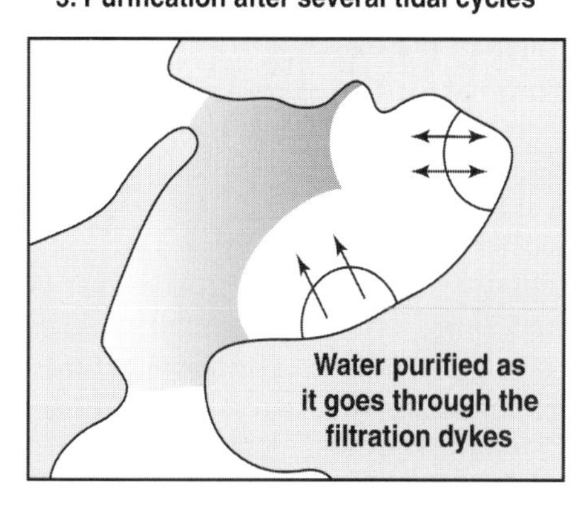

6. Final situation after many cycles

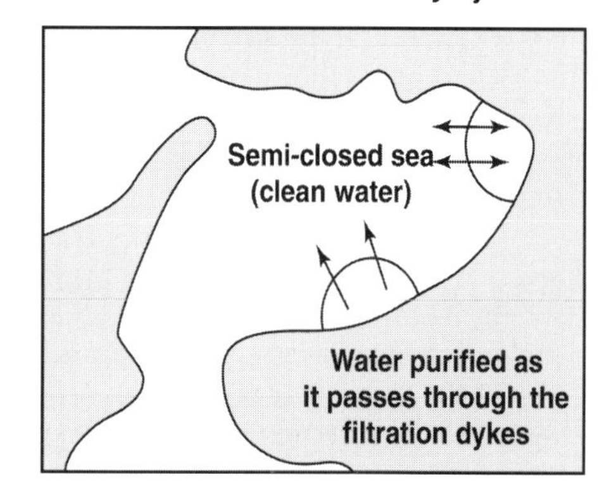

Fig. 14.18. Application of water purification process to a closed sea (adapted from Akai *et al.*, 1991).

hundred at Port Cros, France). On some coral areas frequented by divers, feeding by the club leaders substitutes for natural gatherings, and in some reserves managers feed their groupers to maintain population numbers during the winter; it is a business (see Section 9.5).

A detailed description of the creation and planning problems involved by using reefs for sport fishing on the coasts of California can be found in the work of Wilson (1991).

The development of coastal waters using artificial reefs for diving equates to conservation; their use by humans does not affect the biomass. It can even increase (for example when feeding is used to gather fish around diving sites). This input of exogenous food increases the biomass and can be favourable for reproduction. In ethical and ecological terms, feeding is questionable and provides a subject for discussion between divers and ecologists.

14.3.11 Reefs for anchorage and recycling associated with aquaculture

Artificial reefs have been used in the Caspian Sea as mooring sites for fish-rearing cages (Bougrova and Bugrov, 1994). In addition, reefs can also be used for purification purposes. It is known that cage rearing causes pollution since between 5% and 25% of salmon food, for example, falls out of the cage. The combination of culture in cages and artificial reefs could resolve the problem of waste conversion. The reefs below the cages are invaded by filter feeding organisms; the waste from the cages can thus be used as food and to reinforce the filtration capacity and self-cleaning potential of the site, and the density of other groups of animals also increases. The "Sadcoshelf" experiment carried out in the Caspian Sea, showed that the number of gobies increased 30–50 times while the number of shrimps was 1000 times higher than in a neighbouring area. The cages play a leading role in the reef complex.

A new ecosystem is created: its function is similar to that of "living seabeds" of filter feeders, which we shall examine in the following chapter. The authors also emphasise the interest in reefs and cages shown by tourists, owing to their "extra-terrestrial" appearance.

In Israel, Spanier (1989) also noted the abundance of small fish around a sea bream rearing cage. In Monaco, we have reported the presence of pelagic, planktonophagic species around a bass rearing cage, anchored 90 m above the seabed (Barnabé, 1995). Cages and their inhabitants seem to provide an attractive habitat, like an "inhabited FAD".

14.3.12 Electrochemical artificial reefs

A reef design which captures mineral salts using an electrochemical process has been tested at the Calvi Marine Station. A difference in potential of 8–12 volts between a cylinder and a wire mesh (cathode) 1 m long and 50 cm in diameter, with a central, graphite anode and a current of 0.5–3 amperes, gave results which exceeded all expectations. The same equipment tested in the Gulf of Aquaba resulted in the attachment of madreporine corals. The authors considered that it speeded up larval attachment.

14.3.13 Reefs in freshwaters

Artificial reefs are just as effective in freshwaters as in coastal waters (see Seaman and Sprague, 1991). The subject will not be covered here in detail. In African lakes, substrates consisting of piles of branches or stones are sunk into the ground at regular intervals; these are known as "acadjas" and are similar to artificial reefs; acadjas constitute a system of both aggregation and culture, with yields of between 7 and 20 tonnes per hectare (Hem and Avit, 1994).

14.4 FISH AGGREGATION DEVICES

14.4.1 Definition, function and use

Fish aggregation devices or fish attraction devices are artificial structures situated in open water, often close to the surface. Like the reefs, their architecture is extremely diverse and descriptions can be found in the same publications.

Fish aggregation devices are widely used in many countries because of their efficiency in gathering pelagic fish. As with reefs, various explanatory theories about this attraction have been put forward.

- Fish aggregation devices are covered in algae; these and the plankton that grow on and around them are consumed by small fish which gather round, in turn attracting larger fish. A food chain is set up.
- Larvae and juveniles gather under fish concentration devices to hide from predators.
- Fish have an innate shoaling behaviour in dark areas under objects, such as are found underneath fish aggregation devices.
- The fish are attracted by underwater vibrations created when waves or currents encounter floating objects or mooring lines. We have seen (Section 14.3.6, above) that attraction caused by these phenomena can occur quite far away.

A fish aggregation device is similar to a tiny island with all the accompanying implications with regard to hydrodynamic effects on biological phenomena (Chapter 4). This interpretation allows an understanding of how a fish aggregation device is more efficient, the further it is situated from the seabed and from the coast.

Without doubt, various factors combine to explain the attraction of fish aggregation devices and the different reactions depending on fish species. Algae and invertebrates as well as juvenile coral reef fish, are also found on the moorings of fish aggregation devices, when they are placed in the littoral zone; fish aggregation devices and their moorings are used as a benthic substrate (see colour section).

The attracting part is a simple substrate forming shade and/or floating shelter (e.g. bamboo, wire mesh, cloth), equipped with buoys of various kinds (Fig. 14.19). The simplest fish aggregation devices are coconut leaves suspended from 200 litre barrels or bamboo canes. The more modern ones are made using trawler buoys in a strong net or wire mesh, but there are also very sophisticated ones constituting platforms several metres in diameter, with traps and automatic data transmission. A long mooring line of synthetic rope 12–20 mm in diameter and then a chain attach them to their anchorage which consists of concrete or anchors.

These structures can show on the surface or be situated at depths of up to 30–40 m. The effect of structure and size on fish abundance has mainly been studied by Rountree (1994). This author has also studied (Rountree, 1990) the trophic relationships around a fish aggregation device: the food source providing for the system is mainly planktonic.

Fish aggregation devices are more lightweight in structure than artificial reefs (Fig. 14.19) and can be relocated according to season. Their first widespread use was in south-east Asia and their Philippine name, "payao", was adopted in the 1970s. Fish aggregation devices are anchored on seabeds, from more than 1000 m in coastal waters to more than 100 km from the coast.

14.4.2 Some examples of fish aggregation devices

Fish aggregation devices are mainly used in the Pacific (e.g. Hawaii, Polynesia, Philippines).

- In Hawaii, fish aggregation devices have been monitored and improved for more than 10 years (Higashi, 1994). Emphasis has been placed on the construction of long-lasting equipment which increases sport fishing catches. A system of 78 fish aggregation devices, 56 at the surface and 22 in mid-water, has clearly increased the catch statistics.
- In the Philippines, 3000 fish aggregation devices have been installed at depths from 3000 m and up to 60 km out into the open sea. They concentrate individuals which were not previously fished because they were so widely dispersed. Fish aggregation occurs three weeks after installation. Fishing is carried out using a seine net: a lightship is put in position before midnight to provide a sufficient period of light and catches start from 04.00h. This takes place every 5–6 days and can produce catches of up to 100 tonnes of albacore and skipjack tuna (*Katsuwonus pelamis*). The fish caught are between 16 and 32 cm in length and no fish greater than 60 cm long are caught in the Philippines. Other species represent 5–40% of the catch. Line fishing is possible in the absence of seine nets. Payaos increase cannibalism. Regulations to set the mesh size of the seine nets and to limit the distance of the payaos to at least 7 miles from the coast have been proposed to overcome these problems. Despite these inconveniences, the increase in catches allows exports, a better standard of living for the fishermen and the elimination of conflict regarding secrecy over fishing ground locations.

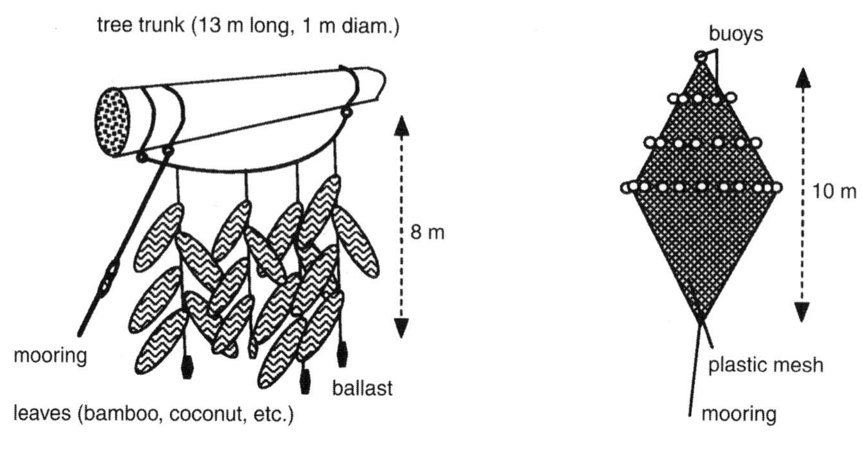

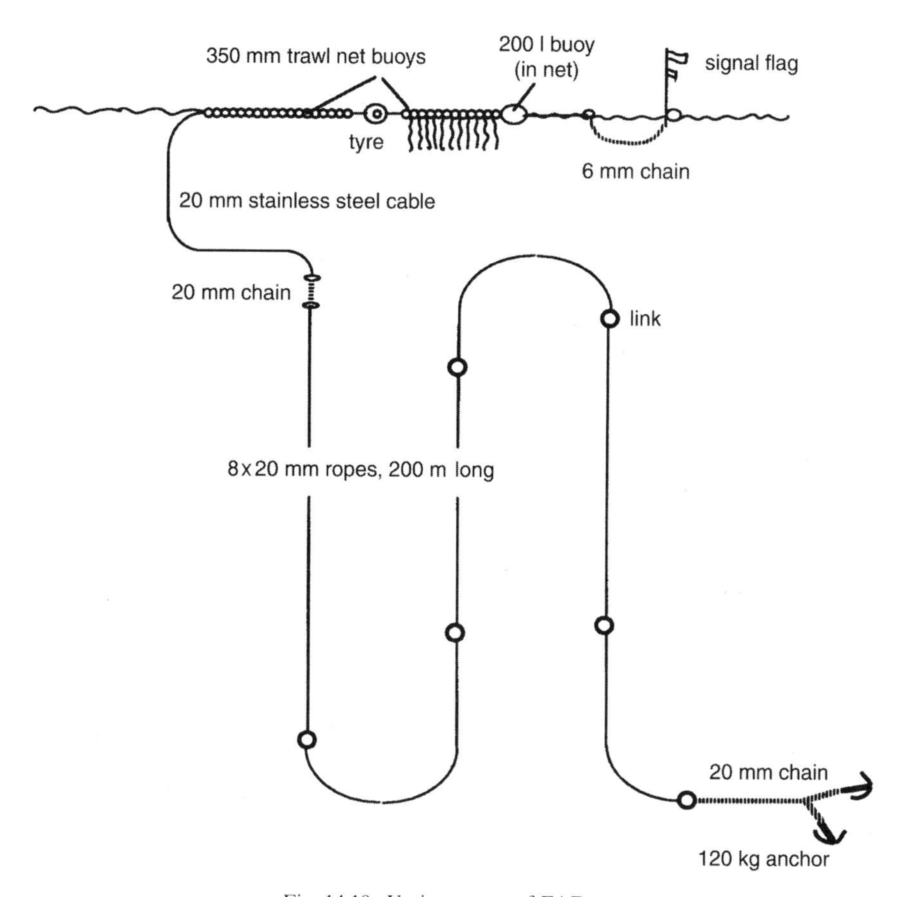

Fig. 14.19. Various types of FAD.

Consumption of these fish has increased and they represent 70% of the protein consumed in the country. This has allowed this country to enter into competition with the large fleets of nations which operate far from home.

- In the Comoro Islands, fish aggregation devices have increased the catch per unit effort of line fishing by 86% and that of dragnet fishing by 29%. They have increased the catches by boats without engines, which are of great social importance (there are 3745 oared boats in contrast with 373 boats with motors).
- There are very many fish aggregation devices in Japan, with 177 around the island of Okinawa alone (Fig. 14.19). The fishery increased from 904 tonnes in 1982 to 2682 tonnes in 1991 (with a value of about €14.8 million and an increase of 310%), due to the presence of fish aggregation devices (Anon., 1994b). The models installed by the Japanese within the framework of the Marinovation programme (see Chapter 16) are more elaborate; they include linked sonars which relay real-time fish concentrations to land-based stations. Catches are monitored continuously. Fish aggregation devices, like reefs, are an essential component in integrated management programmes, to which we shall return.
- The exclusive economic zone in Tahiti corresponds to 5 million km^2. Between 1981 and 1990, 113 fish aggregation devices have been deployed there. They are mostly under-used and the efficiency of fishing methods (line fishing) is limited. The objectives assigned to fish aggregation devices have not been reached since there is insufficient knowledge about the way they work in gathering fish, and other problems concerning their durability in the sea. Fishing at fish aggregation devices can give 230 times the yield of tuna in comparison to other fishing methods. Despite this, fish aggregation devices are not viable in economic terms, since they constitute an indirect subsidy to the fisheries sector (however, in Japan or elsewhere, fisheries cooperatives pay for the fish aggregation devices).

A fish aggregation device can be paid for very quickly, despite its high price, by the resultant catches. Mottet (1985) reported the capture of 32 tonnes of various pelagic fish one week after the installation of such a device in the middle of a sandy, desert area.

14.5 SOCIO-ECONOMIC IMPACTS OF ARTIFICIAL REEFS AND FISH AGGREGATION DEVICES

Many data concerning the socio-economic impacts of these structures have been reported throughout this chapter. These can be summarised here, while stating that, each time that there has been an improvement in fisheries catches, this has resulted in the setting up of a certain type of management (or restocking, discussed in Chapter 13).

Some specific studies are cited below.

- An exhaustive survey of the socio-economic perception of artificial reefs has been carried out in the southern United States (Murray and Betz, 1994).
- The economic role that can be played by reefs within the framework of a professional or sport fishery has been examined in the Gulf of Mexico (Auyong *et al.*, 1986). Pleasure or fishing boats frequenting 164 of the 910 platforms in this gulf were monitored; platforms closest to urban centres (18 km) attracted sport fishermen or divers, while fishing boats were found more often in the open sea (20–55 km). Sport fishing was more active from May to August and at the weekends.
- In the Adriatic, Bombace *et al.* (1994a) showed that reefs increase catches by 2.5 times and that the cost of the reefs is paid off three times in seven years.
- In the Philippines, the catch value covers the cost of installing bamboo reef modules after the first year (the cost being about €4.4/m^3 and the catch of 8 kg/m^3/year exceeds this value).
- According to Chii (1991), reefs play an essential social role. They attract not only fish, but fishermen, cutting down the fuel and time used searching for fish. They also reduce steaming times and increase catch per unit effort. They make the fisherman's life more regular; thus, this author showed that in Japan the number of days spent at sea fell from 250–260 in 1955 to 150–160 in 1985.
- With regard to fisheries management, reefs can be viewed as the allocation of marine space and resources to a particular category, but they also have a role to play in the protection of juveniles and recruitment.
- In qualitative terms, reefs increase biodiversity and reefs or fish aggregation devices favour the capture of more varied and often more prized species.

14.6 CONCLUSIONS

Studying the Japanese and Far East approach to artificial reefs and fish aggregation devices and coastal waters in general, there seems to be a difference in direction. The Europeans are oriented towards theoretical studies and the modelling of aquatic ecosystems, from a conventional point of view which has not changed for a long time (e.g. studies of stocks, natural recruitment and its determining factors, fisheries management in the form of regulation using quotas and TAC) and which seems to us to be rather narrow-minded and based on privileged position. The results of this management have been the depletion of stocks, problems regulating access to resources, conflicts between fishermen and crisis across the entire sector.

The more pragmatic Asiatic approach is aimed at increasing production of its coastal fisheries, whatever the price, in order to be less susceptible to pressure from foreign governments, by reducing dependence on its long-distance fisheries which are limited by the EEZs. This trend was noted by Mottet in 1981 (Mottet, 1985) and has been confirmed with the planning and then the achievement of the large Marinovation programme. These big projects increasingly provide work for businesses.

The accumulated proof on the effectiveness of artificial reefs in their capacity to increase abundance of exploited species has been demonstrated almost throughout the world. They have become a valuable management tool and a link between biology and engineering.

In Japan, the many experiments undertaken and the large investments made go beyond the area of reefs and allow the development of ecotechnological knowledge of coastal waters towards a real "gardening" of the sea, with its seasons and culture areas, and also its land and its proprietors, the fishermen.

In addition to their fisheries-related role, which often is the only one taken into account, they carry out many functions at the level of coastal ecosystems, from physical protection of coasts or seabeds, to water recycling or purification, to the protection of aquatic populations and, above all, the creation of new, productive, ecosystems on seabeds which were previously unproductive.

Reefs are certainly not a panacea, but they should be considered as an additional means of using coastal resources and they interact with other uses of coastal waters. Ecologically, reefs increase spatial heterogeneity, in a way similar to a forest replacing a moor, allowing colonisation by birds and game.

With regard to fisheries, reefs should not be viewed as a common resource, but as a stage in the privatisation of marine resources. Thus development acts, perhaps, as a "proprietary" investment. They have already been used in this way, in the form of marine parks. As in other situations, regulatory measures for reefs have become necessary—management follows development.

In socio-economic terms, such investments, which are profitable in the long-term for civilised society, still constitute the type of large-scale land management which public bodies have to support.

Recycling of certain types of non-toxic waste can be achieved cheaply in coastal waters. Their use as artificial reefs contributes simultaneously to the protection of the shore line, of reserves or artisanal fishing grounds and especially to creating new biomasses and new aquatic production. We have seen that compacted blocks of ash are already used, as well as worn-out tyres or shipwrecks. Dismantled offshore oil platforms could also be recycled as artificial reefs since they play such a role while they are still functional.

14.7 BIBLIOGRAPHY

Akai K., Ueda S., Wada Y. and Tuda R. The water quality purification system for the enclosed sea area. *Marine Pollution Bulletin*, 1991; 23: 683–685.

Anon. Nouveau récif brise-lames. *Flash Japon*, 1993; 47: 6.

Anon. Domestication type fish farming of striped jack: An experimental marine ranch. *Fish. J.*, 1994a; 43: 7–8.

Anon. Payao fishery in Okinawa. *Fish. J.*, 1994b; 43: 4–6.

Antsulevitch A.E. Artificial reefs project for improvement of water quality and environmental enhancement of Neva Bay (St Petersburg County Region). *Bull. Mar. Sc.*, 1994; 55(2–3): 1189–1192.

Auyong J., Ditton R.B. and Reggio V.C. *Offshore petroleum structures lure fishermen seaward in the central Gulf of Mexico.* Gulf of Mexico OCS Region, Minerals Management Service, PO Box 7944, Métairie, LA 70010; 1986.

Barnabé G. 1995 *Evaluation de la faune ichtyologique dans la réserve sous-marine de Monaco et aux alentours du Labrax.* C.R. Assoc. Monegr. Prot. Nat., 1995 (unpublished).

Barnabé G. and Cantou M. *Suivi de la zone de pêche protégée de Marsellan.* Rapport à la Région Languedon-Roussillon; 1996 (unpublished).

Barnabé G. and Chauvet C. *Evaluation de la faune ichtyologique dans la réserve sous-marine de Monaco.* C.R. Assoc. Monegr. Prot. Nat., 1990–91; 51–59.

Barnabé G., Marinaro Y.Y., Carbonnel E., Francour P. and Ody D. Artificial reef in France. In: Jensen A., Collins K. and Lockwood A. (Eds) *Artificial reefs in European seas.* Kluwer Academic Press, Dordrecht, 2000: 167–174.

Barshaw I. and Spanier P.K. Anti-predator behaviors of the Mediterranean slipper lobster, *Scyllarides latus. Bull. Mar. Sc.* 1994; 55 (2–3): 375–382.

Bartol I.J. Intertidal oyster reef as a tool for enhancing settlement, growth, and survival of the oyster *Crassostrea virginica. Shellfish Res.* 1994; 13(1): 280–281.

Beets J. and Hixon M.A. Distribution persistence and growth of groupers (Pisces, Serranidae) on artificial and natural patch reefs in the Virgin Islands. *Bull. Mar. Sc.* 1994; 55(2–3): 470–483.

Bell M. and Hall J.W. Effect of Hurricane Hugo on South Carolina's marine artificial reef. *Bull. Mar. Sc.* 1994; 55(2–3): 836–847.

Bohnsack J.A. Habitat structure and the design of artificial reefs. In: Bell S., Suzan S. and McCoy E.D. (Eds) *Habitat structure: The physical arrangement of objects in space.* Chapman & Hall, London, 1991; 413–426.

Bohnsack J.A., Johnson D.L. and Ambrose R.F. Ecology of artificial reef habitats and fishes. In Seaman W. and Sprague L.M. (Eds) *Artificial habitat practice in aquatic systems.* Academic Press, San Diego, 1991; 61–107.

Bombace G., Fabi G., Fiorentini L. and Speranza S. Analysis of the efficacity of artificial reefs located in five different areas of the Adriatic Sea. *Bull. Mar. Sci.* 1994a; 55 (2–3): 559–580.

Bombace G., Fabi G. and Fiorentini L. *Théorie et expériences sur les récifs artificiels.* C.R. Colloque Océanos, Montpellier, Maison de l'Environnement 1994b; 68–72.

Boudouresque C.F. *Impact de l'homme et conservation du milieu marin en Méditerranée.* Gis Posidonie Éd, Marseille, 1995.

Bougrova L.A. and Bugrov L.Y. Artificial reefs as fish cage anchors. *Bull. Mar. Sc.* 1994, 55 (2–3): 1122–1136.

Branden K., Pollard D. and Reimers H. A review of recent artificial reef developments in Australia. *Bull Mar. Sc.* 1994 55 (2–3): 982–994.

Brock R.E. Beyond fisheries enhancement: Artificial reefs and ecotourism. *Bull. Mar. Sc.* 1994; 55(2–3): 1181–1188.

Buckley R.M. Recruitment of juvenile rockfish (Sebastes) to artificial reef habitats in Puget Sound, Washington. In: *Proceedings of Japan–US Symposium on Artificial Habitats for Fisheries* (JUS 91). Japan Intern. Mar. Sc. Technol. Feder. 1991; 77–81.

Ceccaldi H. Récifs artificiels: Le Japon donne le ton. *Océanorama* 1988; 12: 20–26.

Cheney D., Oestman R., Volkhardt G. and Getz J. Creation of rocky intertidal and shallow subtidal habitats to mitigate for the construction of a large marina in Puget Sound, Washington. *Bull. Mar. Sc.* 1994; 55(2–3): 772–782.

Chii A. The socio-economic role of artificial reefs seen as social capital. In: *Proceedings of Japan-US Symposium on Artificial Habitats for Fisheries (JUS 91).* Japan Intern. Mar. Sc. Technol. Feder. 1991; 27–32.

Clark S. and Edwards A.S. Use of artificial reef structures to rehabilitate reef flats degraded by coral mining in the Maldives. *Bull. Mar. Sc.* 1994; 55(2–3): 724–744.

Collins K.J., Jensen A.C., Lockwood A.P. and Lockwood S.J. Coastal structures, waste material and fishery enhancement. *Bull. Mar. Sc.* 1994a; 55(2–3) 1240–1250.

Collins K.J., Jensen A.C., Lockwood A.P. and Turnpenny A.W. Evaluation of stabilized coal-fired power station waste for artificial reef construction. *Bull. Mar. Sc.* 1994b; 55(2–3): 1251–1262.

Cruz T., Crech S. and Fernandez J. Comparison of catch rates and species composition from artificial and natural reefs in Kerala, India. *Bull. Mar. Sc.* 1994; 55(2–3): 1029–1037.

D'Itri F.M. (Ed). *Artificial reefs. Marine and freshwater applications.* Lewis Publishers Inc., Chelsea, MI; 1985.

Doumenge F. Aquaculture in Japan. In: Barnabé G. (Ed.) *Aquaculture*, Vol. 2. Ellis Horwood, Chichester, 1990; 849–944.

Duclerc J. and Bertrand J. Variabilité spatiale et temporelle d'une pêcherie au filet dans le Golfe du Lion. Essai d'évaluation de l'impact d'un récif artificiel. *Rapport IFREMER*, DRV 93.003/RH/Sète, 1993: 1–42.

Dufour V., Planes S. and Doherty P. Les poissons des récifs coralliens. *La Recherche* 1995; 277: 640–647.

Ellis E. Spiny lobster: A mariculture candidate for the Caribbean. *World Aquaculture* 1991; 22(1): 60–63.

Fage L. Recherche sur la biologie de la sardine (*Clupea pilchardus*). *Arch. Zool. Exp. et Gen.* 1913; 52: 305–341.

FAO. *Rapport du groupe de travail sur les récifs artificiels et la mariculture du CG PM.* Rapport sur les pêches. No 428. FAO, Rome, 1990.

Foster K.L., Steimle F.W., Muir W.C., Kropp R.K. and Conlin B.E. Mitigation potential of habitat replacement: Concrete artificial reef in Delaware Bay – Preliminary results. *Bull. Mar. Sc.* 1994; 55(2–3): 783–795.

Govorin I., Adobovskij V. and Katkov V. Sanitary-bacteriological aspects of the use of mussel mariculture for marine environment bioamelioration. *Gidobiol. Zh. Hydrobiol. J.* 1994; 30(1): 44–53 (in Russian, with English résumé).

Grove R.S. Fishing reefs planning in Japan. In: D'Itri F.M. (Ed) *Artificial reefs. Marine and freshwater applications.* Lewis Publishers Inc., Chelsea, MI; 1985; 187–251.

Harmelin-Vivien M.L., Harmelin J.G., Chauvet C., Duval C., Galzin R., Lejeune P., Barnabé G., Blanc F., Chevalier R., Duclerc J. and Lasserre G. Évaluation visuelle des peuplements et populations de poissons: Méthodes et problèmes. *Rev. Ecol. (Terre et Vie)*, 1985; 40: 467–549 (English translation 1988).

Hem S. and Avit M. First results on "acadja-enclos" as an extensive aquaculture system (West Africa). *Bull. Mar. Sc.* 1994; 55(2–3): 1038–1049.

Héral M. Traditional oyster culture in France. In: Barnabé G. (Ed) *Aquaculture Vol 1*. Ellis Horwood, Chichester, 1990; 342–380.

Higashi G.R. Ten years of fish aggegrating devices (FAD) design development in Hawaii. *Bull. Mar. Sc.* 1994; 55(2–3): 651–666.

Jensen A.C., Collins K.J., Lockwood A.P., Mallinson J.J., Turnpenny A.W. Colonization and fishery potential of a coal-ash artificial reef, Poole Bay, United Kingdom. *Bull. Mar. Sc.* 1994; 55(2–3) 1263–1276.

Johnson T.D., Barnett A.M., De Martini E.E., Craft L.L., Ambrose R.F. and Purcell L.J. Fish production and habitat utilisation on a southern California artificial reef. *Bull. Mar. Sc.* 1994; 55(2–3) 709–723.

Love M., Hyland J., Ebeling A., Herlinger R., Brooks A. and Imamura E. A pilot study of the distribution and abundances of rockfishes in relation to natural environmental factors and an offshore oil and gas production platform off the coast of Southern California. Fish production and habitat utilisation on a southern California artificial reef. *Bull. Mar. Sc.* 1994; 55(2–3) 1062–1085.

Marumo R. and Kamada K. Oil globules and their attached organisms in the East China Sea and the Kuroshio area. *J. Oceanogr. Soc. of Japan* 1973; 29: 155–158.

Moffit R.B., Parrish F.A. and Polovina J.L. Community structure, biomass and productivity of deep water artificial reefs in Hawaii. Fish production and habitat utilisation on a southern California artificial reef. *Bull. Mar. Sc.* 1989; 44: 616–630.

Mottet M.G. Enhancement of the marine environment for fisheries and aquaculture in Japan. In: D'Itri F.M. (Ed). *Artificial reefs. Marine and freshwater applications.* Lewis Publishers Inc., Chelsea, MI, 1985; 13–94.

Murray J.D. and Betz C.J. User views of artificial reef management in the Southeastern US. *Bull. Mar. Sc.* 1994; 55(2–3): 970–981.

Nakamura M. Evolution of artificial reef concept in Japan. *Bull. Mar. Sc.* 1985; 37: 271–278.

Nelson W.G., Savercool D.M., Neth T.E. and Rodda J.R. A comparison of the fouling community development on stabilized oil-ash and concrete reefs. *Bull. Mar. Sc.* 1994; 55(2–3): 1303–1315.

Ody D. *Les peuplements ichtyologiques des récifs artificiels de Provence (France, Méditerranée Nord occidentale).* Thesis, University Aix Marseille 11, 1987.

Polovina J.J. Fisheries applications and biological impact of artificial habitats. In: Seaman W. and Sprague L.M. (Eds) *Artificial habitat practice in aquatic systems.* Academic Press, San Diego, 1991; 153–176.

Pondella D.J. and Stephens J.S. Factors affecting the abundance of juvenile fish species on a temperate artificial reef. *Bull. Mar. Sc.* 1994; 55(2–3): 1216–1223.

Potts T.A. and Hulbert A.W. Structural influences of artificial and natural habitat on fish aggregations in Onslow Bay, North Carolina. *Bull. Mar. Sc.* 1994; 55(2–3): 609–622.

Rountree R.A. Community structure of fishes attracted to shallow fish aggregation devices off South Carolina, USA. *Environmental Biology of Fishes* 1990; 29: 241–262.

Rountree R.A. Association of fishes with fish aggregation devices: Effects of structure size on fish abundance. *Bull. Mar. Sc.* 1994; 55(2–3): 960–972.

Sale, P.F. and Ferell D.J. Early survivorship of juvenile coral reef fishes. *Mar. Ecol. Prog. Ser.* 1988; 7: 117–124.

Sampaola A. and Relini G. Coal ash for artificial habitats in Italy. *Bull. Mar. Sc.* 1994; 55(2–3): 1277–1294.

Scarborough M. and Kendall J.C. An indication of the process: Offshore platforms as artificial reefs in the Gulf of Mexico. *Bull. Mar. Sc.* 1994; 55(2–3): 1086–1098

Seaman W. and Sprague L.M. Artificial habitat practice in aquatic systems. In: Seaman W. and Sprague L.M. (Eds) *Artificial habitat practice in aquatic systems.* Academic Press, San Diego, 1991; 1–29

Shieh C.S. and Duedall I.W. Chemical behavior of stabilized oil-ash artificial reef at sea. *Bull. Mar. Sc.* 1994; 55(2–3): 1295–1302.

Shpigel M. and Blaylock R.A. The Pacific oyster *Crassostrea gigas* as a biological filter for marine fish aquaculture pond. *Aquaculture* 1991; 92: 187–197.

Shulman M.J. Resource limitation and recruitment patterns in a coral reef fish assemblage. *J. Exp. Mar. Biol.* 1984; 74: 85–109.

Spanier E. How to increase the fisheries yield in low productive environments. In: *The global ocean. Vol 1: Fisheries.* MTS/IEEE 1989; 297–301.

Stephens J.S., Morris P.A., Pondella D.J., Koonce T.A. and Jordan G.A. Overview of the dynamics of an urban artificial reef at King Harbour, California, USA, 1974–1991: A recruitment driven system. *Bull. Mar. Sc.* 1994; 55(2–3): 1224–1239.

Titus J.G. Greenhouse effect, sea level rise and coastal zone management. *Coastal Zone Management* 1986; 14(3):147–171.

Tocci C. Champs récifaux formant obstacle aux arts traînants. *Comm. Colloque: De l'étude à l'aménagement des eaux lagunaires et cotières*, Sète, March 1994; 29–31.

Tokuda J. (Translator) *Seaweeds of Japan. A photographic guide.* Midori Shobo Co. Ed., Tokyo, 1994.

Wilson K.C. Artificial reef plan for sport fishery enhancement. In: *Proceedings of Japan–US Symposium on Artificial Habitats for Fisheries (JUS 91)*. Japan Intern. Mar. Sc. Technol. Feder. 1991.

Yorouchi H., Yamamoto R. and Ishizaki Y. Construction of artificial seaweed bed accompanied with the reclamation for unit no. 3 of Ikata power station. *Mar. Poll. Bull.* 1991; 23: 719–722.

Zhuykov A. and Panyushkin S. Use of conditioned-reflex concentration of fish for fishing artificial reefs. *Voprosy ikhtiologii* 1990; 30(6): 1044–1046. *Journal of Ichthyology* 1991, 31(3): 50–54.

15

Filter feeding molluscs and managed ecosystems

15.1 RECYCLING OF PLANKTONIC PRODUCTION BY THE BENTHOS

15.1.1 Benthos of solid substrates

The living organisms found on solid substrates on the shore are the same ones that colonise artificial reefs. Algae are most abundant close to the surface but, below a depth of a few metres, they give way progressively to animals. Of the latter, bivalve molluscs (mussels and oysters) are predominant, but sponges, echinoderms (ophiurids) and many other invertebrates are also found. This fauna is thus characteristic of artificial reefs, because of their immersion depth, which is often greater than about 10 m.

Almost all filter feeders that attach and grow on a solid substrate take their food from living organisms or particles suspended in the water (these species could be called suspensivores). Thus, the plankton provides food for attached animals. Thus an artificial reef covered in benthic organisms functions like a plankton or particle filter; hence their use for filtration (see Chapter 14).

At the same time, these benthic organisms discharge their soluble (CO_2 and ammonia) or solid (faeces) waste into the surrounding water and these wastes provide the starting point for new primary production (Chapter 3). Thus recycling of organic matter occurs in areas where benthic life is rich. The functioning of an artificial reef as such a "reef machine" (Bombace *et al.*, 1994) is shown in Fig. 14.14 in the preceding chapter. The richness of the fixed fauna is thus determined by the richness of the waters; in oligotrophic waters, benthic biomass is very low and mainly consists of encrusting algae; animal life is rare.

The presence of filter feeding fauna on all submerged solid substrates in littoral waters shifts the food chain away from planktonic life forms (small and dispersed) towards larger, benthic organisms. Man cannot exploit biomass which is particulate in form; in contrast, these larger benthic species can be exploited (bivalve molluscs such as mussels or oysters). Interactions between a mussel bed and several other components of the food web are shown in Fig. 15.1. Mussel density can reach 10

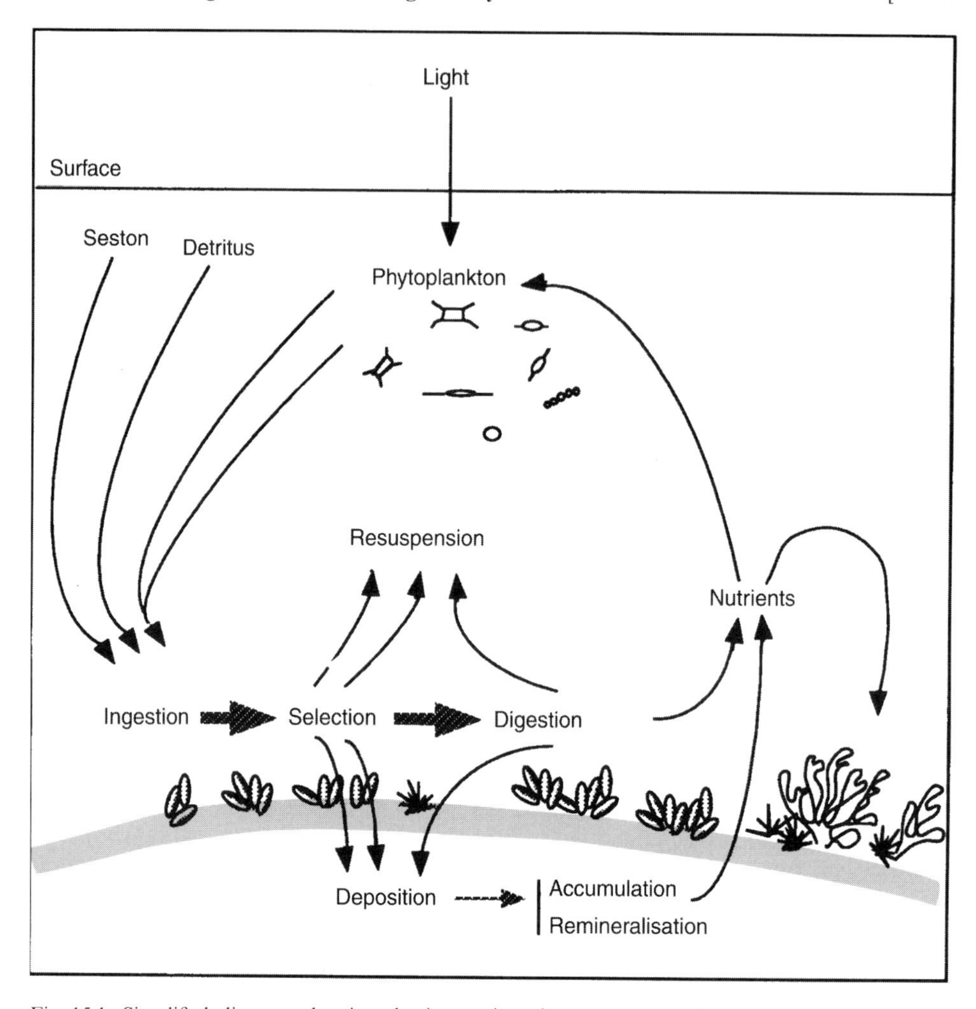

Fig. 15.1. Simplified diagram showing the interactions between a mussel bed, the inorganic seston, detritus and the phytoplankton (adapted from Asmus, 1993).

thousand per m^2 and they are found at depths of over 30 m (Marine Forest Society, 1994).

These fixed organisms form niches and micro-habitats that are colonised by an entirely interstitial fauna (e.g. small crustaceans such as shrimps). Many authors have emphasised that the species diversity of macro- and micro-fauna associated with mussel populations is huge. This fauna provides food for several species of fish and macro-invertebrates that can also be exploited by man (crustacean predators of bivalves, octopus and various fish). There is thus a shift from the pelagic food chain towards benthic production, with a much shorter chain (e.g. phytoplankton–mussel–benthic fish). The yield from a pelagic food chain is low, one tonne of phytoplankton producing less than a kilogram of adult fish, while the same quantity results in the production of more than 100 kg of mussels.

An example of this type of enrichment has been reported by Foster *et al.* (1994). To compensate for the loss of habitats in estuaries and coastal zones, artificial reefs were installed in the inland part of Delaware Bay (USA) and monitored for five years. The reefs increased the total benthic biomass by between 147 and 895 times the initial biomass of the sandy site. These figures do not necessarily mean that soft sea beds are deserts; it is useful, from a management point of view, to describe their differences.

15.1.2 Benthos of soft substrates

In order to remain sandy, the seabed must keep moving and special adaptations are required by the organisms living in it; some molluscs and crabs have developed methods of burrowing to avoid being swept away.

When waves and currents become weak, matter suspended in the water sediments and creates sand-mud or silt sea beds (Chapter 2). These sea beds are typical sedimentation zones for fine organic material, such as dead plankton, produced by productive, mid-water ecosystems. In ecological terms, much of this detritus is treated like additional plankton and is thus grazed on or filtered out along with bacterial or fungal fauna, as has been noted in a mesocosm (Adey and Loveland, 1991). Many molluscs and worms, which live in soft sea beds, feed on plankton extracted from the water above them, but there are also transitional species, from plankton filter feeders to feeders on particles in the sediment.

Bacterial activity on organic particles is significant and leads to high levels of oxygen consumption. When water becomes stagnant (lagoons, closed bays), anaerobic conditions can occur in the sediment, which becomes black. Hydrogen sulphide (H_2S) is produced and very few animals can survive in the seabed, since the organic material is completely broken down into CO_2, water and minerals by the large numbers of bacteria. In contrast, these bacteria can be used as a food source by many inhabitants of the seabed sediment. There are numerous interactions between aerobic organisms situated above the seabed and the buried anaerobic biomass (see Adey and Loveland, 1991). If there is not excessive agitation, these sediments can become the place for storage of organic material, sometimes called a "carbon sink".

Hundreds of species of worms, amphipods and molluscs are hidden under the surface of these uniform seabeds; the number of macro-invertebrates found in soft substrates in estuaries varied from 722–30,000 individuals/m^2 (mean = 7400). Despite this abundance, the total biomass remains low, since it often includes small sized species (see Table 15.1, from Adey and Loveland, 1991). Few species are adapted to live in very mobile sandy substrates.

When these sedimentary seabeds are more stable (e.g. lagoons, bays), some species, which require calmer waters, abound—examples of these are worms used as fishing bait and burrowing molluscs such as the clam. Such species are exploited in sheltered littoral zones since their presence can be detected by their siphons, which they use for respiration or excretion. Various methods are used to dig them out of the substrate (e.g. hook, tube or dredge). In Italy, in the Adriatic close to Sacca di Goro, where restocking with clams has been very successful, population densities can

Table 15.1. Mean number and wet weight of organisms in 1 m^2 of muddy seabed (5 mm thick, superficial layer).

Group	Number	Biomass (g)
Large macrobenthos	2.8	3.75
Small macrobenthos	230	3.30
Meiobenthos	146×10^3	1.15
Protozoa	283×10^6	0.02
Diatoms	590×10^6	0.06
Bacteria	355×10^6	0.07

reach 2000–3000/m^2. In the Étang de Thau, the fishery for the European clam by diving or dredging is about 1500 tonnes/year and exceeds the turnover for shellfish culture in this lagoon. These densely-populated areas are unstable and can require water circulation to avoid oxygen deficits (Chapter 12).

15.2 DOMINANT FILTER FEEDING BIVALVES IN COASTAL WATERS

15.2.1 Huge capacity for filtration

Mussels are very often the most abundant, with the highest biomass, of all animals populating marine shores. From land, they can be seen blackening exposed rocks, colonising beacons at sea and buoys in harbours. Underwater they extend much further, wherever the substrate is solid enough for them to attach: breakwaters, jetties and poorly-maintained ships' hulls are thus uniformly covered in a carpet of mussels whose size and composition vary depending on the time of year and where they are found. In Europe, they are present almost everywhere, from north to south. Mussels are also amongst the most efficient filters of small particles, retaining objects 1–2 μm in size with maximum efficiency (Conover, 1976).

The filtering capacities of mussels and oysters are as follows, according to the review by Lubet (1994):

- a mussel, *Mytilus edulis* (18 months old) filters between 0.3 and 2 l/h per gram dry weight;
- a Japanese oyster, *Crassostrea gigas* (2 years old) filters between 2 and 6 l/h per gram dry weight;
- a flat (European) oyster, *Ostrea edulis* (2 years old) filters between 1 and 3 l/h per gram dry weight.

(It is estimated that the dry weight corresponds to approximately one-fifth wet weight, and that flesh weight is 25–30% of the total weight of the mollusc.)

In the Galician Rias, a mussel can filter between 2–5 litres of water/hour, i.e. 90,000 litres per day for one rope; a raft of mussels can filter 70,000 m^3/day. Plankton and detritus retention has been estimated at 35–40%. A raft of mussels ingests around 180 tonnes of organic material per year, from which 100 tonnes of detritus are returned to the sea.

On littoral seabeds populated by mussels (and, to a lesser extent, by oysters or other bivalves), or in rearing areas, tonnes of these species are found and the volumes of water which they filter amount to thousands of m^3, which is why they have been used for depuration (see Chapter 14). Thus, these bivalves are sometimes called "harvesters of the sea" as they are capable of collecting and converting a biomass inaccessible to man. These species are edible and are at a low position in the food chain, increasing their value. Development that allows the shifting of primary production by water circulation and fertilisation, etc., also allows the production of species useful to man, mainly that of bivalves.

In Liverpool mussels are used to clean the waters in the docks (a series of 3–10 m deep linked docks) within the framework of a rehabilitation plan. Naturally-attached or introduced spat are used, and from a permanently eutrophic and turbid state, with sometimes toxic phytoplankton blooms* as was the case prior to the mussels' installation, the water has become permanently clear (Allen and Hawkins, 1993). In the USA, the quays of a marina have been used for the nursery culture of mollusc spat (Rheault, 1989).

Smaal and Prins (1993) described several distinct coastal ecosystems (e.g. area, mean depth, residence time of water) and went on to estimate the numbers of bivalves they would sustain and that their water masses would be filtered in times of between 0.3 and 13 days by these same bivalves. Certainly, this does involve the estimation of filtration of volumes of water equivalent to that of the ecosystem, but the data from a large number of sites do correspond. Below are some "filtration" times for waters in large ecosystems:

San Francisco Bay: 0.7 days (various species)
Asko area (Baltic Sea): 0.3 days (mussels)
East part of the Wadden Sea: 0.5 days (mussels)
Brest Harbour: 2.8 days (various species)
Étang de Thau: 2.8 days (reared suspended oysters)
Ria de Arosa: 12.4 days (reared suspended mussels)

15.2.2 Exploitation through fisheries

Mussels that attach and grow naturally at sites in the open sea where the seabed is sufficiently stable to resist storms, are subjected to fisheries using dredges; this occurs mainly in the coastal waters of Barfleur in the English Channel where this harvest produces 5000–18,000 tonnes, depending on the year and fishing intensity. Currently 92% of the site is exploited.

Deep water oyster fisheries use the same method; these bivalves are attached to old shells, other members of the same genus, gravel, etc. Unfortunately, the over-exploitation of these natural beds often leads to their disappearance—this has happened with the flat oyster, *Ostrea edulis* and also the scallop in many coastal European waters. During storms, waves carry individuals which have attached in too shallow water up onto the shore; large numbers of little oysters can be found on the beaches of the Golfe du Lion after the first autumnal storms.

15.2.3 Exploitation through aquaculture

Shellfish culture exploits this potential of both mussels and oysters, to the benefit of the human consumer. We have described several examples of shellfish culture in sheltered waters (Chapter 10) and refer the reader to aquaculture publications for more details (Korringa, 1976; Dardignac-Corbeil, 1990; Héral, 1991). At 75% of total, mollusc production by aquaculture is much greater than that by fisheries.

With a few exceptions, most shellfish culture methods are complex and very different from simple management; the rearing of mussels on the seabed in Holland is a typical example of a "living seabed".

15.3 BIVALVES AND MANAGED COASTAL ECOSYSTEMS

15.3.1 Simplified exploitation of natural production in Holland

Bivalves such as the mussel or oyster can live on seabeds which are not solid, such as soft bottoms in still waters. This is the situation in the waters of the Wadden Sea in northern Holland which connects with the North Sea. Juvenile mussels (spat) are found on a sandy seabed, but attached to shells or gravel; using their byssus threads, the spat also attach to congeners, since a mixture of sand and mud predominates in this area. With a tide of 1.4–1.5 m amplitude, 2 billion m^3 of water is displaced with each tide and the current can reach 220 cm/s, although the waves remain less than 3 m high. Clusters of mussels on the soft mud are subjected to erosion by the strong tidal movements and mud in suspension can reach 50 mg/l. Observation by divers shows that the mussels do not make up a carpet but form groups, piled on top of and attached to each other, providing security against waves and currents. There are 4000 ha of plots exploited for shellfish culture; their average depth is less than 10 m. Spat is collected at a size of 25 mm using dredges and is spread over other seabeds to grow. Production after growing on is less than 8 kg/m^2, since unexplained mortality can reach 85%.

The starting point of the culture is the installation of the larvae on the seabed where they will grow, taking their food from the water, ultimately producing spat for the mussel farmers. Nature provides the initial creation of this living seabed which is subsequently exploited by man. It can also have many other uses.

Mussel beds are not the only example of living seabeds. Ophiurid beds are found on sand-mud beds and oyster or other bivalve beds can also be developed.

15.3.2 Management of the seabed for oyster production in China

The techniques used for oyster culture have been described by Cai and Li (1991); the Chinese produce 50,000 tonnes of flesh from 250,000 tonnes of oysters in culture zones that extend over 37,000 ha.

The oldest method consists of placing flat stones, about 90 × 25 cm by 10 cm thick, in the intertidal zone. Often coated with lime, these slabs are placed in groups of five about 70 cm apart; oyster larvae attach to them and grow there until they are harvested.

Another method uses irregularly-shaped 4–5 kg rocks distributed uniformly or grouped in "blossom" shapes; about 60,000 rocks are used per hectare. Predation by sea stars (starfish) is high and production of the order of 1–3 t/ha. This management technique has developed towards the culture of "stone bridges", with immersion of traps using the techniques of traditional shellfish culture.

Bamboo stakes are also used on softer seabeds; they are planted either in lines or in groups of aligned bundles. Per hectare 150,000–200,000 stakes are used, and they can be moved several times during rearing. These methods require more manpower, like the culture of mussels on bouchots in France.

15.3.3 Management of oyster beds in Louisiana (USA)

In Louisiana, production of oyster flesh is a tradition giving approximately 5000 tonnes, or 20,000–30,000 tonnes of oysters with their shells per annum. The American oyster (*Crassostrea virginica*) prefers the estuarine and brackish waters of this coast, which is over 1 million hectares in area.

Production is carried out on public or private concessions and proper management has been undertaken in order to maximise attachment and growth of oysters. On its concessions (356,000 hectares), the state produces spat that it provides to private concessions, which spread it in order to grow-on the oysters. The growing-on beds are 133,000 ha in area.

To facilitate attachment (capture of larvae), the state tips suitable material into the water, consisting of oyster shells, but mainly cheaply-available clam (*Rangia cuneata*) shells, which produce a well-formed oyster. Since 1926, 746,000 m^3 of material has been immersed (Perret *et al.*, 1991). The aim of this dumping is to create new solid substrates.

The dumps must be well synchronised with the maximal larval density and carried out on suitable sites; studies of plankton, currents, sediments and water temperature are used to provide information for site selection. The oyster fishermen also have their say in site selection.

Specialised barges are used to spread the shellfish, using pressure jets in order to form a thin layer of shellfish on the sediment. Between 38 and 75 m^3 of shellfish are used per hectare.

The cost benefit ratio is 12 : 1 (Perret *et al.*, 1991), which speaks for itself.

15.3.4 Management of a whole bay: Mutsu (Japan)

Mutsu Bay in Japan extends over 1660 km^2. With a mean depth of 38 m, it is 57 km long and about 40 km wide. It is very sheltered and is linked to the open sea by an 11 km wide, 60 m deep channel. An oceanic current enters and circulates in this bay so the waters are turned over about every 50 days. Water temperature varies from

3°C in February to 23°C in August, and salinity is between 32 and 34‰. The management of shellfish in this bay has been described by Ozaki *et al.* (1991).

Fisheries production in 1986 was about 69,000 tonnes, representing about €119 million. The scallop (*Patinopecten yesoensis*) is the main catch (52,180 tonnes and about €76 million), followed by the sardine (5010 tonnes) and the Pacific cod (1798 tonnes). Scallops represent 72% of the volume and 67% of the value of production. It involves a managed fishery:

- The fishermen collect spat on suspended traps which is then released on the fishing grounds where it will reach commercial size in 2–3 years. Owing to the form of the biological cycle of this species, there is capture on suspended structures, followed by a growing-on period.
- To these ends, coastal waters are controlled by the fishing rights of community fisheries, with capture zones and exclusion zones (culture beds), for scallops. The culture beds occupy 498 km^2, or 90% of the total culture beds in the bay. The fishermen are organised into cooperatives which have the fishing rights.
- Variations in production have occurred, owing to the quality of spat collected, which led to selection of spat retained, and also management of the culture beds and control of the density of seeding, etc.
- Normal mortality is between 40% and 60% (in contrast to 5–10% for suspended culture).
- Production capacity of the bay has been estimated from primary production (70–100 g C/m^2/year), corresponding to 88,000–178,000 t/year. Production from the seabed is about 20,000–23,000 t/year and could reach 49,000–63,000 tonnes, based on extensions planned for the future. Total production (beds + suspended) expected from the total extended area is 69,000–86,000 t/year, which is still well below the potential of the environment.

Fishing ports have been developed that allow all the operations required by this activity (slipways, areas for grading, cleaning, storage and waste treatment). The coastal population and the fishermen preserve the bay while it remains a fishing zone.

The Japanese have thus moved towards an intensive, managed fishery, while maintaining respect for the productive capacities of the environment by controlling recruitment. The concern for the future is the preservation of the quality of the environment in order to increase production and to manage hatchery and nursery zones for fish such as salmon, cod, flatfish and squid.

The second objective is a complete management system for the bay. Already five constantly-operating communication buoys relay temperature, salinity and current speed to a land-based station, which provides daily and monthly figures to a computer. This is very useful for predicting capture dates, larval distribution, catch levels and danger periods (temperature >23°C), for this cold-water species.

We have seen (Chapter 13) a similar example to that of the scallop fishery in Japan, of fisheries management in the open sea. However, it was based on restocking with juveniles produced in a hatchery and not from capture *in situ*.

15.4 NEW SYSTEMS OF SUSPENDED CULTURE IN THE OPEN SEA

In terms of biomass production per unit surface area, figures for mussel production cultivated in suspension are the highest known for any animal species: for example, nearly 250 tonnes/ha/year in the Galician Rias (Spain). Certainly, it involves production in a volume of water rather than on a surface, and the plankton used are produced elsewhere, but this is very advantageous in practical terms.

Suspended culture methods are quite complex and rely more on marine engineering than straightforward management, but they allow the exploitation of water volumes and can be used in exposed areas in the open sea. They thus constitute a means of polyvalent management. We shall not discuss traditional suspended cultures in sheltered sites here (see Chapter 10), but the method of managing the shellfish remains the same.

When coastal waters in Japan became saturated with aquaculture, the Japanese turned to the open sea and they were the first to set up shellfish culture after the Second World War. The Pacific oyster, *Crassostrea gigas* and the scallop, *Patinopecten yesoensis* are the main species reared. Culture methods have been described by Lucien-Brun and Lâchais (1983), Doumenge (1989) and Anon. (1989).

The structure of a long line system is relatively simple. Depending on its length, the number of individuals it supports and sea conditions, the dimensions of the materials used may vary, but the design remains the same: the main line is moored to the seabed by moorings, stakes planted in the bottom, or by anchors. Submerged or floating buoys carry this single line under the surface or at the surface and rearing baskets or ropes are suspended from this line (Fig. 15.2). The molluscs or algae are thus in contact with the open water which brings them both food and oxygen and takes away waste products. Currents renew this nourishing water around the reared individuals.

End to end, these lines can be up to 1800 m long and are often arranged in networks of parallel lines spaced several tens of metres apart and occupying hundreds of hectares on coastal seabeds up to 50 m in depth. Japan is one of the main producers of shellfish in the world, with a production of 300,000 tonnes of oysters and scallops.

Japanese long line cultures have been copied throughout the world, with varying degrees of success. In the Golfe du Lion (France), mussel culture in the open sea started in the 1980s and production is about 5000 tonnes. A diagram of the type of long line used is shown in Fig. 15.3 and the technology used has been described (Barnabé, 1990). The coastal administrators have created four shellfish culture sites, several km^2 in area, in the open sea, on soft seabeds between 15 and 35 m deep; this activity has created 300 new jobs. There have been some difficulties with reliability of anchor points and profitability remains uncertain; this activity has been burdened with technical constraints and regulations which have limited its development. It should be made simpler and less costly.

In the Sagres region of Portugal, which is exposed to great Atlantic storms, a series of long lines for rearing were installed, each about 30 m apart and linked together to form a square framework 300 m × 300 m (Podeur, personal communication, 1993).

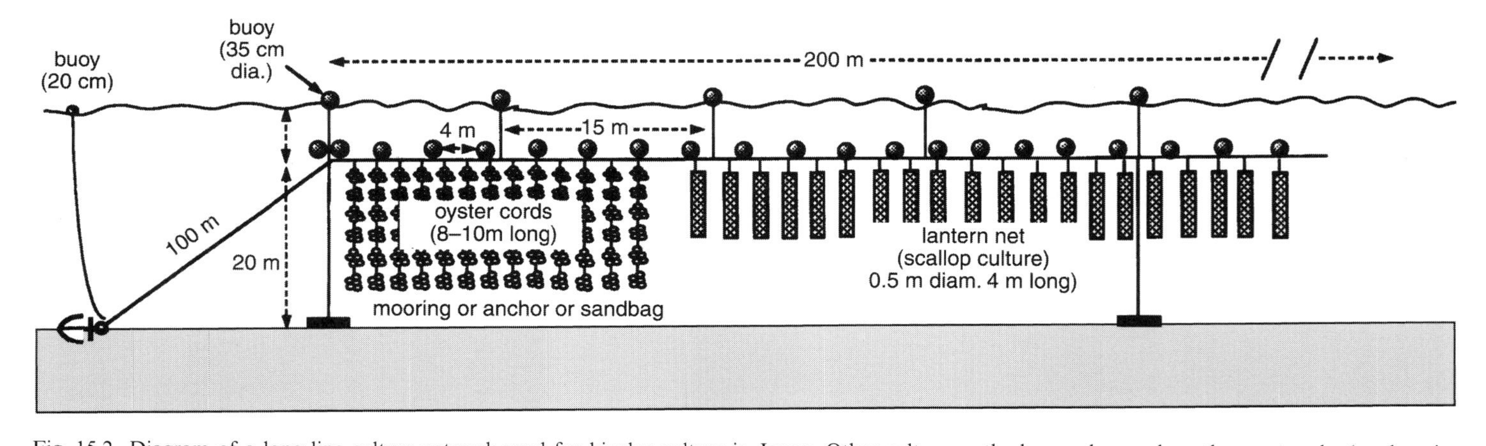

Fig. 15.2. Diagram of a long line culture network used for bivalve culture in Japan. Other culture methods are also used on these networks (ear-hanging method for scallop culture, baskets for pearl oysters, and harvesting of spat in traps, etc.). Unlike in the diagram, one network is devoted to a single culture type.

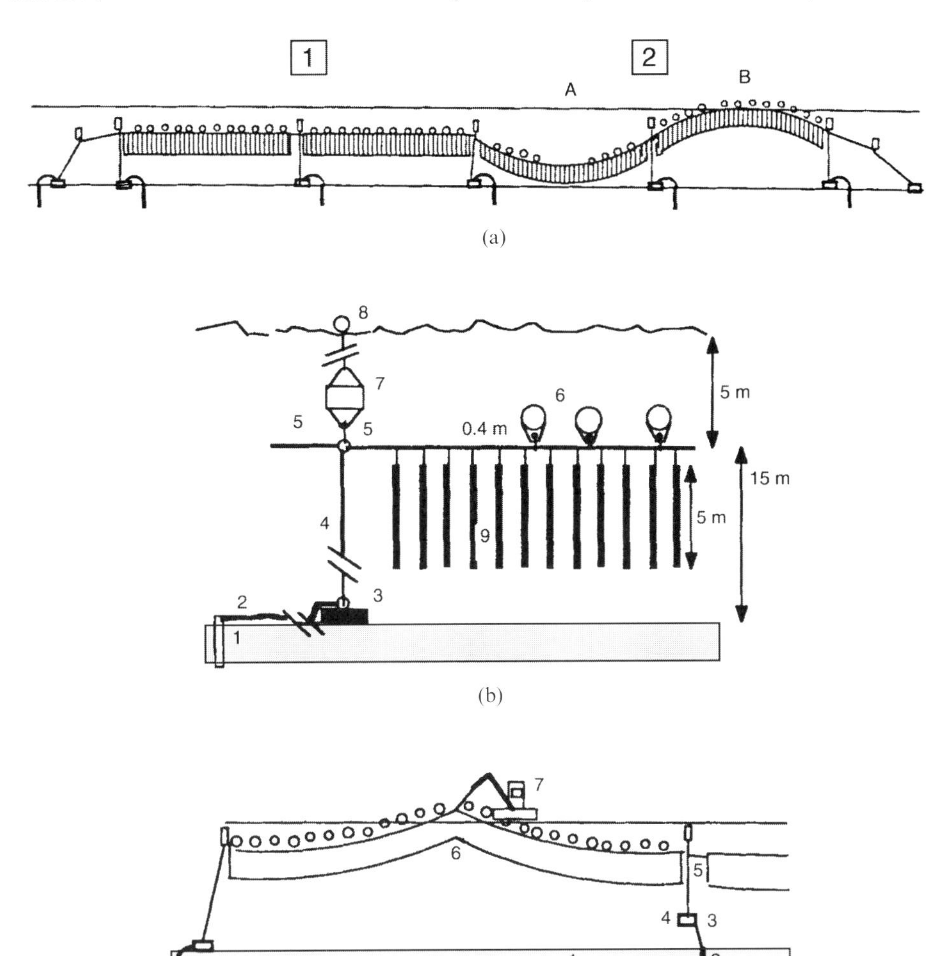

Fig. 15.3a. Diagram of the "Espuna" sub-surface network.
1. Theoretical location of network.
2. Observed locations.
   A. Too little flotation.
   B. Too much flotation.

Fig. 15.3b. Structure details of the "Espuna" network.
1. post
2. chain
4. stay
5. main line
6. 501 buoys
8. beacon
9. mussel cords

Fig. 15.3c. "Espuna" network is brought to the surface (for adding floats or spat, or harvesting mussels).
1. soft seabed
2. anchoring post
3. chain
4. mooring
5. stay
6. mussel cords
7. barge

This framework withstood storms without any problems for three years. This farmer also succeeded in using a rearing method for flat oysters in the sea on this type of structure and managed to grow-on this highly-valued species. Other installations of this type, used for rearing scallops, have been used in Chile and long lines are also used in other places, including Great Britain and Eastern Europe.

The flexibility of the long lines allows them to withstand cyclonic waves and thus to colonise the open sea. This fact, and their moderate cost, have led to their success: long lines constitute an irreplaceable management tool for exposed coastal waters. They are now used for algae, oysters, scallops and mussels, but the system could doubtless be adapted for rearing many other species.

15.5 CREATION OF NEW PRODUCTIVE ECOSYSTEMS IN COASTAL WATERS

The natural abundance of bivalves on rocky shores is common and in most countries where shellfish culture is carried out, rocky zones provide the spat which are the starting point for all rearing activities (see colour section). These living seabeds or mollusc reefs have been the subject of a recent collaborative work (Dame, 1993a). According to this author, they fulfil the same role as large herbivorous mammals on land (Dame, 1993b).

These rich littoral ecosystems have received little attention from ecologists who have been focussed on mangroves, coral reefs and rainforests. Nevertheless, mussel beds are just as productive, more resistant to pollution (which they are capable of eliminating) and to climatic variations, are not endangered and are dependent on a resource which is not directly exploited, i.e. marine primary production and matter suspended in the water.

The biomass of mussels found on a hard littoral substrate is commonly between 10 and 100 kg/m^2, which makes an interesting comparison with figures for soft substrates at the same depth (see Table 15.1). Similar biomasses of mussels are found on artificial reefs installed for fish and Fabi *et al.* (1989) proposed the association of artificial reefs and shellfish culture during trial experiments.

In contrast to mangrove swamps or coral reefs, living mussel (or oyster) beds can be created by man very rapidly. It is sufficient to submerge hard substrates, rocks or artificial reefs made of various materials in order to have an abundance of mussels the following year, without any other action being necessary. Attached mussels can be found at depths of about 30 m, but they can survive at much greater depths if they are installed there.

15.5.1 "Mussel reefs" in the USA

The practical solutions which have been proposed to create new ecosystems are very simple. In California, the Marine Forest Society proposed the creation of mussel reefs as ecosystems of the future, using old tyres arranged on the seabed. We shall not repeat the advantages of attached filter feeding ecosystems, but the

Fig. 15.4. Reefs of tyres being colonised by mussels (adapted from a drawing by the Marine Forest Society).

environmental and economic benefits of the tyre-mussel association are difficult to dispute. Figure 15.4 shows a diagram of this proposal.

These reefs can be used either directly (mussels are harvested) or indirectly—they could also be used for professional or sport fishing or diving.

15.5.2 Development of European coastal waters

In France, the coastal waters of the Golfe du Lion, which used to be rich in oysters, mussels and scallops, could be developed and reserved for culture of these species. While mussels cover all hard substrates submerged in shallow water, many little flat oysters, 2–5 cm in size, are found in shallow waters in summer and autumn, on sandy, soft seabeds and on small stones or empty shells. Their presence indicates the unexploited potential of the environment. In fact, this spat is very rare and widely sought after. In uneven areas in the open sea (artificial reefs, rocks), which escape trawling, flat oysters, queen scallops and scallops are found. It is similar for many other areas in the Atlantic (reefs of tyres, installed experimentally, also become covered in mussels and oysters; Casamajor, 1992). The majority of European coastal waters, characterised by vast continental shelves, could be developed in such a way.

The technical solutions are numerous and include stakes or immovable bouchots, rocks, nets and specific artificial reefs. Attachment of mussels or oysters can be improved, depending on the depth and duration of immersion. Exploitation can be aimed at these shellfish, or at the associated fauna, i.e. octopus, crustaceans (shrimps, crabs, lobsters, Mediterranean slipper lobster) and fish.

The use of old tyres allows the disposal of a waste product. If they are burnt, they pollute the air; stored on land, they retain rain water and favour the proliferation of mosquitoes, which are disease vectors. On the other hand, they constitute an excellent substrate for the capture of mussels or construction of artificial reefs, especially when they are used in conjunction with concrete as ballast. Millions of tyres have already been used throughout the world for the construction of artificial reefs and they have been tested on a small scale on the French Atlantic coast (Mimizan) and on the Côte d'Azur. It appears that their use does not pose any health problems (see Seaman and Sprague, 1991). Many other materials could be used, but this reconciles waste recycling and development.

These examples in Louisiana, China and Mutsu Bay should be carefully considered. They show that very basic developments, which can be very simply achieved on a large scale, can form the basis of very high production; these developments are often the fruit of research carried out by professionals.

While fisheries collapse because of overexploitation or poor management, many littoral seabeds on which they operate could be managed using the very simple procedures outlined above.

15.5.3 Indirect effects

We have seen above how mussels on soft seabeds in Holland do not aggregate to form a carpet, but form little clusters which make micro-reefs that can provide shelter for fish larvae or create a backwash favouring attachment of larval benthic species (Chapters 2 and 4). Living seabeds thus play a vital role; by disturbing the hydrodynamics of the limiting layer, they favour recruitment of benthic species. In terms of food, the installation of a living seabed provides a food reservoir to other links in the food web, since it can cause a hundredfold increase the benthic biomass.

It is known that the filter feeding ecosystem has an impact on the phytoplankton and the particulate material in suspension, but in structural terms, this type of ecosystem also affects the geomorphology of inlets and the sediment texture (Dame, 1993a).

With regard to health, filter feeding molluscs can be used for the purification of coastal waters polluted by man (as described in preceding chapters), but they can also reduce the phenomena of coloured waters which is widespread in these waters. This double function means that molluscs destined for human consumption must be kept separate from those used for other purposes and that production areas must be carefully supervised. Molluscs concentrate toxic substances (carcinogens), such as okadaic acid, from coloured waters. In fact, their sale is only prohibited when the concentration of toxic phytoplankton exceeds 200 individuals per litre; regular consumption of molluscs contaminated to a lesser degree is also dangerous—economic interest must not take priority over public health.

15.6 CONCLUSIONS

Living seabeds of filter feeders, whether they are natural or managed, fulfil many functions, including water filtration, production and recycling of living matter in the ecosystem and the production of edible biomass for man.

Because of their advantages both to man and the environment and their low cost, their use could be extended in future large-scale development of coastal waters. An abundant and free source of substrate can be used in the form of old tyres, a waste product typical of our society. Its use for the creation of beds of filter feeders links waste recycling to new production of living material and water purification—what other material or equipment simultaneously assumes three ecological functions, free? It would be better to bequeath coastal waters populated with living seabeds than mountains of old tyres on land to our descendants. Dismantled oil platforms could also be put to use in this way.

In the sea, there is no equivalent to the tree that holds up each of its leaves into the air and the light, far above the ground; in the sea, water masses are inhabited by small life-forms, the only ones which are able to survive in these conditions. It is here, in the surface layers of water, that very "dilute" plant production flourishes in the sunlight. Benthic filter feeders exploit very efficiently this small amount of production, which is carried to the seabed by water movements; but this is not always true, rendering overexploitation impossible: the organisation of the food chains does not allow it.

15.7 BIBLIOGRAPHY

Adey W.H. and Loveland K. *Dynamic aquaria: Building living ecosystems*. Academic Press, San Diego, 1991.

Allen J. and Hawkins S.J. *Mytilus edulis* populations and water quality control in the South Docks. In: Dame R.F. (Ed) *Bivalve filter feeders in estuarine and coastal ecosystem processes*. Springer-Verlag, Heidelberg, 1993.

Anon. Oyster culture fishing. *J. of Fisheries* 1989; 28.

Asmus J. and Asmus R. Phytoplankton-mussel bed interaction in intertidal ecosystem. In: Dame R.F. (Ed) *Bivalve filter feeders in estuarine and coastal ecosystem processes*. Springer-Verlag, Heidelberg, 1993.

Barnabé G. Open sea culture of molluscs in the Mediterranean. In: Barnabé G. (Ed) *Aquaculture Vol 1*. Ellis Horwood Ltd, Chichester, 1990; 429–442.

Bombace G., Gabi G. and Fiorentini L. Théorie et expériences sur les récifs artificiels. *C.R. Colloque Océanos*, Montpellier, Maison de l'Environnement, 1994; 68–72.

Cai Y. and Li X. Oyster culture in the People's Republic of China. *World Aquaculture* 1991; 21(4): 67–72.

de Casamajor M.N. *Suivi biologique du récif artificiel de Porto*. Mémoire de Maîtrise des Sciences et Techniques, University of Bordeaux III, Talence, 1992.

Conovet R.J. The role of filter feeders in stabilizing phytoplankton communities with some considerations for aquaculture. In: Devik O. (Ed) *Harvesting polluted waters*. Plenum Publishing, New York, 1976; 67–85.

Dame R.F. (Ed). *Bivalve filter feeders in estuarine and coastal ecosystem processes*. Springer-Verlag, Heidelberg, 1993a.

Dame R.F. The role of bivalves filter feeder material fluxes in estuarine ecosystems. In: Dame R.F. (Ed) *Bivalve filter feeders in estuarine and coastal ecosystem processes*. Springer-Verlag, Heidelberg, 1993b.

Dardignac-Corbeil, M.J. Traditional mussel culture. In: Barnabé G. (Ed) *Aquaculture Vol 1*. Ellis Horwood Ltd, Chichester, 1990; 285–338.

Doumenge F. Aquaculture in Japan. In: Barnabé G. (Ed) *Aquaculture Vol 2*. Ellis Horwood Ltd, Chichester, 1990; 849–944.

Fabi G., Fiorentini L. and Giannini S. Experimental shellfish culture on an artificial reef in the Adriatic Sea. *Bull. Mar. Sc.* 1989; 44(2): 923–933.

Foster K.L., Steimle F.W., Muir W.C., Kropp R.K. and Conlin B.E. Mitigation potential of habitat replacement: Concrete artificial reef in Delaware Bay – Preliminary results. *Bull. Mar. Sc.* 1994; 55(2–3): 783–795.

Héral M. Traditional oyster culture in France. In: Barnabé G. (Ed) *Aquaculture*, Vol. 1. Ellis Horwood, Chichester, 1990; 342–380.

Korringa P. *Farming marine organisms low in the food chain*. Elsevier Science, Amsterdam, 1976.

Lubet P. Mollusc culture. In: Barnabé G. (Ed) *Aquaculture: Biology and Ecology of Cultured Species*. Ellis Horwood Ltd, Chichester, 1994.

Lucien-Brun H. and Lâchais A. Évolution de la pectiniculture au Japon. *La Pêche Maritime*, July 1983; 388–396.

Marine Forest Society. *Mussel reefs: Ecosystems of the future*. Marine Forest Society, Balboa Island, CA, 1994.

Ozaki T., Mitsuashi K. and Tanka M. Scallop culture and its supporting system in Mutsu Bay. *Mar. Poll. Bull.* EMECS, 1991; 23: 297–303.

Perret W.S., Dugas R.J. and Chatry M.F. Louisiana oyster: Enhancing the resource through shell planting. *World Aquaculture* 1991; 22(4): 42–45.

Rheault R. Nursery culture of shellfish seed in marinas. *Maritime* 1989; 33(3): 4–6.

Seaman W. and Sprague L.M. (Eds) *Artificial habitat practice in aquatic systems*. Academic Press, San Diego, 1991.

Smaal A.C., and Prins T.C. The uptake of organic matter and the release of inorganic nutrients by bivalve suspension feeder beds. In: Dame R.F. (Ed). *Bivalve filter feeders in estuarine and coastal ecosystem processes*. Springer-Verlag, Heidelberg, 1993.

16

Examples of integrated development and management

16.1 DEFINITION

We have seen throughout this book that, very often, development can initiate a cascade of consequences. For example, it is impossible to separate the circulation of waters from their richness in nutrient salts, their dissolved oxygen concentration, the presence of substrates and filter feeders, etc.

Management thus presents itself as an assemblage of interdependent, interwoven processes, destined to affect the functioning of ecosystems. When several management methods are used together to achieve an overall goal, this is termed integrated management.

We have tried to separate the principal lines of development into their main categories within successive chapters, but in many cases the combinations and their interference with each other has led us to consider groups of developments, or integrated developments. Although we shall limit ourselves here to the categories of development corresponding to Chapters 9 to 15, the possible combinations are innumerable.

16.2 MARINOVATION – A MAJOR JAPANESE PROGRAMME

Rather than repeating and adjusting our remarks, we have chosen to illustrate integrated development of coastal waters as it has been devised in Japan within the framework of the large Marinovation programme and its satellite projects. This has been made possible thanks to F. Simard (of the Monaco Museum, and former resident in the Franco-Japanese household) who adapted the Japanese plans into French and gave us authorisation to use them (Figures 16.1–16.4). This assistance is gratefully acknowledged.

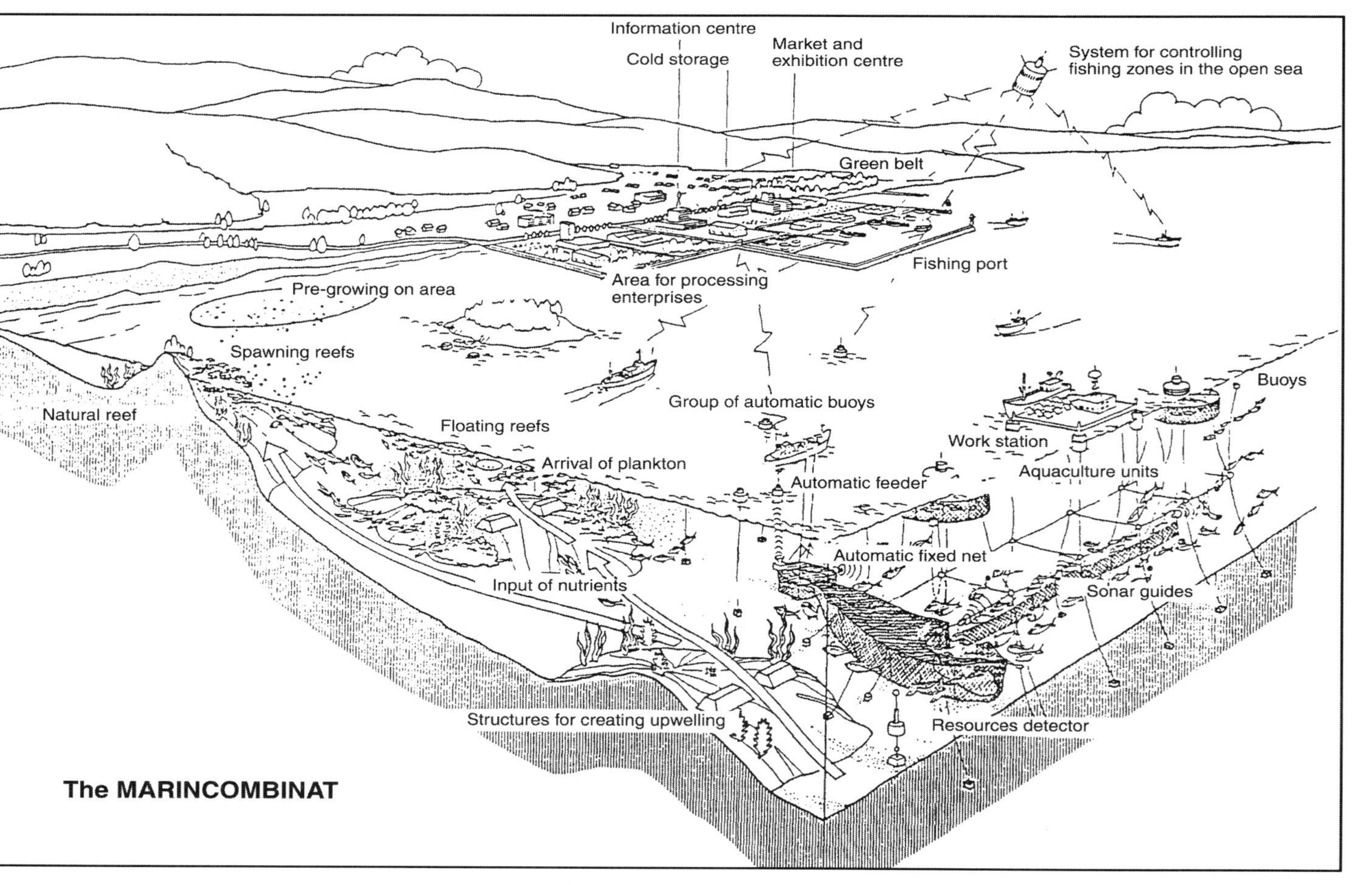

Fig. 16.1. Reproduced with kind permision of F. Simard.

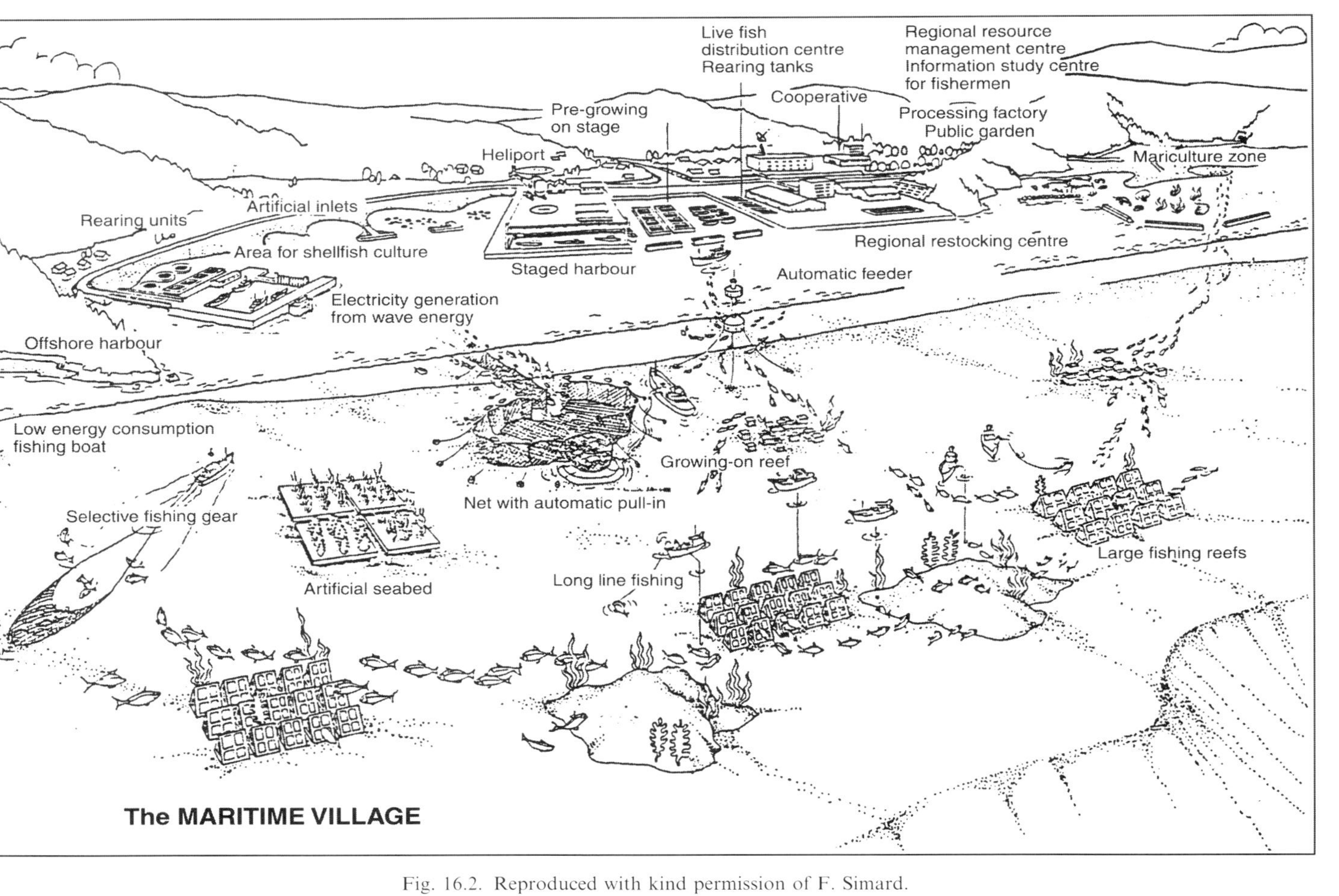

Fig. 16.2. Reproduced with kind permission of F. Simard.

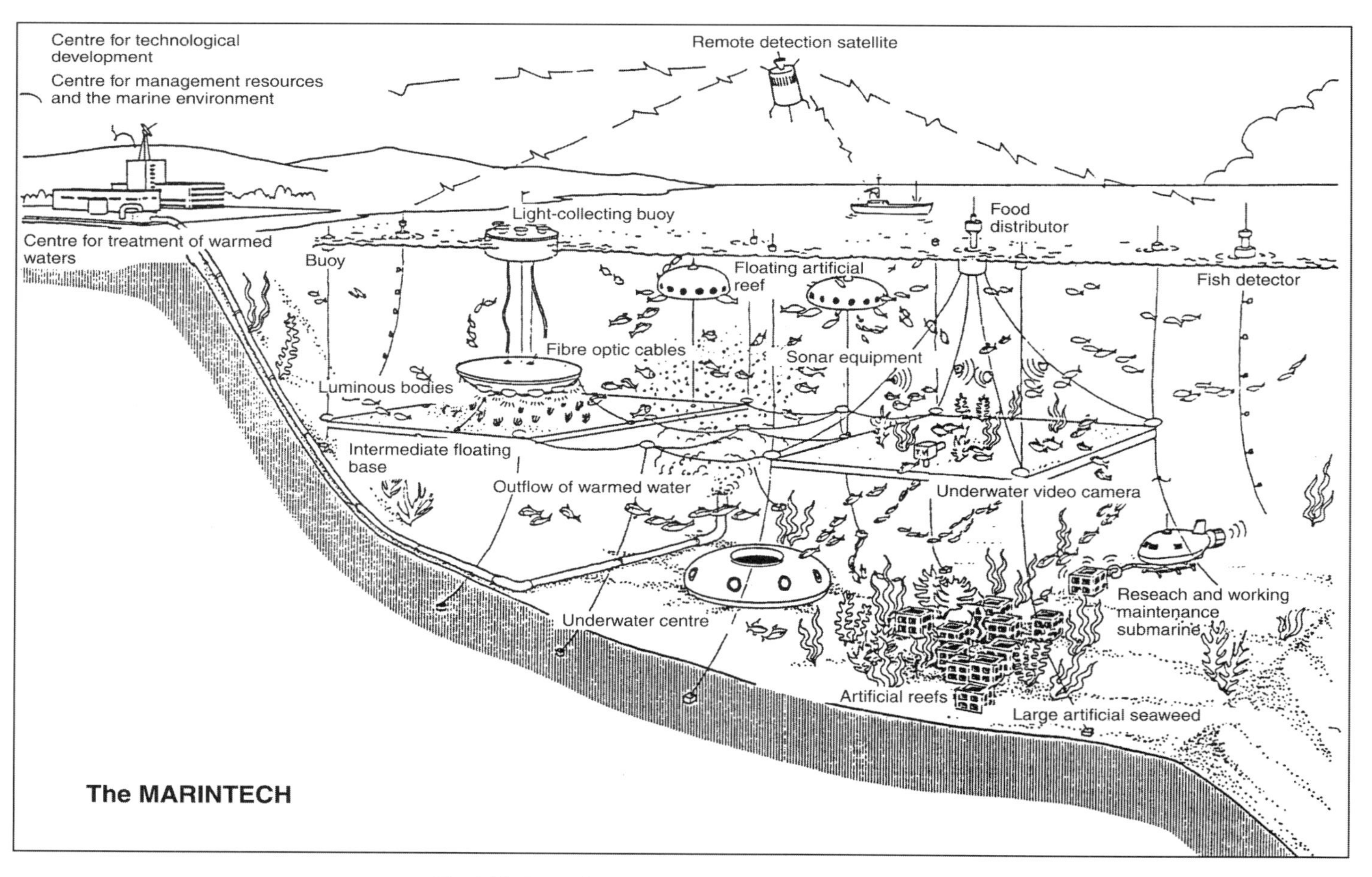

Fig. 16.3. Reproduced with kind permission of F. Simard.

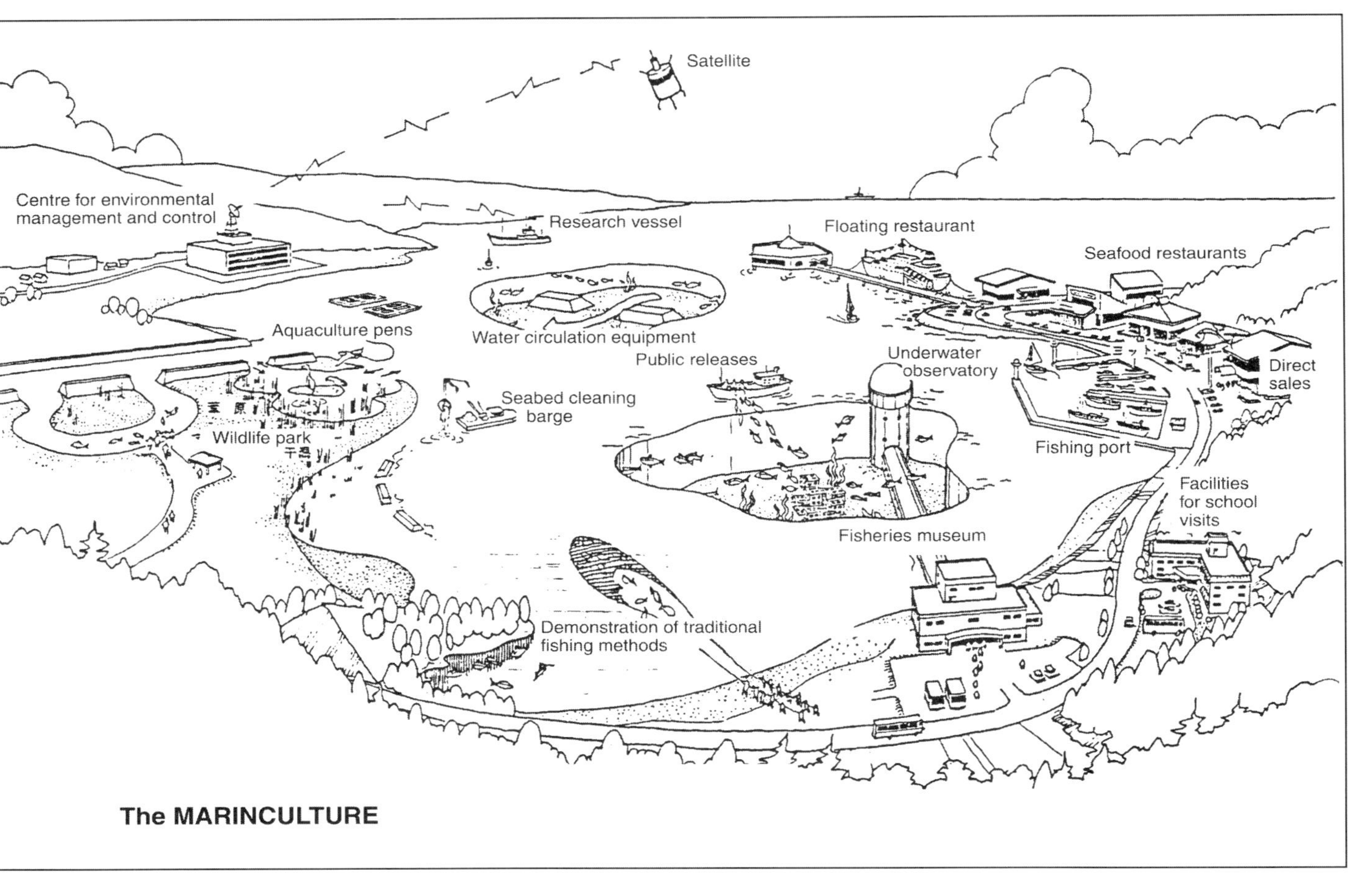

Fig. 16.4. Reproduced with kind permission of F. Simard.

The Marinovation programme, according to the word created by the Japanese Fisheries Agency, marks out the long-term objectives of the government to transform the coastal and off-shore environment into a system that will be completely at the service of economic, social, cultural and recreational requirements of the Japanese nation in the 21st century.

Much of our descriptive information and, above all, its combination and integration, can be found illustrated in these figures.

These illustrations, which were published several years ago (Simard, 1986), have another quality. They show that this conceptualisation has not gone unheeded. It presents a short summary and a concentration of the achievements that we have described in preceding chapters: artificial reefs and fish concentration devices, which were only experimental, have become a reality, in the same way as have submerged obstacles to create upwellings. Activity in hatcheries and restocking is not as easy to represent as these objects; in each case, the achievements that we have presented from recent work carried out on all these subjects prove that the development of coastal waters is widespread, with transferable results. In contrast to many of our programmes of technological research and development, it should also be emphasised that this involves long-term programmes.

16.3 RESULTS

Some results from the Marinovation programme are already known (Simard, 1991). In 1989, a fish aggregation device (7 m diameter buoy) was placed in Tosa Bay, allowing the daily capture, in autumn, of 700 kg of pelagic fish per boat for a fleet of 5–19 boats.

Thierry (1989) described another of the achievements of this programme (in French)—the Oita marine ranch off Kyushu Island, a diagram of which is shown in Figs 16.5 and 16.6. Sea bream were being restocked and were conditioned by an acoustic signal to come and feed; a collection of surveillance equipment allowed this

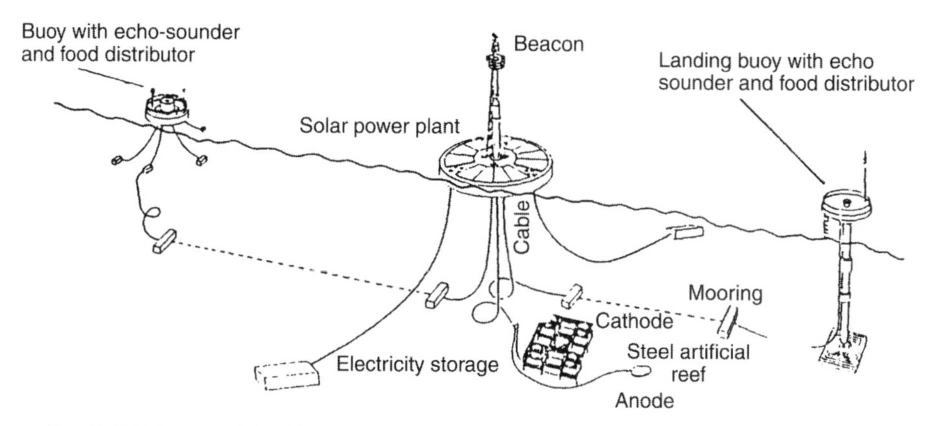

Fig. 16.5. Diagram of the Oita "marine ranch" with its solar power plant (from Thierry, 1989).

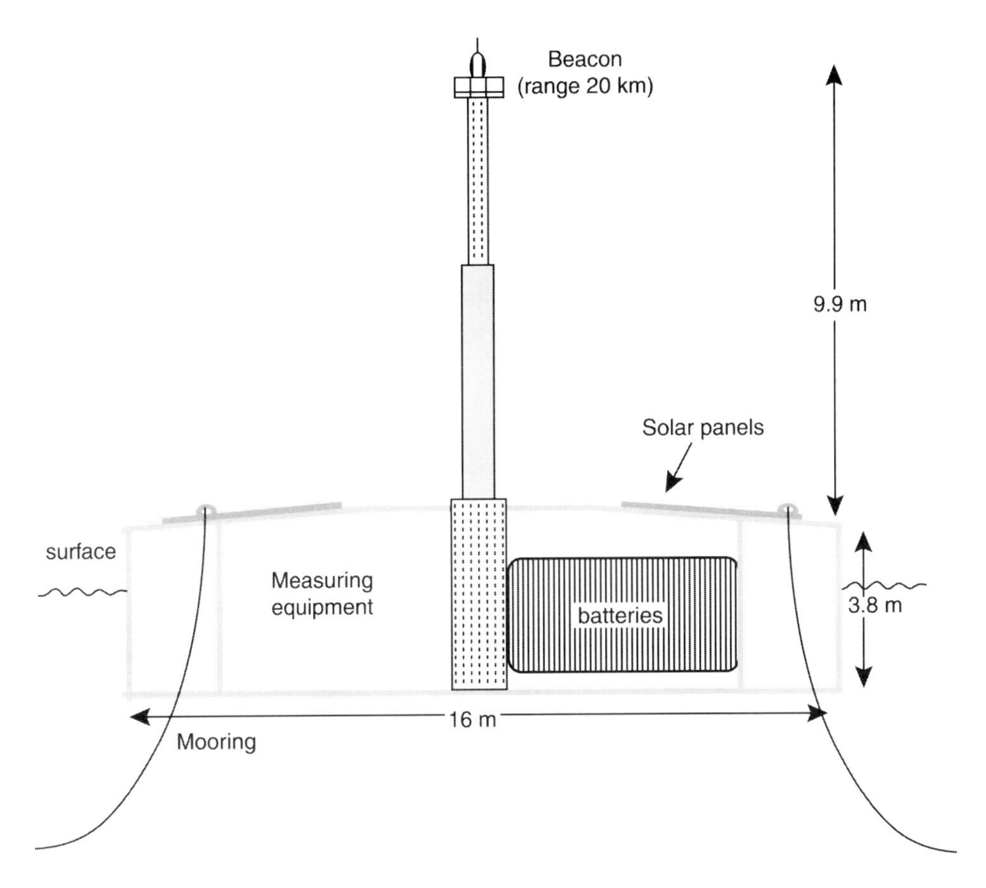

Fig. 16.6. The floating power plant of the Oita marine ranch.

conditioning to be monitored (echo-sounders and traps). In order to allow the group of food distributors and surveillance apparatus to work and to transmit messages to land, a solar power generator was constructed and installed in the sea (Fig. 16.6). This 16 m diameter circular platform was anchored by loose cables and has been designed to resist cyclones.

Descriptions of other achievements of the Marinovation programme can be found in a report by Grove *et al.* (1994).

Plant feeds for yellowtails have been studied within the framework of the subsidiary programme, "Marineforum 21". An artificial feed incorporating between 23% and 70% plant meal (soy, maize and kidney bean) has been made and tested. The growth results obtained were almost identical to those using traditional feed: fish weighing 1.5 kg at the start reached 2.5 kg five months later, whatever the percentage of plant material, and the fish's flavour was not altered (Yoshushu, 1995).

Within the same programme, in 1993 a pilot offshore fish farm was being installed on the west coast of Hokkaido; a tidal wave in July 1993 led to 200 deaths and the

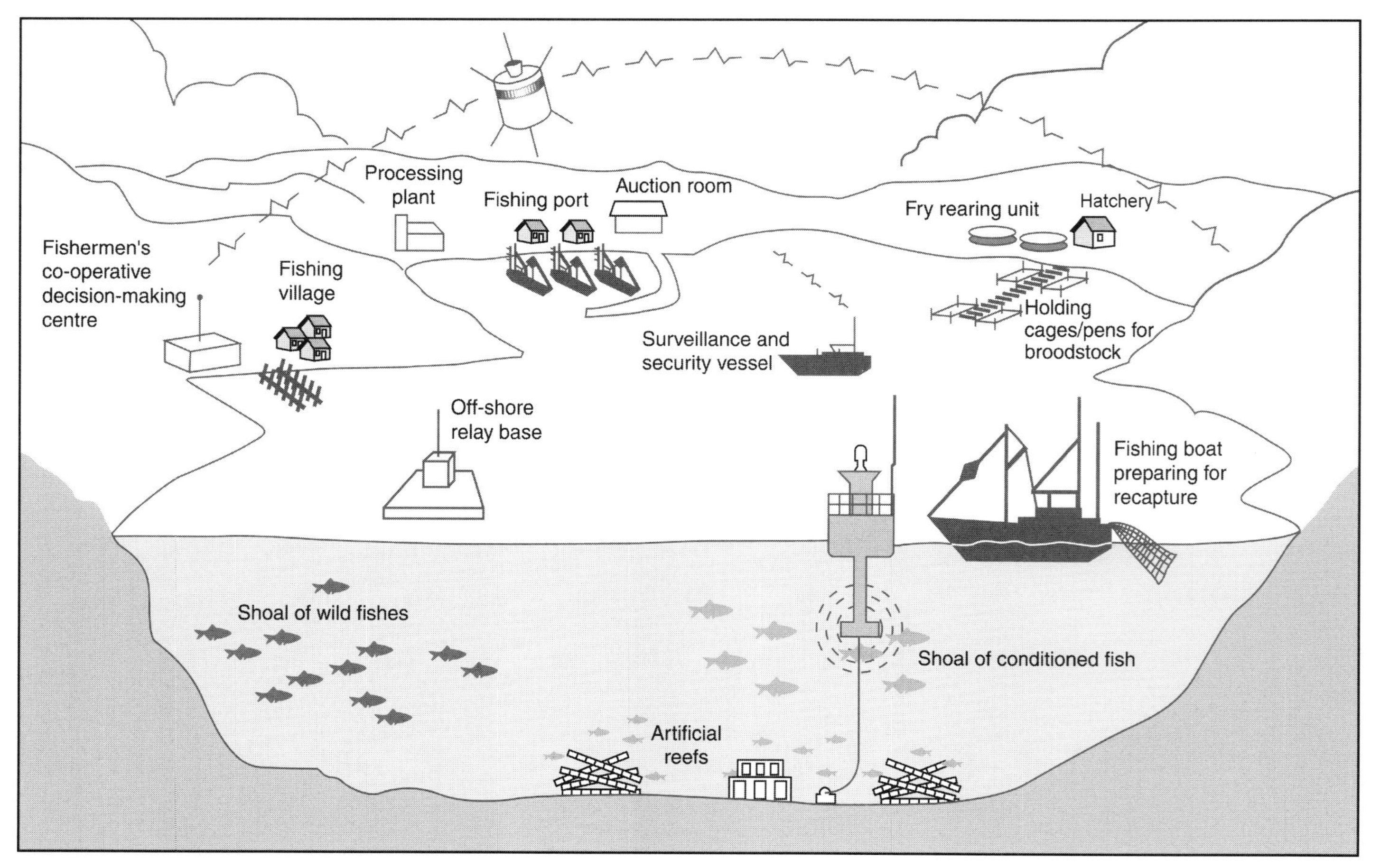

Fig. 16.7. Aquacapture, project of the Arcachon marine ranch. Reproduced by kind permission of GIE Aquacapture.

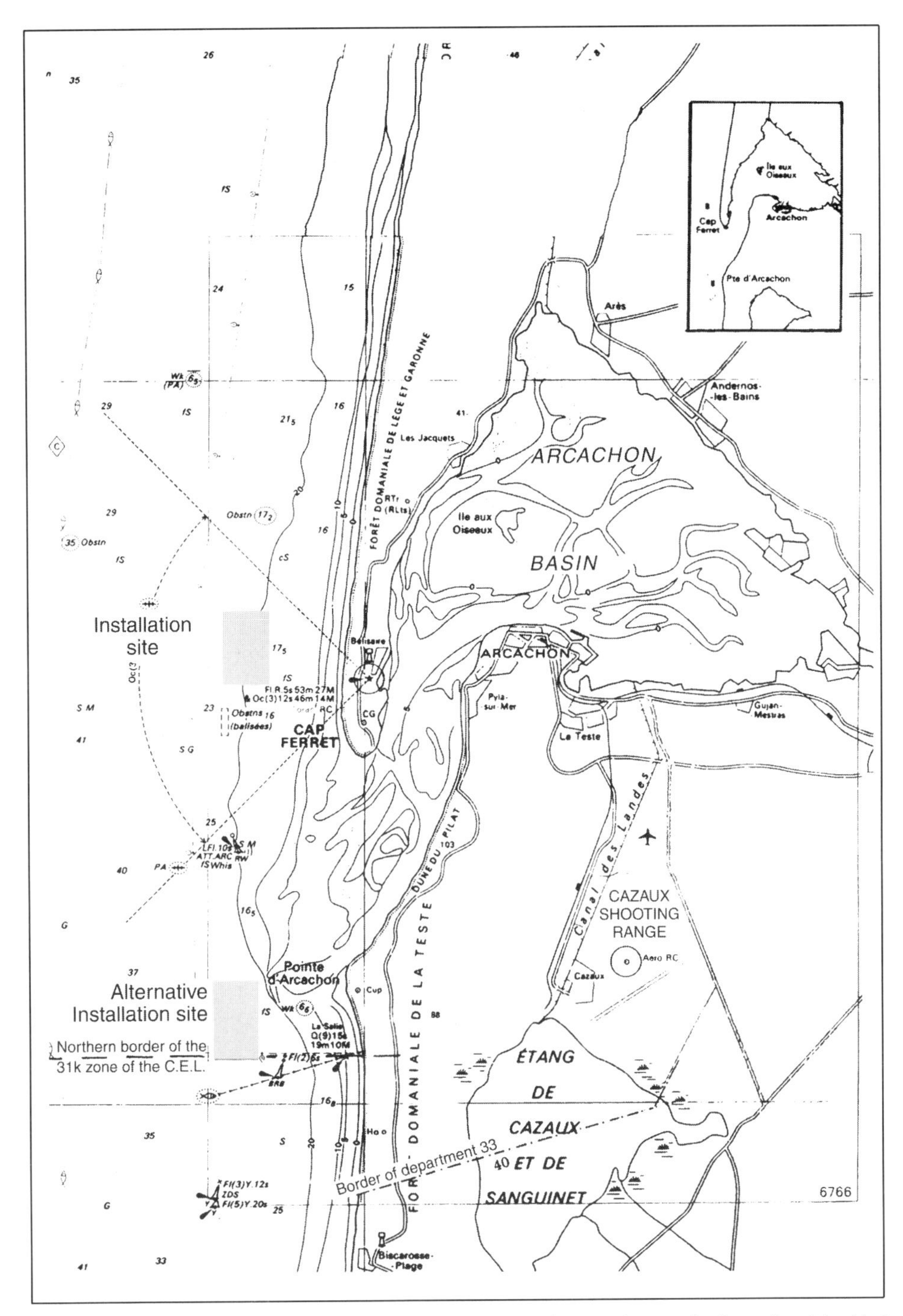

Fig. 16.8. Diagram of alternative setting-up site for the Arcachon marine ranch. Reproduced by kind permission of GIE Aquacapture.

destruction of 66% of the boats. The farm, however, withstood the disaster: there was no damage to cables or platform, nor to the automatic feeding system (source, *Flash Japan*, no. 53, May 1994). One single, 6000 m^3 cage was installed, containing 40,000 cherry salmon (*Onchorhynchus masou*). The maximum capacity predicted for 1996 was 120,000 fish.

One of the most recent Japanese programmes, "Marine Ecotopia 21" involves the installation of observatories on the coastal fringes, with the objective of improving water quality by installing depuration systems, managing reefs and algal zones. Not only the exploited resource is taken into account, but also the biotope (Minato, 1995).

An integrated development programme called "Aquacapture, projet de Ranch Marin" was started in 1996 in France (Figs 16.7 and 16.8). It was similar in many ways to the Marinovation programme.

16.4 CONCLUSIONS

Integrated development of coastal waters is not a future possibility, but a current reality. Japan is certainly the most advanced country in this development, following lengthy work on conceptualisation, a large amount of experimental work and long term programmes. Nearer to home, some interesting achievements, often with European funding, have been made; these have mainly been in southern countries, with development of coastal lagoons (hydraulic and ecological management of lagoons and forced water circulation in Italy), the installation of huge artificial reefs (Italy and Spain) and the development of rearing fish in cages (Greece and Spain).

16.5 BIBLIOGRAPHY

Grove R.S., Sonu C.J. and Nakamura M. Recent Japanese trends in fishing reef design and planning. *Bull. Mar. Sci.* 1944; 44(2); 984–996.

"Minato". Plan gouvernemental: Marine ecotopic. *Flash Japon* 1995; 60: 1.

Simard F. Un nouveau plan de développement pour les pêches au Japon: Le "Marinovation". *La Pêche Maritime*, April 1986; 258–270.

Simard F. Two examples of artificial floating reefs in Japan. In: *Symposium on artificial reefs and fish aggregating devices as tools for the management and the enhancement of marine fisheries resources.* FAO RAPA Report 1991/11, Bangkok, 1991; 314–339.

Thierry J.M. Les technologies avancées au service du développement des pêches au Japon: Le "ranch marin d'Oita". *Aqua-Revue* 1989; 24: 28–42.

"Yoshoshu". Alimentation végétale des sérioles. *Flash Japon* 1995; 60: 3.

17

General conclusion

17.1 SPECIFIC CHARACTERISTICS OF AQUATIC ECOSYSTEMS

17.1.1 The aquatic environment

The aquatic environment is not the natural environment for a terrestrial vertebrate such as man. It is both hostile and impenetrable, without the accessibility of a field or a forest; it has no borders or physical frontiers and involves a three-dimensional space, i.e. a volume. The properties of water mean that its mass acts in a certain way but, in addition, all other physical and chemical characteristics manifest themselves in specific ways. This environment, therefore, functions in a different way to a land-based environment, on a physical, chemical and biological level; life is widely dispersed. These differences to terrestrial life and the particular perception of man for the world of water might thus appear to provide as many difficulties for ecological management as could be conceived for any environment. Fortunately, this is not so.

17.1.2 New dimensions in marine ecology

Over the past two decades, coastal marine ecology has been marked by significant progress at various levels.

- An understanding of the role of physical phenomena in biological processes has led to the phrase "physical factors which feed the fish". In other words, biology can no longer be considered separately.
- On the small scale, there is a boundary layer around organisms and viscosity governs exchanges over distances less than a millimetre.
- Above a millimetre, the phenomena are dependent on water turbulence: eddies transport nutrients and wastes. This water turbulence is regulated by the tides and major marine currents, i.e. by large-scale hydrological phenomena.

17.1.3 Hydrology and biological production: unpredictability

Hydrodynamic phenomena have a determining influence on biological phenomena, but the former are themselves dependent on atmospheric or climatic changes. Another achievement of contemporary science is to have shown that hydrodynamics, like the climate, are controlled by the deterministic law of chaos—currently they remain unpredictable. Since hydrodynamics affect biological production, there is no way that the latter can be predicted, even if the functioning of food chains is precisely understood.

17.1.4 Primary production is dependent on mixing of waters

Initial production of living matter in the aquatic environment is carried out by many tiny organisms which are spread, like a "living dust" throughout this three-dimensional world, which has no physical barriers. The surface areas available for exchange between living organisms and environment are enormous and this can occur very rapidly, creating a daily production which can represent several times the biomass.

The simultaneous variation of hydrological and biological factors in both space and time is significant. This synchrony therefore constitutes the key to biological success: surface waters that receive solar energy are often nutrient-poor and warm, while deep waters are richer and cold, and also denser. Breaking down this stratification allows productive mixing. In this way, water movements constitute a form of energy which is indispensable to the ecosystem.

These data have led to a major revision of knowledge acquired in still waters and in the laboratory and have turned traditional oceanography upside down. They are also of major importance in terms of exploitation of marine living resources.

17.1.5 Huge production, still far from exploitation of the full potential

Despite the fact that global primary production of the oceans is enormous and is estimated at 200–500 billion tonnes per annum, it involves a photoreactor which covers two-thirds of the planet; "average" primary production is much lower than terrestrial plant production. When conditions are optimal, which is often the case in coastal waters, aquatic biological production is the highest known: 50 tonnes/ha (dry weight) for lagoons in the tropics and 250 tonnes/ha (wet weight) for mussels in the Galician Rias. These exceptional figures describe particular ecosystems which are limited to certain areas, underlining the fact that, in other waters, biological processes function very slowly.

17.2 POPULATION EXPLOSION AND ENVIRONMENTAL PROBLEMS

17.2.1 Demographic explosion

The opinion of many ecologists has been expressed by Ramade (1994). The human population explosion is in fact the most gigantic ecological catastrophe to be confronted by our species; in practice, it is the origin of many others.

Two-thirds of the human population currently lives close to the sea and it is predicted that, in the future, double this number will wish to live there. As a result, coastal waters are subjected to degradations which are probably just as significant as those found in tropical rainforests, and their biodiversity decreases continuously. However, all this occurs in silence under the surface, without any signs appearing above to trouble the movements of the waves (Carleton-Ray, 1988).

17.2.2 Climatic threats

Whether this involves global warming (greenhouse effect due to industrial pollution which has caused a 0.5°C increase in temperature), the chemical pollution of the air or the breakdown of the ozone layer, all these problems have repercussions for the oceans whose role in the climate is primordial (11% of precipitation on land is due to terrestrial evaporation, the remaining 89% from evaporation from the oceans). The increase in ultraviolet radiation may have unknown effects on oceanic plant production, which regulates the levels of oxygen and carbon dioxide in the atmosphere, and coral bleaching has already been attributed to increased water temperature. Other phenomena such as the strong intensification of the depth of convection currents in the Labrador Sea, or their attenuation in the Greenland Sea could lead to significant climatic changes (e.g. more severe depressions, more frequent storms). The reality of the greenhouse effect is recognised.

17.2.3 Chemical threats and the concentration of pollutants

As stated by Lavoisier many years ago, sea water may be considered as the result of the washing of the entire surface of the globe (and of humankind). Man has synthesised molecules which did not exist before in nature and which end up in the sea; the majority of the oceans are now changed by human activities. More than a million different chemical compounds exist and their long-term effects are not known. For example, a decrease in human sperm quality has been witnessed in Europe.

Black tides and oil or other pollution of beaches have become commonplace but only form the obvious part of a state of affairs which is much more serious, but less visible—widespread chemical pollution of most of the seas, due largely to agriculture and industry. Traces of radioactivity from the nuclear accident at Chernobyl were found in collectors submerged in the sea less than a week later.

The sensitivity of aquatic organisms to tiny concentrations of various substances has already modified the subtle equilibria that regulate the processes of aquatic production, but toxic products are also concentrated along food chains, which can sometimes render consumption of molluscs or fish dangerous, or even lethal. Tuna with mercury is no longer the speciality of Minamata and phytoplankton toxins wreak havoc in many coastal areas. Alongside acute episodes of illness with obvious symptoms, a latent toxicity is suspected (carcinogenic substances); regular consumption of mussels may not be without danger.

It is possible that we are witnessing an ecological drift of the marine littoral environment as a result of man's inputs. This is a very serious threat.

17.3 THE RESTRICTIVE VIEW OF SCIENCE AND TECHNOLOGY

17.3.1 The coincident increase in knowledge and deterioration of water quality

The coastal environment is increasingly well understood, but also increasingly degraded by man, with no solutions being put forward by science (Rougerie, 1995). According to some authors (Goldsmith, 1994), this irresponsible attitude is the basis of all our troubles. When geographers, agronomists, biologists and economists return to look at the data that their research has produced, they realise that overpopulation, extinction or pollution have already reduced our standard of living and placed our future in jeopardy (Arms, 1990). According to the Director General of the Japanese Environment Agency, Seiji Ikkatai, if our civilisation, at present based on mass production and consumption, does not radically change its objectives, it will soon come to know the fate of ancient civilisations which have disappeared—it will die (agency White Paper).

In spite of subtle differences, all authors agree in establishing a concerned account of environmental degradation, but this does not improve matters involving the exploitation of aquatic resources through fishing.

17.3.2 Crisis in fisheries management

Exploitation of oceanic production has been very poorly managed by fisheries and has reached a ceiling of 90 million tonnes, since the food chains leading to fisheries production are long and complex. Man removes less than one two-thousandth of the vast primary production of the oceans (200–500 billion tonnes).

In order to catch €53 million worth of fish, the world's fisheries industry spends €95 million per annum and most of the €40 million deficit is made up by various subsidies (Safina, 1996). It has been estimated that 70% of stocks are overexploited: between 60 and 90% of hake, cod, haddock and Greenland halibut are juveniles.

The decline in major fisheries, e.g., presented as models of fisheries management and the recovery in fisheries whose decline was predicted, such as the lobster in Canada, removes credibility from fisheries scientists and accentuates the rift between scientists and professional fishermen. The links between climate, hydrology, planktonic production and fisheries are well known in the way they develop, but they remain unpredictable. An ultimate objective is to understand these mechanisms and, in particular, the way in which recruitment is determined, but is it feasible? It is not only the level of complexity or the absence of data which limit understanding of recruitment determinism, but the nature of the phenomena on which it depends.

The question must therefore be posed, as it has been by some international organisations, as to whether it would not be simpler to change the paradigm. Fisheries theory no longer corresponds to the facts and has become outdated. If it were possible to predict recruitment levels, of what practical use would this knowledge be? Understanding and modelling a state of penury would not provide a solution to overexploitation or the fisheries crisis.

17.3.3 Increasing distance between scientists and fishermen

In France, research work has been devoted to the analysis of the interface between scientific knowledge and other issues of a more practical nature (Van Tilbeurgh, 1994). This study has shown that scientists are preoccupied with aquaculture and their relationships with the producers are not in accord with the producers' problems. Such misunderstanding is not the exception, but has become the rule. It is also found in agriculture but, in the marine domain, this state of affairs leads to violent demonstrations (as for example, by the shellfish farmers in Sète in 1989 and fishermen in Rennes in 1994) or confrontations at sea (for example the Spanish and the French in 1995). The problem is not purely intellectual; it has become socio-economic.

This situation is in contrast to what has happened in Asiatic countries such as Japan: the synergy between "Fisheries Universities", "Fisheries Prefectures" and professional fishermen's cooperatives results in an extraordinary dynamism of the productive sector, which is constantly progressing. Japanese artisanal fisheries and aquaculture exceeds 2 million tonnes. Aquaculture is expanding rapidly throughout Asia and more than compensates for the decline in fisheries.

In terms of pure ecology, these concepts result in one single, very strict form of management: conservation. Although it is not useless, conservation is limited by rules, regulations and prohibitions which are enforced to varying degrees and are thus relatively inefficient. This restrictive step is not very innovative and only constitutes a tiny part of the response to the problems posed in coastal waters.

17.4 A NEW STRATEGIC VISION

17.4.1 The role of the scientist

It has been remarked upon by some authors that the most densely populated areas, such as south-east Asia, are also the most productive. Rice culture (semi-aquatic culture) has transformed the countryside into a garden and even the mountainsides have been developed into meticulously maintained aquatic terraces. Ninety per cent of aquaculture production is from Asia.

Despite the degradation of coastal waters and their overexploitation, the full production potential of living material in coastal ecosystems is far from being attained; we are not inevitably doomed to failure since things can change very rapidly. In this context, the scientists' role should not finish with reports; they should compare their results with reality and society's demands, becoming a force in proposals and players in their own right in political decision-making. This is a new definition of research, a new strategic vision which is being put forward to be exploited. A complete change in models should be made; in addition to searching and finding, one must exploit (Salvat, 1995). Ecology and fisheries should come together with engineering and technological sciences; the separation between research and its application is only artificial.

17.4.2 Management based more on ecology than technology

Ecological management is integrated with the natural and reversible functioning of ecosystems. Major engineering works consisting of blocking or re-routing rivers, damming straits and changing the course of large marine currents irreversibly changes natural ecosystems This does not belong to ecological management; it involves what might be termed "mismanagement", while the complete opposite must be done in order to manage nature.

Ecological management of waters is often easier than the management of land. While the large terrestrial areas of desert are hardly comparable to the vastly larger, low-production areas considered as oceanic deserts, everything is lacking to transform them into a fertile plain: soil, water, fertilisers and seed. In contrast, almost everything is present in the sea to fertilise the surface waters which are lacking in life; inputs of fertiliser or the addition of trace elements in very small quantities are sufficient, as has been shown by the positive results of fertilisation experiments carried out in the mid-Pacific. Forced circulation or fertilisation of waters have no equivalent in terms of the energy expenditure of ploughing. Lagoon systems such as the valli, tambaks, salt marshes and oyster beds constitute developments of aquatic environments which are so well integrated into our horizons and littoral ecology, that it has now been proposed that they should be conserved as "wild" ecosystems. This demonstrates the ecological validity of water management.

There are many management methods, each of which has been broadly considered in each of the chapters in Part C of this book; combining these possibilities results in a multitude of solutions which are also forms of integrated management.

At present, this type of action is no longer conceived in the form of large-scale, offshore technical installations, as was the case in the 1970s, but more in the form of gentle manipulation of food webs in coastal waters. Similarly, the concepts of exploitation and management which then prevailed have given way to "sustainable development".

Given that these types of management become integrated with natural ecosystem functioning, an approach requiring cooperation between humans and nature becomes possible, through regulations or shifts in natural production and application of ecological principles. In contrast to agriculture, there is no simplification of the aquatic ecosystem and biological diversity is not altered. Ecotechnology, in contrast to biotechnology, proposes reversible processes.

17.4.3 Management as a weapon against pollution

One of the paradoxes of aquatic ecosystem function is that new biomass is created at the same time as pollutants present in the water are transformed or concentrated there; these self-purifying processes form an integral part of natural aquatic food chains. Management fits into the functioning of aquatic ecosystems. Fish culture in ponds, traditional shellfish culture and water purification using lagoon systems all constitute well-known forms of management, and their potential could be extended to marine coastal waters (with a surface area equivalent to that of Africa).

Ecotechnological development is thus in a unique and extraordinary situation: it can simultaneously generate production of living matter that can be used by man, and carry out recycling or large-scale concentration of dispersed pollutants in the aquatic environment.

Management can thus play a major role in the protection and rehabilitation of coastal waters that have been severely polluted by man. This type of activity can integrate the often divergent aspirations of the many users of coastal waters, since they all agree that respect for the sea and natural processes come high on their agenda.

If ecology is to be combined with economics, the management of coastal waters constitutes a rampart for the defence of the environment. Whether it involves the conservation of the natural character of tourist areas or encourages production of marine species for financial ends, it is vital that there is no pollution. While this is a difficult goal to achieve in the absence of economic justification, it can be done when such factors come into play. This is a new phenomenon which contrasts strongly with the trends of the recent past (fisheries overexploitation, oil industry pollution and use of the sea as the ultimate dustbin).

In Europe, the law of the sea implicitly confers a national or supranational (EU) right of ownership, constituting an appropriate framework for such activities. Unfortunately, in these countries, coastal waters have no electoral constituency and thus "have no society living there to defend their house and garden". The problem is therefore that of knowing whether our society is capable of imposing a set of constraints and, especially, of conceiving a group of solutions leading to sustainable development, overcoming the tendency of politicians for short-term action and economic objectives.

17.4.4 New tools for aquatic ecotechnology

With a global production of 25 million tonnes each, artisanal fisheries and aquaculture together represent more than half of the world's aquatic production. There are many other assets: lower intermediate consumption, increased gross and net added value, lower cost of tied-up capital and more direct employment. The riches created return to the national economies and production from coastal waters is destined for human consumption.

The developments of many aquatic ecosystems throughout the world and the jobs that they create, the 50 million tonnes produced by artisanal fisheries and aquaculture, provide an example of widespread ecological management of coastal waters. New possibilities extend these prospects even further:

- The hatchery-nursery revolution: The hatchery-nursery set up is the basis for the production of tens of thousands of tonnes of numerous marine species. This is a new stage in man's history, in which we have the power to reproduce and rear these species, and it is also the realisation of work attempted many times since the start of the 20th century. The hatchery-nursery may now revolutionise fisheries' perspectives and this is already true for salmonids in the sea.

- Artificial reefs; protective and productive substrates: Evidence has accumulated on the efficiency of artificial reefs, and their capacity for increasing the abundance of exploited populations, discouraging trawling and recycling primary production or purifying waters has been demonstrated. They result in the creation of productive new ecosystems on seabeds which were previously deserts and are capable of ensuring the physical protection of coasts or seabeds. They have therefore become a very useful management tool and result from collaboration between biology and engineering. In Japan, many experiments have been undertaken and vast investments made, allowing the passage from the domain of reefs to evolving ecotechnological knowledge of coastal waters towards a real "gardening of the sea". On these bases, conceptual reflection resulted in the large "Marinovation" programme, whose instigation resulted in new production and the creation of jobs. This is an example worth thinking about.
- Utilisation of benthic filter feeders—the choice between purification and production: Bivalve filter feeders, such as mussels or oysters, are associated with hard substrates or reefs and form the basis of the recycling of living material, production of exploitable benthic biomass and water filtration in many ecosystems. This type of development can fulfil various functions and its extension to the large scale is one of the least costly; it can multiply the production from soft desert seabeds of European coastal waters several hundreds of times.
- Induced water circulation: As a form of artificial auxiliary energy introduced by man to aquatic ecosystems, induced water circulation mixes, oxygenates, fertilises and renews—in short "gives life". Already widely used for purification (of beaches, harbours, aquaculture zones and stagnant waters), this technique has a good future in coastal waters for production and depuration (e.g. shellfish culture and limitation of coloured waters).
- Management of waters and sustainable development: These prospects correspond to the requirements for sustainable development and constitute an exceptional opportunity for the development of this area, as the future of the shore lies on its seaward side.

Many routes exist, therefore, to reinforce ecology and stimulate the economy. May the scientists shed light on the dreams of the decision-makers, by embodying them in the blue horizon of coastal waters.

17.5 BIBLIOGRAPHY

Arms K. *Environmental science*. Saunders College Publishing, PA, 1990.

Carleton-Ray G. Ecological diversity in coastal waters and oceans. In Wilson E.O. (Ed) *Biodiversity*. National Academic Press, Washington, 1988.

Goldsmith E. *Le défi du XXI siècle. Une vision écologique du monde*. Éd du Rocher, Monaco, 1994.

Ramade F. Écologie, démographie et développement. *Plein Sud* 1994, 15: 6–7.

Rougerie F. Réflexions sur l'océanologie hauturière et récifale dans le Pacifique Sud (1965–1994). In: *Quelle recherche environnementale dans le Pacifique sud: Bilans thématiques (compléments)*. CR Colloque, MESR, Paris, 28–31 March 1995.

Safina C. Les excès de la pêche en mer. *Pour la Science* 1996; 219: 28–36.

Salvat B. La biodiversité et son maintien en Polynésie française. In: *Quelle recherche environnementale dans le Pacifique sud: Bilans thématiques (compléments)*. CR Colloque, MESR, Paris, 28–31 March 1995; 37–46.

Van Tilbeurgh V. *L'huître, le biologiste et l'hostréiculteur: Lecture entrecroisée d'un milieu naturel*. Coll. Logiques Sociales, L'Harmattan Éd., Paris, 1994.

Glossary

anoxia	total absence of oxygen
*autoecology	the ecology of individual organisms or species
*benthic	pertaining to the bottom terrain of water bodies; animals living on or near the bottom of a body of water
*biocoenosis	a community or natural assemblage of organisms inhabiting a biotope, specifically excluding physical aspects of the ecosystem
biodiversity	variety and variability of living organisms and the ecological complexes to which they belong. Also the assemblage of biological and genetic resources or biological capital
bioindicator	organism or group of organisms that, in relation to biochemical, cytological, physiological, ethological and ecological variables, allow the practical and precise description of the state of an ecosystem or an ecocomplex and the accurate demonstration of natural or induced changes (Blandin, 1986)
biomass	quantitative estimate of total mass of plant or animal organisms in a given area at a given time. It is most often expressed as weight of dry matter
*bloom	rapid and localised increase of one or more planktonic species, often in spring or summer, causing large temporary concentrations of these organisms to dominate the community
*bouchots	method of culturing mussels primarily used along the

	French Atlantic coast; ropes with young mussels attached are transported to the on-growing area and wrapped around wooden poles (i.e. bouchots) which are embedded in the intertidal zone
*byssus	a mass of thread-like fibres secreted by the byssus gland situated in the foot of certain bivalve molluscs and which serves to fix the animal to the substrate
ciguatera	human disease caused by the ingestion of marine animals (especially fish) which concentrate toxins from unicellular algae. Its symptoms are nausea, vomiting and diarrhoea
*claire	type of earthen pond used in the final stages of oyster culture in France
coastal zone	area in which the terrestrial environment influences the marine or lacustrine environment, and vice versa (Carte, 1989); area including marshes, bogs, reclaimed land or natural, artificial, permanent or temporary areas of water, whether stagnant or flowing, fresh, brackish or salt, including areas of marine water whose depth at low tide does not exceed 6 m (definition of the RAMSAR convention)
*conservation	the planned management of natural resources; the preservation of natural balance, diversity and evolutionary change in the environment
*convection	a mode of heat transfer within a fluid, involving the movement of substantial volumes of the fluid from one place to another as a result of changes in the density of heated and non-heated areas
cryptic (cryptic fauna)	hidden fauna
current	in oceanography, water movement characterised by a precise direction, speed and flow
*demersal	living on or close to the bottom of a pond, lake or the sea
*detergent	tensio-active surfactant compound that features a non-polar, hydrophobic hydrocarbon chain with a polar end. Detergents are often included in cleaning solutions as they emulsify dirt and oil. Detergents (with or without phosphate) have a marked inhibitory effect on *Scenedesmus quadricaudata* (Chlorophycae) and *Phaeodactylum tricornutum* (Diatom), according to Crouzet (1991)

downstream migration	migration of salmonid fry, or those of other species, from the river in which they were spawned, towards the sea
dry matter	it is assumed that 8–10 g of living matter corresponds to 2.5 g of dry matter; this contains 40–50% organic carbon
*ecological niche	the conceptual space occupied by a species, which includes both the physical space and the functional role of the species
ecology	scientific study of the interrelationships between living organisms and their environments; extended to the study of mechanisms and processes which explain the distribution and abundance of organisms
*ecosystem	an integrated biological system which results from the interaction of all or a limited number of biotic or abiotic factors within a defined section of the biosphere
ecotone	intermediate zone between two ecosystems
embryo	organism in process of development; this is the developmental stage which terminates in hatching of marine animals
endogenous phase	in larvae, the period of larval life during which an endogenous food source is assured from reserves inherited from the parent; there is no requirement to obtain food from the external environment
endotrophy, endotrophic	feeding on yolk reserves
*environment	all the external or internal factors or conditions supporting or influencing the existence or development of an organism or assemblages of organisms
*epibenthos	the benthic organisms living on the surface of the sea bed or lake floor
*epizootic	a disease attacking many animals in a population at the same time, widely diffused and rapidly spreading
*estuary	that part of a river where the current meets ocean waters and is subject to its effects
exotrophy (exotrophic phase)	the animal's food is taken from the external environment, in contrast to the yolk sac phase
foreshore	the zone between mean high water and mean low water

genetic diversity	see biodiversity
IFREMER	Institut Français pour l'Exploitation de la Mer (France)
inhabitant equivalent	estimation of the mean quantity of pollutants or waste water produced per human inhabitant
*inoculum	the individual or group of individuals comprising the founders of a colony or newly-established population
INRA	Institut National de la Recherche Agronomique (France)
intertidal zone	the shore zone between the highest and lowest tides
*lagoon	a shallow body of salt or brackish water, which may have a shallow, restricted outlet to the sea. Offshore bars may temporarily block this channel and isolate the lagoon from the sea
lagunage	a system of interconnecting lagoons used typically in France for water purification
larva	an organism from hatching to metamorphosis into a juvenile, which is morphologically similar to the adult
*long line culture	form of open-water suspended culture in which cultured species are on-grown on ropes or diverse containers (baskets, stacked trays, lantern nets) suspended from anchored and buoyed surface or subsurface ropes (long lines)
*lux	the SI unit of illumination equivalent to 1 lumen/m^2. Light levels in a public place, such as a lecture theatre, range between 330 and 600 lux; in full sunlight, light intensity exceeds 60,000 lux
m	milli = thousandth; metre
M or Molarity (mole/litre of solution)	a mole or a gram molecule of water (H_2O) contains 6.02 $\times 10^{23}$ molecules and "weighs" 18 g (molecular weight of oxygen is 16, that of hydrogen, 1). A mole of sodium chloride NaCl contains 6.02×10^{23} Cl^- ions and 6.02×10^{23} Na ions and "weighs" 58.5 g; a molar solution (1M) of NaCl thus contains 58.5 g of NaCl per litre
*macro-algae	algae visible to the naked eye: non-specific collective term referring to the larger algae
*macrophyte	a large macroscopic plant, used especially of aquatic forms such as kelp, and tall plants such as sea grasses

mangrove	a tidal salt marsh characterised by the presence of mangroves
mariculture	cultivation or rearing of marine organisms
marine pollution (UN definition)	the introduction by human activities, directly or indirectly, of substances or energy into the environment resulting in a deleterious effect on living organisms. While aquaculture often suffers from specific impairments of water quality by sewage, pesticides and industrial wastes, it can, as an industry, also negatively affect the environment, if not properly managed
*mesh size	length of mesh side; the distance between two sequential knots or joints, measured from centre to centre when the yarn between those points is fully extended
*mesocosm	enclosed experimental facility large enough ($>10\,m^3$) to allow natural ecosystems to be studied or exploited
mg	milligram
mixed layer	surface layer of marine waters which is mixed by waves and surface currents
ml	millilitre
mm	millimetre
mM	millimole
μ	(micro = millionth)
μg	microgram
μl	microlitre
μm	micrometre or micron
*nekton	those actively swimming pelagic organisms able to move independently of water currents
*nutrient	a substance which provides nourishment; the term is often used to refer to dietary compounds
OECD	Organisation for Economic and Co-operative Development; 2, rue A. Pascal, 75775 Paris Cedex 16, France
*oligotrophic	pertaining to waters having low levels of the mineral nutrients required by green plants
organic carbon or organic C	carbon entering the makeup of living organisms; dry matter contains 40–50% organic carbon

paradigm	an example or model
*pelagic	relating to, living, or occurring in open water areas of lakes or oceans (cf. benthic)
*photosynthesis	synthesis by plant cells of organic compounds (mainly carbohydrates) in the presence of light, from carbon dioxide and water, with simultaneous production of oxygen
phylum	a main division of the animal or plant kingdom
phytoplanktonophage	an organism which feeds on plant plankton (microalgae or phytoplankton)
*plankton	passively drifting or weakly swimming organisms, including many microscopic plants and animals, but also some large jellyfish (cf. nekton)
*poikilotherm	refers to animals that lack a mechanism for internal temperature regulation; body temperature and metabolic rate are influenced by external temperature
ppm	parts per million (often comparable, but not equivalent to, milligrams per litre: mg/l)
*production	the assimilation of energy into organic materials by an organism, population or community
*productivity	rate of production of biomass; expressed as production during a specific time period
recruitment	numbers in each cohort which succeed in passing through the successive stages of egg, larva and juvenile, to reach the phase exploited by a fishery and contribute to stock renewal; results in the arrival of young adults in a population, the characteristics of which are acquired
reserve	marine zone within which marine flora and fauna are protected
*sea ranching	the release of juvenile fish, crustaceans or molluscs from culture facilities for growth to a harvestable size in a natural habitat
*sessile	fixed or attached; not free-moving
*seston	collective term for the particulate material suspended within the water column; it thus includes the plankton
*spat	young bivalve molluscs just past the veliger stage, which have settled and become attached to some hard object

stratification	formation of distinct horizontal layers within a water mass
*suspended solids	term used to describe the presence of sediment in culture water
sustainable development	according to the OECD, the assemblage of activities and processes allowing the current requirements of man and other species to be met, while preserving the biosphere so that it can respond to and provide for the reasonable and predictable requirements of man and all other species in the future
*synecology	ecology of communities as opposed to that of individual species (q.v. autecology)
T 90	the period of time after which 90% of bacteria have disappeared from a given environment
telemediator	chemical substance produced into the water by a species, capable of acting on the development of other species at a distance
*thermocline	the zone of rapid temperature change in a thermally-stratified body of water
*thigmotropism	a directed response of a motile organism to continuous contact with a solid surface
*tidal range	the difference in height between consecutive high and low waters
tourist	according to the World Tourist Organisation, a temporary visitor who spends at least 24 hours in the visited area, for either leisure or work purposes
UNESCO	United Nations Educational Scientific and Cultural Organisation, 7 place de Fontenoy, 75700 Paris, France
*upwelling	the process by which deep water masses rise, usually as a result of divergence and offshore currents
valli	private lagoon system developed and managed for the extensive ongrowing of marine fish (the term is Italian since this type of rearing is specific to Italy)
valliculture	extensive fish culture in valli
*viscosity	the resistance of liquids, semi-solids and gases to movement or flow

wet weight: equivalent to fresh weight; 1 g wet weight of organic matter represents about 1 kcal of energy

* indicates definition partly, or entirely, from Eleftheriou M. (Ed.) *Aqualex: A glossary of aquaculture terms*. Wiley-Praxis Series in Aquaculture and Fisheries. John Wiley & Sons, Chichester, 1997.

Index